Practical Machinery Management for Process Plants
Volume 3, Third Edition

Machinery Component Maintenance and Repair

Practical Machinery Management for Process Plants
Volume 3, Third Edition

Machinery Component Maintenance and Repair

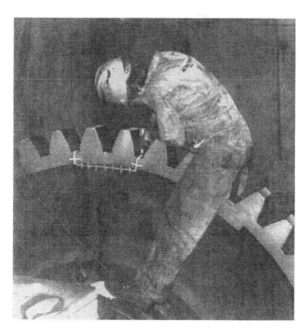

Heinz P. Bloch and
Fred K. Geitner

ELSEVIER

AMSTERDAM • BOSTON • HEIDELBERG • LONDON • NEW YORK
OXFORD • PARIS • SAN DIEGO • SAN FRANCISCO • SINGAPORE
SYDNEY • TOKYO

Gulf Professional Publishing is an imprint of Elsevier

Gulf Professional Publishing is an imprint of Elsevier
30 Corporate Drive, Suite 400, Burlington, MA 01803, USA
Linacre House, Jordan Hill, Oxford OX2 8DP, UK

 Recognizing the importance of preserving what has been written, Elsevier prints its books on acid-
free paper whenever possible.

Library of Congress Cataloging-in-Publication Data
Application submitted

British Library Cataloguing-in-Publication Data
A catalogue record for this book is available from the British Library.

ISBN-13:978-0-7506-7726-4
ISBN-10: 0-7506-7726-0

For information on all Gulf Professional Publishing publications visit our Web site at
www.books.elsevier.com

Transferred to Digital Printing 2009

Contents

Type of Operation. Manager's Role. Maintenance. Central Control System. Incentives for Computer Systems. Setting Up an Effective System. Machinery Maintenance on the Plant Level. Assignment of Qualified Personnel. Timing and Basic Definition of Critical Pre-Turnaround Tasks. Specific Preparation and Planning. Documenting What You've Done.

What's an Epoxy? Epoxy Grouts. Proper Grout Mixing Is Important. Job Planning. Conventional Grouting. Methods of Installing Machinery. Pressure-Injection Regrouting. Prefilled Equipment Baseplates: How to Get a Superior Equipment Installation for Less Money. Appendix 3-A—Detailed Checklist for Rotating Equipment: Horizontal Pump Baseplate Checklist. Appendix 3-B—Specification for Portland Cement Grouting of Rotating Equipment. Appendix 3-C—Detailed Checklist for Rotating Equipment: Baseplate Grouting. Appendix 3-D—Specifications for Epoxy Grouting of Rotating Equipment. Appendix 3-E—Specification and Installation of Pregrouted Pump Baseplates.

Bearing. Shaft and Housing Preparation. Checking Shaft and Housing Measurements. Basic Mounting Methods. Hints on Mounting Duplex Bearings. Preloading of Duplex Bearings. Importance of the Correct Amount of Preload. Assembly of Bearings on Shaft. Cautions to Observe During Assembly of Bearings into Units. Mounting with Heat. Checking Bearings and Shaft After Installation. Assembly of Shaft and Bearings into Housing. Testing of Finished Spindle. Maintain Service Records on All Spindles.

Foreword

A machinery engineer's job was accurately described by this ad, which appeared in the classified section of the New York *Times* on January 2, 1972:

> Personable, well-educated, literate individual with college degree in any form of engineering or physics to work . . . Job requires wide knowledge and experience in physical sciences, materials, construction techniques, mathematics and drafting. Competence in the use of spoken and written English is required. Must be willing to suffer personal indignities from clients, professional derision from peers in more conventional jobs, and slanderous insults from colleagues.
>
> Job involves frequent physical danger, trips to inaccessible locations throughout the world, manual labor and extreme frustration from lack of data on which to base decisions.
>
> Applicant must be willing to risk personal and professional future on decisions based on inadequate information and complete lack of control over acceptance of recommendations . . .

The situation has not changed. As this third edition goes to press, there is an even greater need to seek guidelines, procedures, and techniques that have worked for our colleagues elsewhere. Collecting these guidelines for every machinery category, size, type, or model would be almost impossible, and the resulting encyclopedia would be voluminous and outrageously expensive. Therefore, the only reasonable course of action has been to be selective and assemble the most important, most frequently misapplied or perhaps even some of the most cost-effective maintenance, repair, installation, and field verification procedures needed by machinery engineers serving the refining and petrochemical process industries.

This is what my colleagues, Heinz P. Bloch and Fred K. Geitner, have succeeded in doing. Volume 3 of this series on machinery management brings us the know-how of some of the most knowledgeable individuals in the field. Engineers and supervisors concerned with machinery and component selection, installation, and maintenance will find this an indispensable guide.

Here, then, is an updated source of practical reference information which the reader can readily adapt to similar machinery or installations in his particular plant environment.

Uri Sela
Walnut Creek, California

Acknowledgments

It would have been quite impossible to write this text without the help and cooperation of many individuals and companies. These contributors have earned our respect and gratitude for allowing us to use, adapt, paraphrase, or otherwise incorporate *their* work in Volume 3: W. J. Scharle (Multi-Plant Maintenance), J. D. Houghton (Planning Turbomachinery Overhauls), E. M. Renfro/Adhesive Services Company (Major Machinery Grouting and Foundation Repair), M. G. Murray (Grouting Checklists, Machinery Alignment), Prueftechnik A. G. (Laser Alignment), P. C. and Todd Monroe (Machinery Installation Checklists and Pre-Grouted Baseplates), J. W. Dufour (Machinery Installation Guidelines), W. Schmidt (Piping Connection Guidelines), Garlock Sealing Technologies and Flexitallic, Inc. (Gasket Selection and Flange Torque Requirements), D. C. Stadelbauer, Schenk Trebel Corporation (Balancing of Machinery Components), MRC Division of SKF Industries (Bearing Installation and Maintenance), Flowserve Corporation (Metallic Seal Installation, Repair, Maintenance), H. A. Scheller (Pump Packing), T. Doody (Welded Repairs to Pump Shafts, etc.), H. A. Erb (Repair Techniques for Machinery Rotor and Case Damage), Byron Jackson, Division of Flowserve Corporation (Field Machining Procedures), Terry Washington, In-Place Machining Company (Metal Stitching Techniques), Tony Casillo (OEM vs. NON-OEM Machinery Repairs), Barney McLaughlin, Hickham Industries, Inc., and W. E. Nelson (Compressor Rotor and Component Repairs, Sealing Compounds, etc.), M. Calistrat/Koppers Company (Mounting Hydraulically Fitted Hubs), Larry Ross, C. R. McKinsey and K. G. Budinski (Hard Surfacing), C. R. Cooper, Van Der Horst Corporation (Chrome Plating), Turbine Metal Technology (Diffusion Alloys) and National O-Ring Company (O-Ring Selection and Application).

We also appreciate our close personal friend Uri Sela who devoted so much of his personal time to a detailed review of the entire draft, galleys, and page proofs. Uri counseled us on technical relevance, spelling, syntax, and other concerns.

More than ever before, we are reminded of some important remarks made by Exxon Chemical Technology Vice President W. J. Porter, Jr. in early 1984. Mr. Porter expressed the belief that through judicious use of outside contacts, participation in relevant activities of technical societies, and publication of pertinent material, we can be sure that our technical productivity will continue to improve. The technical person will thus be updated on the availability of "state-of-the-art" tools and individual creativity encouraged.

We hope this revised text will allow readers to find new and better ways to do their jobs, broaden their perspective as engineers, and contribute to a fund of knowledge which—if properly tapped—will bring benefits to everyone.

Heinz P. Bloch
Fred K. Geitner

ix

Part I

Background to Process Machinery Maintenance Programming

Chapter 1
Machinery Maintenance: An Overview

Maintenance and repair of machinery in a petrochemical process plant was defined in a preceding volume as simply "defending machinery equipment against deterioration."[1] Four strategies within the failure-fighting role of maintenance were defined:

- Preventive
- Predictive
- Breakdown or demand based
- "Bad actor" or weak spot management

Machinery maintenance can often be quite costly in a petrochemical plant operation. Prior to the publication of the first two volumes of this series, very few studies were available describing quantitative or objective methods for arriving at the optimization of the four strategies[2]. Though our readers should not expect detailed contributions to those subjects in this volume, we did opt to include an overview section describing the maintenance philosophy practiced in a large multi-plant corporation which makes effective use of centralized staff and computerized planning and tracking methods.

What, then, can our readers expect? After a short definition of the machinery maintenance problem we will highlight centralized maintenance planning. We will then guide our readers through the world of machinery maintenance procedures by identifying the What, When, Where, Why, How—and sometimes Who—of most maintenance and repair activities around petrochemical process machinery. We ask, however, that our readers never lose sight of the total picture. What, then, is the total picture?

3

It is the awareness that true cost savings and profitability can only be achieved by combining machinery reliability, safety, availability, and maintainability into a cost-effective total—consistent with the intent of our series of volumes on process machinery management. Figure 1-1 illustrates this concept. Consequently, machinery maintenance cannot be looked at in isolation. It will have to be governed by equipment failure experience, by our effectiveness in failure analysis and troubleshooting[1], and by built-in reliability[3].

Maintenance in a broad definition is concerned with controlling the condition of equipment. Figure 1-2 is a classification of most machinery maintenance problems.

Deterministic or predictive component life problems are those where no uncertainty is associated with the timing or consequence of the maintenance action. For example, we may have equipment whose components are not subject to actual failure but whose operating cost increases with time. A good illustration would be labyrinths in a centrifugal process compressor. To reduce operating cost caused by increasing leakage rate, some form of maintenance work can be done—usually in the form of replacement or overhaul. After maintenance the future trend in operating cost is known or at least anticipated. Such a deterministic situation is illustrated in Figure 1-3.

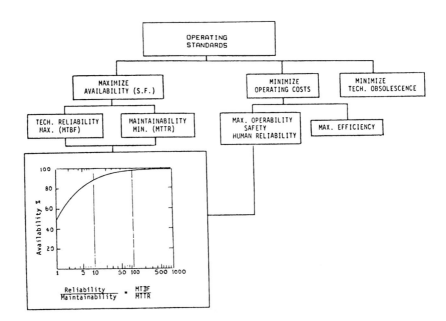

Figure 1-1. The total picture: Possible goals of process machinery management.

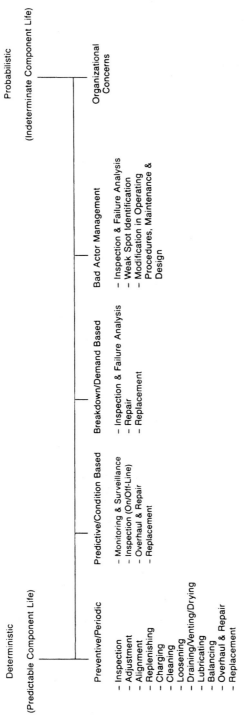

Figure 1-2. Classification of machinery maintenance problems.

Figure 1-3. Deterministic trend in costs.

In probabilistic or indeterminate component life problems, the timing and result of maintenance may depend on chance. In the simplest situation a piece of machinery can be described as being "good" or "failed." From a frequency distribution of the time elapsed between maintenance activity and failure it is possible to determine the variations in the probability of failure with elapsed time. These relationships are thoroughly dealt with in Reference 1.

We saw from Figure 1-2 that *inspection, overhaul, repair* and finally *replacement* are common to all maintenance strategies. The basic purpose of inspection is to determine the condition of our equipment. All machinery inspection should be based on these considerations:

1. Expected failure experience:
   - Deterministic
   - Probabilistic
2. Inspection cost.
3. Probability and risk of failure.
4. Probable consequences of failure, i.e., safety-health and business loss.
5. The risk that inspection will *cause* a problem[4].
6. The quality of on-stream condition monitoring results.

The terms *overhaul* and *repair* are often reserved for maintenance actions that improve the conditions of an item, but may or may not establish "good as new" condition. In fact, *overhaul* is often interpreted as a preventive maintenance action while *repair* is strictly reserved for maintenance of an item that has reached a defined failed state or defect limit[5]. *Replacement* should be understood in our context as a broad term that includes the replacement of components, operating fluids and charges, as well as of complex machinery and systems. Finally, we understand *organizational structure* problems in machinery maintenance as those concerns that deal with maintainability parameters such as facilities, manpower, training, and tools. Figure 1-4 illustrates this point.

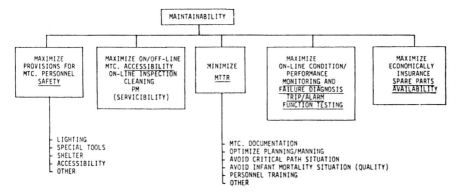

Figure 1-4. Process machinery maintainability components.

Most petrochemical process plants have a preventive maintenance (PM) system. The authors know of a plant where 95 percent of the maintenance work orders are turned in by the PM crews and not the operators. While this is an extreme—and probably not a very cost effective—way of failure fighting, we can support a moderate approach to machinery PM. This moderate approach begins with an attempt to plan all PM actions by following this pattern:

1. Determine what defect, failure, or deterioration mode it is you want to prevent from occurring.
2. Determine whether the defect, failure, or deterioration mode can be prevented by periodic actions. If not, determine how it can be predicted and its consequence reduced by perhaps continuous or daily surveillance.
3. Select PM task.
4. Determine normal life span before defect, failure, or deterioration mode will develop.
5. Choose PM interval within normal life span.
6. Determine who should do the job—operating crew or maintenance personnel.

More often than not we will find that machinery failure modes are probabilistic and indeterminate. PM will therefore not help and predictive strategies are indicated: By continuously looking for problems, we expect not to reduce the deterioration rate of machinery components, but to control the consequences of unexpected defect or failure. This maintenance strategy is often referred to as predictive- or condition-based maintenance. Together with "post mortem" failure analysis, this strategy is the most powerful weapon in the arsenal of the machinery maintenance person. Figure 1-5 shows how predictive maintenance works in connection with large petrochemical process machinery such as turbocompressors, reciprocating compressors, and their drivers.

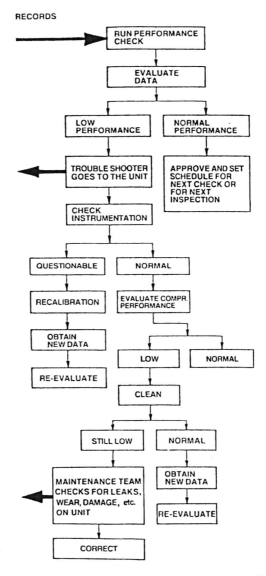

Figure 1-5. Machinery predictive maintenance routine (adapted from Ref. 6).

Table 1-1
State-of-the-Art Instrumentation and Monitoring Methods

MACHINERY COMPONENTS (columns): Structure · Piping · Foundation · Flange joints · Casing joints · Gear couplings · Diaphragm cplgs. · Packing · Seals · Shafts · Fan wheels · Impellers · Motor/bearings · Diffusers · Volutes · Anti-friction · Gearings · Gears · Pistons/cylinders · Recip. comp. · Valves · Combustors · Electric motor · Windings · Cooling systems · Lubrication systems

SENSORY PERCEPTION — DIRECT: Sound (Change) · Resonance · Vibration · Liquid leakage · Discoloration · Gas/vapor leakage · Temperature (Touch) · Wear debris

SENSORY PERCEPTION — INDIRECT: Pressure · Temperature · Flow · Position · Speed · Torque · Vibration · Pressure ratios · Temperature ratios · Power demand · Recycle rates · Efficiency

MONITORING/INSPECTION METHOD AND PARAMETER: Visual inspection · SPG test · Alignment check · Dye penetrants · Magnaflux · Radiographic inspection · Vibration survey · Vib. spectrum analysis · Wear analysis · Shock pulse level · Spike energy level · Ultrasonic measurement · Acoustic emission · Peak firing pressure · Gas path condition · Thermographic survey · Process fluid analysis · Oil analysis/off-line · Oil analysis/on-line · Oil filter inspection

INSTRUMENTATION REQUIRED: Process instrumentation · Engine analyzer · On-line alignment instr. · On-line torque meter · Proximity probe system · Velocity probe system · Accelerometer system · Electronic stethoscope · Sound spectrum analyzer · Ultrasonic meter · Ultrasonic imaging equipment · Acoustic valve leak detector · Portable X-ray analyzer · Gamma ray equipment · Stroboscope: portscope · Portable vibration meter · Vibration spectrum analyzer · Shock pulse meter · Vibration spike energy analyzer · Acoustic emission detector · Acoustic flaw analyzer · Pocket digital thermometer · Infrared thermometer · Spectroscope · Magnetic plugs · Contaminant monitor

The fundamental difference between preventive maintenance and predictive- or condition-based maintenance strategies is that PM is carried out as soon as a predetermined interval has elapsed, while condition-based maintenance requires checking at predetermined intervals, with the maintenance action carried out only if inspection shows that it is required. The main factors in a predictive machinery maintenance program are:

- State-of-the-art instrumentation and monitoring methods as shown in Table 1-1
- Skilled analysts
- Information system allowing easy data retrieval
- Flexible maintenance organization allowing for an easy operations/ maintenance interface
- Ability to perform on-line analysis[7]

In the following chapters we will further deal with predictive maintenance tools.

References

1. Bloch, H. P and Geitner, F. K., *Machinery Failure Analysis and Troubleshooting*, Gulf Publishing Company, Houston, Texas, third edition, 1997, Pages 484–488.
2. Jardine, A. K. S., *The Use of Mathematical Models in Industrial Maintenance*, The Institute of Mathematics and its Applications, U.K., August/September 1976, Pages 232–235.
3. Bloch, H. P., *Improving Machinery Reliability*, Gulf Publishing Company, Houston, Texas, third edition, 1998, Pages 1–667.
4. Grothus, H., *Die Total Vorbeugende Instandhaltung*, Grothus Verlag, Dorsten, W. Germany, 1974, Pages 63–66.
5. Reference 2, Page 232.
6. Fucini, G. M., *Maintenance Shops*, Quaderni Pignone, Nuovo Pignone, Firenze, Italy, Number 27, 1983, Page 30.
7. Baldin, A. E., "Condition-Based Maintenance," *Chemical Engineering*, August 10, 1981, Pages 89–95.

Bibliography

Whittaker, G. A., Shives, T. R., and Philips, *Technology Advances in Engineering and Their Impact on Detection, Diagnosis and Prognosis Methods*, Cambridge University Press, New York, New York, 1983, Pages 11–286.

Chapter 2

Maintenance Organization and Control for Multi-Plant Corporations*

There are many approaches to performing maintenance and engineering activities at an operating facility. The type of process, plant size, location, and business conditions at a particular time are all variables that can affect this approach. The system must fit the basic overall corporate goals. The final evaluation of success, however, for whichever system selected, is achieving the lowest possible product cost over extended periods of time at varying business conditions.

This segment of our text will concentrate on plant maintenance and engineering service in a multi-plant corporation operated on a combination centralized-decentralized basis. However, the reader will quickly discern the applicability of this approach to many aspects of equipment maintenance in "stand-alone" plants. Organizational control methods are all planned for an optimum approach to cost economy. Basically, then, we are presenting corporate management's approach to an overall maintenance strategy. This approach is as valid in 2004 as it has been in the 1965–1970 time period.

---

*Based on articles by W. J. Scharle ("Multi-Plant Maintenance and Engineering Control," *Chemical Engineering Progress*, January 1969) and J. A. Trotter ("Reduce Maintenance Cost with Computers," *Hydrocarbon Processing*, January 1979). By permission of the authors. Updated in 2004.

Type of Operation

To understand the organizational approach to maintenance and engineering described here, it is first necessary to understand the size and type of operations involved. We should assume that the facilities would fall into virtually all size categories. The plants are quite autonomous and may select maintenance organizations to fit their particular needs. Through their own unique experiences of plant maintenance and engineering problems and studying alternative approaches used by others, a mature organization will have gradually formulated an operations control system, including plant maintenance and engineering services, which best serves its type of operation and is flexible for future needs.

This implies that large plants, which have the technical and maintenance support resources to be totally self-sufficient, may opt to deviate from the organizational and implementation-oriented setups we are about to describe. However, for best results, the deviation should not be very drastic because the basic principles of effective maintenance organization and control hold true for any plant environment.

Before discussing plant maintenance specifics and engineering functions, we will discuss why this multi-plant corporation went to the present approach. Like many companies, the corporation started with an approach wherein the plant manager was autonomous in his responsibility for production, maintenance, and most engineering services. He depended largely on the equipment manufacturer to help solve problems.

As more plants were added to the network and more significant operational and mechanical problems were encountered, it was gradually recognized that the most economical solution to critical problems was to quickly interject the best technical specialists within the company, regardless of location. However, it was not possible or economical to have these highly skilled specialists at each facility or to adequately train the plant manager in all areas when the facility normally operated at an extremely high onstream factor. Again, as a higher degree of technical knowledge was gained, equipment improvements made, and sophisticated process and machinery monitoring devices introduced, it was found that the periods between major equipment maintenance could be significantly extended without risking costly equipment failures. The use of a relatively small group of mobile, technical specialists from within the company was the key to better plant performance and lower costs. Equipment manufacturers and vendors' representatives have neither the incentive nor the responsibility to provide the prompt technical services required.

Manager's Role

Yet, it was strongly desired to have these specialists report to and work solely under the guidance of the individual plant manager in order not to confuse the chain of command. Thus, a decentralized system of giving the plant manager responsibility for general operations, cost performance, and maintenance performance, but with a strong centralized approach to all aspects of monitoring plant performance and providing specific maintenance and engineering services as required, was evolved as the fundamental organizational concept. Once this basic concept was reached, efforts were then devoted to understanding and establishing specific methods of accomplishing plant maintenance and engineering services under the general system concept.

Since the plant managers' responsibilities on a decentralized basis represented a rather conventional approach to day-to-day operation, we will dwell on considerations relative to the centralized aspect of plant maintenance and engineering services and the monitoring function. These centralized services were provided by a group of specialists located for the most part at the home office or at the location of the largest affiliated plant. Some advantages of this centralized approach to plant maintenance and engineering services are:

1. Better solutions to important technical problems. With the varied plant problems, the ability to use key specialists will normally result in the best technical solution.
2. More efficient use of talent. With extremely high onstream factors, chemical and mechanical engineering specialists at each facility cannot be fully justified, since the rate of problems and/or severity would not normally warrant their continuous presence. Minimum staffing at each plant to handle normal day-to-day problems, plus a mobile technical and maintenance organization will result in lower overall costs. The question of overstaffing at a particular facility to take care of "first year" startup problems is a very real one. The ability to have this same mobile specialist group help in quickly solving first year operation problems allows a flexible and easy method of reducing a facility to its minimum labor cost at the earliest time.
3. Better communication. Technical solutions, procedures, and other important factors which have a direct and immediate effect on on-stream factors and costs can be more readily transmitted from one plant to another. The use of plant shutdown and maintenance reports prepared by the plant manager allows the central technical

organization to evaluate and disseminate information pertinent to other facilities.

4. Better response to management and business outlook. Constantly varying market conditions change product demand and value. These important factors often become the overriding consideration in scheduling maintenance work and turnarounds. Centralized overall maintenance planning can more readily assimilate these factors in considering a large number of plants at different locations. This is an important consideration in minimizing peaks and valleys in major maintenance work and allowing a smaller specialist group to handle a broader scope of activities.

5. More consistent organizational policies, procedures, and better methods of making comparisons on general performance, cost, production, prompt action, and managerial talent.

To keep the centralized organization current on the facts of life at plant facilities, a program of specialist and management visits to each facility must be established. These visits, coupled with careful production monitoring, normal maintenance, and general cost performance are necessary prerequisites for the system discussed herein. The extra travel and communication costs are far outweighed by better personnel utilization.

Maintenance

Total plant profitability is obviously affected both by onstream factors and maintenance costs. One cannot be separated from the other. Any system, therefore, must account for how cheaply maintenance can be performed from an organizational setup, and also what must be done and how often. The ability to update maintenance requirements and improved planning based on experience at a group of plants has a large bearing on overall maintenance costs.

Other than breakdown maintenance, all maintenance work is planned. Some can be done while the plant is operating and the rest during shutdown. The effectiveness of this planned or preventive maintenance (PM) program to reduce breakdowns and the organizational methods used to accomplish the planned major maintenance work will determine maintenance costs. Preventive maintenance as discussed here covers all planned maintenance work, whether major or minor, regardless of whether the plant is running or shut down. The selection of what shall be done as part of the PM program and how often it shall be done is one of the most important factors affecting corporate maintenance costs and the realization of an optimum onstream factor.

It is a generally accepted practice to let each plant manager handle the PM program for his facility. In some plants, this is being done with individual check sheets or production boards using equipment manufacturers' recommendations and the limited experience of plant personnel. However, the demand for plant operation attention often prevents timely maintenance performance. Another defect is that it lacks uniformity and does not provide compliance reports to home office management. And, there is often no effective way to compare the PM performance at similar plants or equipment at different locations. Most important of all, equipment failures may occur because proper consideration and judgement is not given to maintenance items whose significance is best understood by qualified specialists.

Central Control System

In view of this, major corporations will frequently opt to incorporate a centrally controlled PM system into the Operations Department. This allows mechanical and process specialists to make the key cost decisions on what kind and how often maintenance should be accomplished at all affiliate plants by coupling it to an electronic data processing monitoring system. This will serve as a management tool in evaluating conformance to the maintenance system. Thus, the plant manager is made responsible for efficiently executing the PM work as outlined by the program, and is monitored for performance by centralized management. The data processing system can be easily adapted to any facility, is inexpensive to install and operate, and lends itself to overall reduced costs as the corporation expands. Some of the system advantages are:

1. The PM performance and frequency program is prepared by the centralized group of qualified engineering specialists based on equipment manufacturers' recommendations, experience, and historical records. The program is reviewed and approved by the plant manager. Program updating to take advantage of new technical knowledge and both good and bad experience is important to ensure continued cost savings.
2. A definite schedule is presented to plant managers so they know what is expected of them.
3. Operations management is advised of system conformance and is made aware of rescheduled tasks.
4. The system identifies overall corporate maintenance requirements so that work can be staggered enabling a minimum mobile group of technical and maintenance specialists to handle the overall program.

5. Historical data are accumulated for analysis.
6. Reduction in clerical work more than offsets the cost of computerization.

Principal Applications Areas for the Maintenance Computer

Conceptual discussions of the past and more recent systems development work have concentrated on six general areas of maintenance support. Systems are, of course, called by different names, according to the company which is developing and implementing them. Systems of any one type may also have differing emphasis, according to the specific company's requirements for maintenance support. The general applications areas are:

1. Materials inventory/stock cataloging.
2. Preventive maintenance/equipment records.
3. Work order costing.
4. Fixed equipment inspection.
5. Planning/scheduling of major maintenance projects.
6. Work order planning and scheduling.

The various computer systems have been developed both separately and as integrated groups through exchange of data between systems. Moreover, maintenance systems generally are designed for data exchange with a conventional accounting system.

Materials inventory/stock catalog systems are designed to support maintenance by making certain that required materials and spare parts are available at the right time, at the right place, and at minimum cost. Well-designed systems in this category provide better availability of parts and materials by supplying up-to-date catalogs, generated in multiple sorts.

Some systems allow stock items to be reserved for future usage in major construction projects or for scheduled plant or unit turnaround projects. A well-designed inventory/stock catalog system also may maintain a history of materials and parts usage. This enables maintenance to evaluate service demand patterns or vendor performance and to adjust inventory levels according to materials/parts usage. Some companies place emphasis on the purchasing function in design of inventory systems. Such systems automatically signal the need for materials or parts reorders on whatever basis the purchasing department wishes to establish (such as order point/order quantity or minimum/maximum quantities). The system may also be designed for automatic purchase order generation and to maintain a file of open purchase orders. It also can report unusual situations such as changes in a manufacturer's parts number, price increases beyond a prescribed limit, or alterations in delivery time requirements.

Ordinarily, an inventory/catalog system produces the majority of its reports on a weekly or monthly basis. Systems also may be run daily for adding new materials or parts, for daily stock status reporting or for processing receipts and issues information.

Maintenance people have long recognized the need for adequate inventory control and cataloging procedures. Without such procedures, the maintenance department runs the risk of having its work planning and scheduling controlled by materials availability. The computerized inventory/catalog system, thus, offers the benefits of improved manpower utilization and unit downtime reduction.

Preventive maintenance and equipment records systems not only bring a highly organized approach to scheduling of periodic inspections and service connected with a preventive maintenance program, but also provide a mechanism for compiling a complete equipment performance and repair history—including costs—for equipment within a processing facility.

The well-designed preventive maintenance and equipment records system is built around failure of equipment description data. Through this file, equipment inspection intervals are assigned according to criticalness or in accordance with laws or safety and environmental protection codes. Service intervals are also assigned—sometimes according to manufacturer's recommendations and sometimes on the basis of experience in extension of equipment life. Overhauls are scheduled in the same way as service intervals. Some types of service and all overhauls must also be backlogged for performance during equipment shutdown periods.

Most preventive maintenance systems produce a periodic listing of PM work to be performed—including specifications, service, and overhauls due. Jobs are entered into the plant's regular work order planning and scheduling system. PM jobs not performed on schedule are then reported—perhaps at a higher priority—for inclusion in the next PM work list.

The equipment records function, a natural extension of the preventive maintenance scheduling function, usually is not limited to equipment covered by the PM program. All equipment may be placed in this system's file. Through feedback cards from the field, the system can compile and maintain a complete repair file on all equipment of interest. Repair history and cost data may be reported in several different ways. Repair history by specific equipment or equipment type, for example, aids maintenance in setting or adjusting inspection, service, or overhaul intervals for equipment. Other reports may aid maintenance in identifying equipment which is costing most to maintain or has the poorest performance history. Some systems also support repair/replace decisions by maintenance or engineering department as well as equipment selection decisions. Equipment interchangeability information, and reports on equipment spare parts, are also available from some systems. The preventive maintenance/equipment records systems is also called the "reliability maintenance system" by some companies and plants.

Work order costing systems are vital for analysis and control of plant maintenance costs. These systems provide a framework for the capture of cost-related information and processing capability for analyzing such information and producing reports required by cost-conscious maintenance management.

Work order cost systems accumulate costs by work order. Usually, cost-related data from time sheets, contractors' invoices, journal vouchers, and spare parts inventory are compiled by the system and analyzed to produce:

1. Detailed and summary listings of costs, by work order.
2. Detailed listings of all current month cost transactions for each work order.
3. Reports which list, for each work order, costs incurred for the current month as well as during the life of the work order.

Using these and other reports, maintenance management can compare actual costs against estimates or budgets and can pinpoint costs which are outside policy guidelines or rising at a rate faster than anticipated.

The work order cost system also may be designed to provide input for other systems. It can generate equipment cost transactions for a preventive maintenance/equipment records system, for example, or can provide summarized accounting entries for a general ledger system.

In summary, work order cost systems provide cost information in a form that is fully usable by maintenance management in identification and definition of cost-related problems within the maintenance function. With this information, control efforts may be concentrated on areas where potential savings exist.

The fixed equipment inspection system adds consistency, comprehension, and effectiveness to a plant's inspection program. It is designed to support the plant inspection department and is structured around a data base of information on equipment critical to a plant's operation, such as piping, pressure vessels, heat exchangers, and furnaces. Fixed equipment may be designated as critical because of its potential for creating safety hazards, its position within the processing train or because of laws or codes governing equipment inspection in certain cases.

The system aids in scheduling inspection activities. Each piece of equipment covered by the system is scheduled for periodic inspection. Inspections that can be performed while the equipment is operating are placed on a monthly schedule for routine execution by the inspection team. Inspections which must await equipment shutdown are placed on a standing work list for coordination with operating and maintenance departments.

Inspection systems also may provide inspection history for particular pieces of equipment, standard inspection procedures for the equipment, forms for recording equipment conditions and thickness measurements, and automatic computation of corrosion rate (based on multiple inspections). The well-designed inspection system also can accommodate thickness measurement data produced by inspection tools such as ultrasonic, infrared, or radiographic devices.

Using results from system computations, inspection groups may report equipment condition to maintenance groups if repair, service, or replacement is required. Maintenance, in turn, would generate a work order consistent with the inspector's requirements. Information also may be routinely provided to engineering personnel to plan equipment replacement or to improve equipment and parts selection as equipment is replaced.

Planning and scheduling major maintenance projects using computer-supported Critical Path Method (CPM) techniques was one of the earliest applications of computers in support of the maintenance function. The central idea behind development and use of such systems was to identify opportunities for parallel execution of tasks associated with a turnaround project so that available manpower and resources may be utilized as efficiently as possible to minimize equipment downtime.

In spite of the CPM system's "head start" in use by maintenance groups, this potentially profitable tool soon was abandoned by a surprisingly large number of plants and companies. Most companies said the available CPM systems were too complex or too cumbersome for effective use in maintenance turnaround projects or small construction jobs.

There is, however, a resurgence of computer-based CPM systems today. Systems currently designed and used for planning and scheduling major maintenance projects are simplified versions of the earlier systems. They are, in fact, designed specifically for use by process industry maintenance personnel. They incorporate terminology readily understood by maintenance people and combine simplicity of operation with flexibility.

Typically, the well-designed CPM system produces reports which show how limited resources may be used to complete a project in the shortest possible time. Alternatively, the system may show the manpower necessary for completion of a project in a given length of time.

Maintenance work order planning and scheduling continues to be a largely manual set of procedures throughout the hydrocarbon processing industry. There are, however, several systems which support daily work planning and scheduling. One such system is a skills inventory file that provides daily information on available personnel for use in manual planning and scheduling of maintenance work. Another is the computer-based file containing standard maintenance procedures that can be retrieved for preparation of work orders and in estimating manpower time requirements.

Additionally, other maintenance-related systems, such as preventive-maintenance systems and inspection-support systems, may generate work orders for inclusion in daily maintenance schedules. Work order planning and scheduling also is supported by materials and parts inventory systems.

The actual computer-based scheduling of daily maintenance manpower resources, however, has remained an elusive goal. Recent systems work has aimed at scheduling shop work where forecasting work requirements is easier than forecasting field work.

Incentives for Computer Systems

The primary incentive for design and implementation of maintenance-related support systems is the potential for reducing maintenance-related costs. The cost of keeping hydrocarbon processing plants running includes maintenance expenditures. These typically range from 1.8 to 2.5 percent of the estimated plant replacement value.

Justification of Systems

Although process industry companies generally agree that maintenance-support systems* are a viable means of reducing maintenance costs, there is no general agreement on the size of benefits available or the source of these benefits.

For this reason, there are probably as many ways to justify computer installation as there are computer applications:

1. Reduced clerical effort.
2. Improved utilization of maintenance work force.
3. Improved equipment reliability.
4. Reduced inventory costs.

---

*Also called "CMMS," for computerized maintenance management system.

Reduction of clerical effort is used when filing, recording, and retrieving become excessive. Sometimes a reduction in clerical staff may even be possible after a computer system is installed. However, the relief of key personnel from clerical responsibilities is usually more important as a justification point. For example, a major oil company partially justified installation of a fixed equipment inspection system at a large refinery on the basis that inspectors could be relieved of the clerical duties of filing and retrieving inspection information. This company also found record keeping on inspection, thickness measurement, and corrosion rates to be more consistent and far more accessible. As a result, information compiled by this refinery's inspection department is far more useful today than when such information was kept mostly in filing cabinets in the individual inspector's office.

Improved utilization of maintenance manpower is widely used as a means for justifying turnaround scheduling systems, planning/scheduling systems, and inventory control systems. Results from a carefully conducted analysis of work delays created by existing manual procedures are compared against improvements expected from computerized systems. Man hours saved—multiplied by hourly rates for maintenance personnel—sometimes provide substantial justification for computer systems.

Improved equipment reliability, with resulting reductions in equipment downtime and improvements in plant throughput, are obvious justifications for preventive and predictive maintenance systems. Some companies have found that benefits from this source alone can provide a payout as quickly as one year from the initial computer system investment. In the complex process environment of the modern refinery or petrochemical plant, monitoring equipment performance, effective diagnostics, and early recognition of equipment problems require computer speed and support.

Improved management reaction to plant equipment problems also has justified computer systems. This is a difficult area to quantify. However, if previous costly equipment failures can be identified as preventable through timely management information, this becomes a very real justification for system installation.

Materials inventory and stock catalog systems have been justified by many companies based on reduced inventory. Computer systems have improved inventory management and control, reduced overall stock requirements, and improved warehouse response to maintenance requirements for materials and spare parts. Identification of obsolete parts and materials is far easier and far more thorough when computer support is available.

Although many quantitative methods exist for justifying computer-based systems in the maintenance area, many such systems are justified by what is called the "faith, hope, and charity" method. Maintenance management simply has faith that maintenance can be made more effective and can be controlled better if maintenance activities and costs can be measured. Through computers, maintenance management also hopes effective record keeping will preserve effective procedures and the maintenance department will be less vulnerable to loss of key personnel because these procedures are recorded within a computer system. The element of charity exists because the accounting or operations departments may have computers which are not fully utilized and are, thus, available for maintenance-related applications.

Unfortunately, the "faith, hope, and charity" justification technique too often has resulted in installation of systems which were thrown together on a part-time basis by data processing personnel and imposed on the maintenance department in the total absence of any obvious maintenance coverage and/or desire for such systems. The result has been immediate rejection of the system by maintenance personnel and a setback in the maintenance department's acceptance of computer support of any type.

Setting Up an Effective System

As previously mentioned, there are a variety of computer systems being installed in processing plants. These systems can be installed either as "stand-alone" systems or as systems which exchange data with other related systems. Just where the first system is installed depends mainly on where help is most needed—or where computerization would produce the most significant benefits.

With any system, however, there are certain "places to start" which are absolutely vital to system success.

The maintenance department which hopes to realize benefits from computer systems must start with a convinced, dedicated management and recognize that system acceptance in the maintenance department must be earned.

The manager who has a system designed and installed as "something we can try to see how it works out" has wasted a lot of company money. If the maintenance manager is not solidly convinced the contemplated system is needed and if he is not dedicated to its success, then the system is likely doomed to failure or to only partial realization of potential benefits before the first computer program statement is written.

Maintenance management has long recognized that certain management techniques must be used to implement any change. Unfortunately, these techniques are not always applied when the change involves a computer. Communication, participation, involvement, and training all must be used to ensure that need for the system is generally recognized throughout the maintenance department and that the system is accepted by maintenance personnel as a problem solver. One of the more effective techniques for implementing a computerized system is to build upon existing, manual systems in order to permit minimal change in the information input activity even though major improvements are effected in available reports and analyses.

A common misconception is that a computer application requires a large volume of additional routine data. If a good manual system exists for preventive maintenance scheduling, inventory control, or other functions, the computer system often requires no more routine input information. As reports are produced, the volume should be carefully limited to necessary information. Report formats should be developed with the ultimate user's participation. Finally, results should be thoroughly communicated throughout the maintenance organization.

A plant also should be careful to allocate the resources necessary to support the system's implementation effort. Computer applications often require a one-time data entry—such as equipment specifications or material descriptions—which imposes a short-term load on available personnel. These tasks may be assigned to existing personnel or contracted to outside firms. The temptation to use existing personnel on a part-time basis has often proven counter-productive to final system success.

After programming, implementation, and training it is also essential that the system be supported. The new maintenance system's "credibility" among maintenance personnel is extremely fragile during the first few months of its existence. Hardware problems, computer priorities and program "bugs" can be disastrous to system acceptance. Parallel operation of existing manual procedures with the computer system for a period of time has been used to prove the computer system and to demonstrate the improvement in information availability and analysis.

Finally, when implementing a computerized maintenance program, it is important to progress from one system to the next at a speed that will not create confusion or misunderstanding. If multi-system maintenance support is a plant's goal, then a long-time strategy for system implementation is necessary to ensure logical growth compatible with needs (and abilities) of plant personnel. To overcome the "too much, too soon" problem, one major chemical company has designed a modular system for eventual installation at all of its plant sites. The modules are made available to the plants—but not forced upon them. Each plant is encouraged to formulate a long-term strategy for use of these systems and to use the techniques of communication and personnel involvement in implementing systems at its own pace. This modular, but preplanned concept of computer system installation at plant sites permits growth into a totally integrated system, even if years separate the installation of individual systems.

Manuals Prepared

To accomplish the preventive maintenance control system, in a large multi-plant environment, manuals are prepared by technical specialists listing the specific maintenance tasks for each equipment item at the operating plants. The manufacturer's recommendations and a plant's own experience are considered in determining the extent of coverage for maintenance procedures and frequency. Differentiation between running maintenance and shutdown maintenance is also made. As operating requirements change, these procedures are improved and updated and revised pages are issued to keep the manuals current. Needless to say, these "manuals" are kept and updated on computers. Paper printouts are produced, as needed.

Maintenance tasks range in frequency from daily shifts to several years, depending on the equipment type, its loading, and serviceability. Maintenance tasks are monitored by the staff at the home office and passed through the data processing equipment that performs the following functions:

1. Prints schedules and feedback cards.
2. Digests feedback information on completed or rescheduled maintenance.
3. Prints reports showing tasks performed or deferred.
4. Calculates percent compliance.
5. Accumulates actions taken and total time expended.
6. Prints addenda to the schedule and addenda feedback cards for uncompleted tasks.

The percent compliance to the schedule for each plant is separated into "normal" and "downtime" categories. This separation permits evaluation of the schedule portion controlled by the plant manager—that portion he can do only during an emergency or planned shutdown. Central management is thus automatically given the opportunity to pass judgement on the desirability of rescheduling "downtime" PM items. Compliance reports are issued monthly and sent to plant managers and the home office.

At the beginning of each month, the computer prints work schedules for all maintenance tasks due in the particular month. These schedules cover machinery and equipment for each plant in the system. Copies are sent to each plant manager and to the home office staff. The schedules list all the PM tasks that must be done during the coming month. An advance schedule of downtime tasks, covering the next three months, is also included. This advance notice assists the plant manager in planning downtime task performance in case an emergency shutdown occurs. The computer schedules are accurate because maintenance task timing is based on the date they were last performed and the frequency assigned. Many international design contractors offer maintenance services that integrate other aspects of asset management (Figure 2-1).

Along with the schedules, the data processing equipment prints out a data log to feed back completion or rescheduling information. This mode of tracking is used by plant maintenance personnel to record actions taken, time expended, date completed, and any pertinent remarks concerning findings when the task was done. The log issued to the plants at the beginning of the month must be answered on the last day of the month.

Performance Reports

The preventive maintenance performance report shows the tasks which are performed on time, performed late, are rescheduled, or remain in a deferred state. It allows the plant manager and home office management to evaluate performance. The number of tasks scheduled, rescheduled, and completed is listed at the end of the report along with the compliance percentages and the total time in hours for normal and downtime categories. Preventive maintenance performance reports are generated by any of the commercially available CMMS software programs.

Preventive maintenance tasks that were not completed as scheduled are summarized in addenda to the schedule and sent to plant managers as reminders. The addenda are printed monthly by the computer, based on noncompliance of tasks previously scheduled. Deferred tasks continue to appear on these addenda until completed. A set of feedback requests accompanies the addenda for the reporting of work completed.

Data reported via the feedback requests are accumulated by the computer. This includes time expended for each maintenance task and the number of times actions such as cleaning, filling, lubricating, overhauling, or testing are performed. A report of accumulated maintenance statistics is produced by the computer and is used by the operations management to make an audit of work done.

Breakdowns Reduced

Since the incorporation of this system at large multi-plant corporations there has been a very definite trend of reductions in breakdowns. This allows nearly all maintenance work to be performed on a planned basis

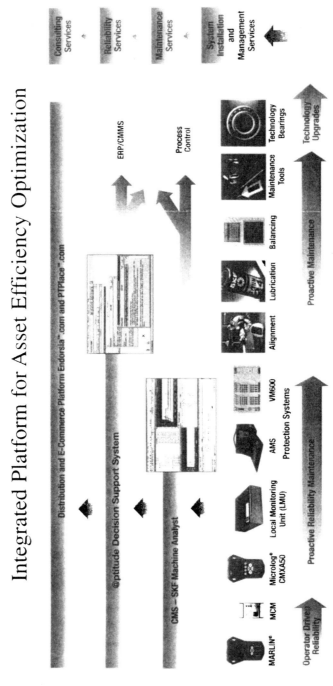

Figure 2-1. Maintenance as part of Asset Management. (Source: SKE Publication 51605 2003)

and on an optimized time schedule to provide the best possible on-stream factor.

In the actual performance of planned maintenance work, there can be several approaches. One approach is to have complete in-house maintenance and supervisory ability at each plant with occasional subcontracting for large peaks. A second is to subcontract all maintenance work, thus eliminating the requirement for maintenance personnel at individual plants. Each system has obvious advantages and disadvantages depending on plant size, location relative to other area plants, etc. Recognizing good planning and skilled supervision as the key elements in low cost major maintenance, an intermediate approach has been taken at some plant locations. Some of the main considerations of this approach are:

1. The plant manager is fully responsible for normal maintenance. Each plant employs an absolute minimum number of resident maintenance people consistent with the day-to-day requirements, plus a normal backlog of work which can be accomplished while the plant is running.
2. The responsibility for planning major maintenance and turnarounds would come under the jurisdiction of a corporate maintenance manager working in close conjunction with the plant managers. His group of mobile planners, technicians and maintenance staff represent a well-trained nucleus for supervising major maintenance work to supplement the normal plant maintenance group. These individuals travel from plant to plant as required. This makes it unnecessary to have skilled supervision at each facility capable of handling planned major maintenance work. By scheduling the total corporate maintenance requirements, this same skilled group can handle a large work volume at a number of facilities at overall lower cost and inject a higher than normal experience factor into the supervision aspect of maintenance. The major maintenance work is performed using standard critical path scheduling, manpower and tooling planning, cost control procedures, inspection reports, etc.
3. Supplementary maintenance manpower is provided by using carefully selected local contractors. However, by having a well-trained nucleus of supervisory and maintenance personnel available from within the company, overall manpower efficiency is kept at a higher level than normal, thus resulting in lower costs and reduced outage time. Operators are used where possible during turnarounds which involve plant shutdown.
4. The travel and living costs for the flexible, rotating group of maintenance technicians and planners is a minor cost factor compared to the more efficient use of personnel and reduced outage time. In many

cases, the technician and central maintenance group are geographically located near key facilities, since this is where they spend most of their time.

Discussion of any maintenance concept is incomplete without including a method of spare parts control. The goal of an effective spare parts program is to keep the investment in capital spares to a minimum without seriously jeopardizing the plant onstream factor, and administering the spare parts program at the lowest possible cost. Only experience, after an extended operating period, will determine the adequacy of decisions made in this regard.

The spare parts program at a multi-plant corporation should most certainly be administered on a centralized basis. The commonality of equipment makes this a prerequisite for low total spare parts investment. The same central mechanical engineering organization responsible for monitoring field mechanical problems is also responsible for the initial selection of spare parts and the approval for reordering major spares. Initial spare selections are based on equipment manufacturer recommendations, operating experiences, and careful analysis of what is in existence. To obtain the best possible price, major spare parts are negotiated as part of the original machinery or equipment purchase.

Central Parts Depot

Specific items not common to other facilities and small, normal spares are maintained at individual facilities. Certain major components common to more than one operating facility and some parts showing high usage are stocked at a centrally located parts depot. This concept allows for a lower total investment in spares. Since spare parts handling, packaging, and long-term storage are so critical and require specialized knowledge, it is necessary to provide this capability at only one location. It is possible to ship spare parts from this depot on a 24-hour, seven day a week basis. Transportation arrangements normally keep the total shipping time to less than eight hours. With most maintenance work performed on a planned basis, actual plant startup delays due to the central stocking depot concept are rare.

By careful analysis on ordering of initial spares and the central depot concept, major corporations have been able to lower the investment in spares (expressed as a percentage of equipment investment) from approximately 5 percent a few years back to under 3 percent on new plants.

To keep the administration of replacement spare parts at a minimum cost, a central data processing system has been established. As parts are used, data are sent to the corporate office for computer input, which automatically generates the parts replacement purchase order. The authorized parts level is periodically and automatically reviewed to prevent reordering of parts with a low turnover. A block diagram showing the spare parts support system is shown in Figure 2-2. A composite listing of all parts in the system is available at the corporate office to facilitate the identification of parts interchangeable with other facilities.

Plant Engineering

Plant engineering referred to here includes those process and mechanical services required for monitoring plant operations, the prompt resolution of special plant problems, normal debottlenecking, and special engineering assistance as required in performing maintenance work. A

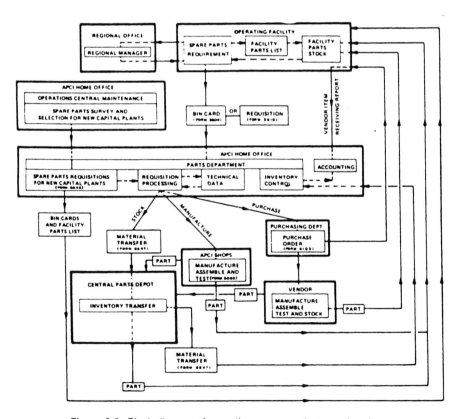

Figure 2-2. Block diagram of operations spare parts support system.

centralized organization of specialists within the operations department is charged with this responsibility for the network of facilities. Major engineering design and construction work related to new plants and plant expansions is handled by a separate corporate engineering department and will not be discussed in this section. The corporate engineering department is also available for special help to the operations department.

As indicated previously, the interjection of technical specialists for the quick and efficient resolution of problems was one of the key points to a centralized system of engineering services, and the engineering staff at individual plants has, therefore, been kept minimal. In some plants, a certain need for minimum on-site staffing of chemical and mechanical engineers is required for day-to-day problems, but here special emphasis is placed on coordinating problem issues with the centralized staff.

The normal day-to-day minor plant and equipment problems are handled by the plant manager with his staff sized on this minimum basis. The centralized engineering services then encompass these major responsibilities:

1. Aid in resolving specific equipment and process problems as they arise based on information gathered through monitoring techniques or through plant manager request. Suitable engineering or technician help is provided, including site visits when necessary.
2. Getting special services from the corporate engineering department to obtain maximum benefits from understanding the design concepts and to provide a valuable source of field problem feedback for future design considerations. This would also include obtaining recommendations from equipment manufacturers and outside consultants.
3. Monitoring process performance of all facilities including overall production, utility efficiency, and gathering sufficient data to generally identify problem areas. Each facility requires a detailed analysis to determine the minimum key data required. Some monitoring is performed on a daily basis and is transmitted to the home office by e-mail. Other monitoring is performed on a weekly or monthly basis. The computer is used to perform routine calculations required for certain evaluations to determine minimum operating costs. Thus, the computer can be economically used and up-to-date process monitoring and reporting allows for prompt management attention to plant problems.
4. Monitoring of machinery and equipment performance. An example would be the periodic collection of data on all large compression equipment to determine stage efficiencies and intercooler perfor-

mance since utility costs represent a significant portion of the total operating costs in most petrochemical or air separation plants. Where possible, the performance evaluations are translated into dollars so that business decisions can be made. Figure 2-3 shows a simplified computer program used for evaluating compressor efficiencies where the results are translated into cost inefficiencies in dollars/day. Of particular interest, also, is the increased use of field monitoring methods as a key element in evaluating equipment performance. In many cases, this represents the actual guidelines for determining frequency of inspections.

5. Monitoring of plant and equipment performance by regular visits of process and mechanical specialists to the facilities. This provides the necessary final tie of the centralized group with plant personnel.
6. Establishing safety, technical, and operating procedures to provide conformity to all plants.
7. The organization of corporate technical and training sessions for plant personnel based on the management evaluation of need.

Summary

The system of plant maintenance and engineering services outlined has been successful in achieving exceptionally low maintenance costs for a nationwide complex of plants.

The overall maintenance system described herein has allowed a steady reduction in plant personnel with corresponding savings. Actual maintenance costs have steadily decreased as a percentage of original investment costs without any allowance for labor and material escalation. With these factors taken into account, the total maintenance cost reductions are indeed significant.

The publishing of monitored data on plant performance and preventive maintenance compliance has allowed for prompt management attention to problems and has stimulated a noticeable spirit of competition among the plants. Needless to say, it also serves as a valuable method of evaluating plant manager performance.

Although the system described herein may not be directly applicable to every large chemical complex for reasons of size or process type, the general trends toward computerization, sophisticated equipment monitoring methods and processes, cheaper transportation and communication costs, are indicative of increasing advantages obtainable in the future from centralized concepts in plant maintenance and engineering for multi-plant corporations.

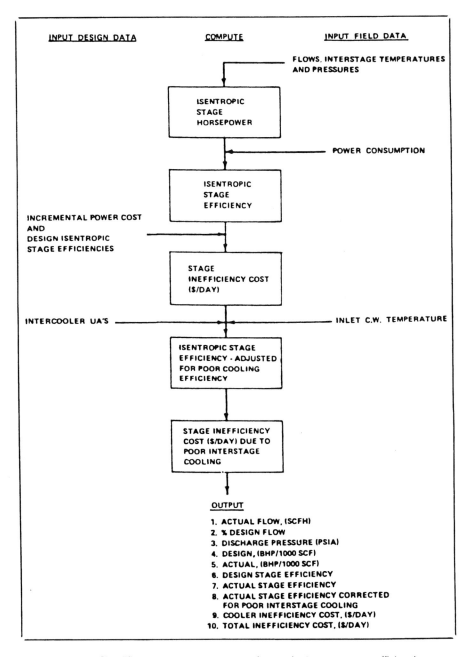

Figure 2-3. Simplified computer program used to evaluate compressor efficiencies.

Machinery Maintenance on the Plant Level

With this overview behind us, let's get back to the machinery engineer's concerns on an individual plant level. As he of course knows, modern turbomachines can run reliably for many years if designed, applied, and operated correctly. As of 2004, the periods between inspection and overhaul, commonly called "turnaround," on machines in clean, noncorrosive service can exceed eight years. It is easy to see how plant personnel may have trouble remembering just how much time and effort are required to successfully plan and execute an overhaul of a particular piece of equipment. A proper turnaround involves preplanning and teamwork among plant technical, warehouse, purchasing, safety, operations and maintenance forces, as well as with the original equipment manufacturer and other noncompany sources. In the case of sophisticated problems, consultants and laboratories may also come into play to restore machinery to a reliable, smooth-running, and efficient operation. Managing these resources and documenting the results presents a real challenge to those assigned the task of heading up the overhaul effort.

This segment of our text deals with turnaround management principles that must be understood and considered by maintenance personnel on the plant level.

Assignment of Qualified Personnel*

Major machinery overhauls require not only early planning input, but also early designation of qualified personnel to execute planning and related tasks. High quality machinery overhauls can be more consistently achieved if machinery expertise is directly applied.

Process plants subscribing to this approach define, in specific outline form, the responsibilities of supervisory and staff personnel involved in turnaround (T/A) of major machinery. The outline explains the various job functions involved in T/A activities and identifies the timing and scheduling requirements which precede the actual shutdown.

The assignment of qualified personnel starts with the designation of an *overall T/A coordinator* no later than nine months before the scheduled shutdown. As of this time, the plant's *senior machinery specialist* is required to maintain formal communications with the T/A coordinator.

---

* Based partly on material prepared by John D. Houghton and originally presented at the Seventh Turbomachinery Symposium, Texas A&M University, College Station, Texas (December 1978). Adapted by permission. Updated in 2004.

Among other duties, the T/A coordinator screens and approves the planning and scheduling efforts of maintenance or contractor personnel involved in major machinery T/A's after the senior machinery specialist has had an opportunity to review these efforts. Working hours and team composition are to be determined jointly by the T/A coordinator and senior machinery specialist.

Experience shows that a *spare parts and materials coordinator* should also be designated no later than nine months before scheduled shutdown. This person will generally be responsible for implementing spare parts procurement requested by the senior machinery specialist and mechanical supervisors. He will be required to forward up-to-date listings of parts on hand to personnel requiring this information.

The *plant senior machinery specialist* is generally charged with responsibility and authority to direct planning and execution of the machinery portion of the T/A. His background and experience should make him uniquely qualified for this job, and as senior resident expert he would be thoroughly familiar with all machinery affected by the planned T/A.

A key job function is to be fulfilled by *turbotrain T/A engineers*. On major machinery T/As it was found essential to have one or more of these engineers assigned the responsibility of verifying the quality of execution of all machinery overhaul tasks. If a given plant does not have enough machinery engineers to man the job around the clock, affiliate loan or contractor engineering personnel should be brought in for the duration. The specific responsibilities of turbotrain T/A engineers have been described as those of a machinery advisor and quality control person who augments the *mechanical supervisor* and reports to the senior machinery specialist for work direction and guidance. His responsibility and authority extends from machine inlet nozzle to machine outlet nozzle and includes lube and seal oil systems. His work begins after all required blinds have been installed and ends after every item of machinery work is complete. He will then turn over the machinery to the mechanical supervisor for removal of blinds.

A turbotrain T/A engineer typically has a degree in an engineering discipline (preferably mechanical engineering) and has had practical machinery engineering experience for a minimum of five years. His past assignments should have included active participation in major machinery erection, commissioning, testing, operation, troubleshooting, and repair. He normally performs work which involves conventional engineering practices but may include a variety of complex features such as resolution of conflicting design requirements, unsuitability of conventional materials, and difficult coordination requirements. His normal sphere of activity requires a broad knowledge of precedents in turbomachinery design and a good knowledge of principles and practices of materials technology, metalworking procedures, instrument-electrical techniques, etc. This person should be a hands-on engineer whose past performance will have established his reputation as a resourceful, highly dependable contributor, a self-starter with sound judgement.

Timing and Basic Definition of Critical Pre-Turnaround Tasks

Senior Machinery Specialist

Immediately following the designation of a T/A coordinator (approximately nine months before T/A), the senior machinery specialist starts to interface with planners, designated turbotrain T/A engineers, maintenance or mechanical supervisors, and the T/A coordinator. From then on, the following action items and timing will be typical of this function:

Nine Months Before T/A:
- Work list items assembled by maintenance are forwarded to the senior machinery specialist for review purposes. A typical work list page is shown in Figure 2-4.
- He receives the most probable work zone outline for review and comment. (For description of work zones, refer to Volume I of this series.)
- The senior machinery specialist requests up-to-date tabulation of spare parts presently on hand for major machinery trains. The spare parts coordinator must provide this tabulation in a format similar to Figure 2-5.

Eight Months Before T/A:
- The senior machinery specialist determines which replacement parts are required for major machinery T/A
- He issues written requests for the spare parts coordinator to place a "hold" on selected parts, locally stocked parts, or to obtain these from the corporate central storage location
- He specifies inspection requirements for existing key spare parts and any additional key spares to be procured

Six Months Before T/A, the senior machinery specialist must:
- Commence refresher training for mechanical supervisors, craftsmen, and designated contract personnel
- Arrange for vendor assistance
- Review the machinery T/A schedule

	IRD WORKLIST
	Model No.: 9EH 1B MH Steam Turbine Package No.: 052-02 Work Order No.: 6305
	Equipment No.: CT-01

20. Check and record clearances as per Machinery Reliability Program. Use forms in INSPECTION CLEARANCE FORM section.

21. Disassemble governor end and exhaust end packing housings (See Procedures-Packing housings). Take top clearances on both upper halves.

22. Check and Record clearances as per Machinery Reliability Program. Use forms in INSPECTION CLEARANCE FORM section.

23. Remove upper halves of Radial bearings and thrust bearing assembly. Visually inspect, take clearance checks on each bearing, then remove to make rotor ready for removal. (See Procedure-Radial Brg. & Thrust Brg. Assembly.)

24. Check and record as per Machinery Reliability Program. Use forms in INSPECTION CLEARANCE FORM section.

25. Remove rotor from lower half of turbine case. (For proper rigging see Procedure-Rotor Assembly.)

26. Remove all upper & lower half packing, oil seals, etc. mark for position, clean and cover until ready for re-installation. (See Procedure-Interstage Packing).

27. Remove upper and lower half diaphragms, clean, inspect/repair non-destructive test and cover until re-installation (See Procedure-Diaphragms).

28. Remove lower half radial bearings, and thrust assembly housing. (See Procedure-Radial Brgs. & Thrust Brg. Assembly).

29. Clean lower and upper halves of turbine casing and prepare for re-installation of diaphragms.

30. Clean inside of relay box, lower half bearing housings .

31. Clean, inspect/repair, and non-destructive test nozzles.

32. Install terminal strips inside of each end bearing housing for new therma-couple wires.

33. Repair 660# steam extraction leak at flange.

34. Repair horizontal joint leak on CT-02.

35. Install new coupling end oil seals.

36. Install new revised ground brushes.

37. Install new governor.

Page 3 of 6	**MACHINERY RELIABILITY PROGRAM**

Figure 2-4. Typical worklist page for major compressor turnaround.

MWO.# 63051	PKG.# 3050-0 6	MATERIAL LIST EC-01		PAGE 1 OF 2	
ITM	QUAN	UNIT	SYMBOL NUMBER	DESCRIPTION OF MATERIAL	PCH/SP ORDER #	STORAGE LOC.
1	2	EA		Seal, Labyrinth, Part #876070-1 Item #2		
2	1	EA		Wind-back Seal C.C.W.Rotation Part #876142-1 Item #4		
3	40	EA		Spring Conical Part #845094-1 Item #5		
4	2	EA		Contact Seal Sleeve Part #876063-1 Item #7		
5	2	EA		Ring, metal-backed carbon Part#876068-3 Item #8		
6	2	EA		Seal Sleeve Outboard Part #876055-3 Item #10		
7	2	EA		"O" Ring Part #P700D459 Item #13		
8	2	EA		"O" Ring Part #P700D282 Item #14		
9	32	EA		"O" Ring Part #P700D115 Item #15		
10	2	EA		"O" Ring Part #P700D449 Item #16		
11	8	EA		"O" Ring Part #P700D021 Item #17		
12	2	EA		"O" Ring Part #P700D279 Item #18		
13	8			Seal Part #P832A045 Item #23		
14	1	EA		Wind-back Seal C.W. Rotation Part #876396-1 Item #25 FOR ITEMS 1-14 REFER TO DRAWING #X-868846		
15	2	EA		"O" Ring Part #700Y855 Item #30		

REMARKS

PREPARED BY: *Jack M. Perkins*   DATE: 9/19/81

Figure 2-5. Typical spare parts tabulation.

Three Months Before T/A:
  • He meets with designated turbotrain T/A engineers for detailed brief-ing and solicitation of additional input

Two Months Before T/A, the senior machinery specialist should:
  • Review final (detailed) T/A plan for each train
  • Verify that work procedures are either available or being produced

One Month Before T/A, his tasks include:
  • Review of detailed information package for each train. This package will have been assembled by the mechanical supervisors and plan-ners, as we will discuss later
  • Review of bar charts prepared by maintenance, and include these in package

Finally, during the actual T/A, the senior machinery specialist must:
  • Participate in daily T/A meeting
  • Verify that work procedures are followed
  • Verify that data are taken and logged in as required
  • Address deviations from plan
  • Review test runs
  • Review updated start-up instructions

Turbotrain Turnaround Engineers

Approximately three months before the scheduled shutdown, desig-nated turbotrain T/A engineers meet with the senior machinery specialist for detailed briefings and reviews of machinery T/A organization, proce-dures, and preparations. Additional responsibilities are as follows:

Three Months Before T/A:
  • Prepare detailed information package for each turbotrain. A typical table of contents for one such package is shown in Figure 2-6. Sample material making up the package is shown in later illustrations.

Two Months Before T/A:
  • Review final (detailed) T/A plan for each train
  • Verify that work procedures are satisfactory

One Month Before T/A:
  • Review detailed information package for each train
  • Review bar charts prepared by and forwarded by the senior machin-ery specialist

```
LIST OF CONTENTS FOR:        LC-01

         Elliott Centrifugal Compressor 70M 6I


MWO#:  6305                    Package No.:  051-02
```

```
         _____  COVER SHEET

         _____  SAFETY REQUIREMENTS

         _____  WORKLIST

         _____  JOB PLAN

         _____  BAR CHART

         _____  MATERIAL LIST & EQUIPMENT CATALOG

         _____  TOOL LIST

         _____  SPECIAL SERVICE & EQUIPMENT CATALOG

         _____  BLIND LIST

         _____  PROCEDURES
                     ALIGNMENT
                     BENDIX FLEXIBLE DIAPHRAGM COUPLING
                     INDIKON MONITOR SYSTEM
                     RADIAL BEARING ASSEMBLY
                     THRUST BEARING ASSEMBLY
                     ISO-CARBON SEAL ASSEMBLY
                     COMPRESSOR CASING
                     ROTOR ASSEMBLY
                     COMPRESSOR INTERNALS
                     INTERSTAGE LABYRINTH SEALS
                     DESIGN CLEARANCES

         _____  FIELD NOTES

         _____  CLEARANCE INSPECTION FORMS

         _____  COST ESTIMATE

         _____  CRITICAL PATH METHOD CHART

         _____  PHOTOGRAPHS
```

MACHINERY RELIABILITY PROGRAM

Figure 2-6. Table of contents for typical compressor turnaround package.

During T/A:
- Participate in daily T/A meeting
- Verify that work procedures are followed
- Review condition of used parts as they are removed from the machine
- Verify that data are taken and logged in as required
- Take photographs and dictate observations into tape recorder
- Verify the installation is mechanically correct

- Resolve deviations from plan
- Verify machinery alignment
- Supervise test runs
- Restart per startup instructions

Mechanical Supervisors/Planners

Planners, and also mechanical and maintenance supervisors provide machinery related data and support to the senior machinery specialist and turbotrain T/A engineers involved in planning and execution of turbotrain turnarounds.

Again, nine months before the scheduled shutdown for a major machinery T/A, planners and mechanical supervisors will be given initial guidance on anticipated duties and responsibilities prior to and during the actual T/A. From then on, typical action and timing would be:

Nine Months Before T/A:
- Maintenance personnel forward machinery-related work lists to senior machinery specialist for review
- The most probable work zone outline (see pages 394–400, Volume I third edition, 1998) is drawn up and forwarded to the senior machinery specialist
- The mechanical supervisors instruct the spare parts coordinator to assemble up-to-date tabulation of spare parts presently on hand for major machinery trains. After review, they forward the tabulation to the senior machinery specialist.

Eight Months Before T/A:
- The spare parts coordinator and maintenance personnel receive the senior machinery specialist's request to:
  1. Place a "hold" on selected parts
  2. Order additional spare parts
  In response, they issue purchase orders for additional replacement parts.
- Next, maintenance supervisors commence dimensional checking of selected (existing) spare parts per request made by the senior machinery specialist. The results should be documented within eight days.
- Dimensional checking has frequently shown serious discrepancies in parts designation, dimensional configurations, and tolerances. These must be identified early if a smooth turnaround is to result.

Six Months Before T/A:

- Maintenance personnel arrange for vendor assistance
- Maintenance personnel also work up a more definitive work zone arrangement and commence tabulation of detailed work list for each zone
- Maintenance or technical department personnel witness check balancing of major turbomachinery rotors
- Assemble tools and identify missing tools
- Arrange for scaffolding, etc.
- Forward data to senior machinery specialist regarding status of spare parts ordered six months earlier
- The planner should now provide final work zone arrangement and detailed work list for each

One month before T/A, maintenance planners or mechanical supervisors provide bar chart diagrams for machinery-related T/A work.

- They participate in a meeting with the senior machinery specialist and designated turbotrain T/A engineers

One week before T/A, mechanical supervisors commence meeting with designated turbotrain T/A engineers for briefings on matters relating to machinery work.

Specific Preparation and Planning

When preparing for an overhaul of a major piece of turbomachinery, it is important to know as much as possible about the machine and why it needs to be taken out of service. There are several obvious sources of information, including the operating and maintenance personnel, the equipment file folder and the vibration history record. If sufficient information is not found in the file folder, which is all too often the case, this fact should reinforce the resolve to do a proper job of documenting the planned overhaul in a special manual or "machinery T/A package."

Before proceeding, one question usually arises: Is a complete overhaul really necessary? To properly answer this question, you will need to evaluate the symptoms. Has the vibration steadily increased over a long period of time or have you witnessed a step change? What does an analysis of the vibration signature reveal? Has the performance gradually fallen off or taken a dramatic drop? Problems such as a locked gear coupling or soluble deposits inside the machine can sometimes be corrected without opening the machine and at a considerable savings of time and effort.

Another obvious source of information is the manufacturer's manual. The good ones provide detailed, step-by-step instructions with clear illustrations; others assume prior knowledge or place undue reliance on the manufacturer's service representative. Consequently, it is prudent to develop procedures, installation instructions, or even detailed commissioning instructions for inclusion in the turnaround package. See Figure 2-6 for typical requirements.

Since a detailed manual is often too bulky for constant reference, we might reduce portions of it to a critical item list. Certain steps, clearances, and methods are vital to doing a good job. These items should be summarized and kept for ready reference during the course of the overhaul. In fact, one complete turnaround package should be on the compressor platform and should be used while the job is in progress.

It is important to assign the responsibility for the overhaul to one person so that conflicting positions do not occur. As indicated earlier, we recommend the appointment of a turbotrain T/A engineer to oversee the job and believe that all decisions and compromises should be made by him. He should be responsible for the engineering coverage, interface with the maintenance and operating departments, interface between user company and original equipment manufacturer, and for documenting the overhaul. It is a responsible assignment, one that requires judgment, maturity, and initiative on the part of the engineer. It is strongly recommended that the turbotrain T/A engineer assume responsibility for the development of data packages and checklists, some of which are shown later in sample form, but which must of course be adapted to fit a specific machine or turnaround situation.

Safety

Work safely. Be sure all power is off, blinds in, purging procedures followed, etc. A prework safety item checklist is strongly recommended, as is a list of all blind locations. The latter item is important at the beginning of a job to ensure all necessary lines are secure, and, at the end of the job, to check off the removal of all installed blinds. Failure to install or remove a blind at the appropriate time could lead to a disaster.

Account for tarps and "welding blankets." These are known to have been left in piping, only to later be ingested into equipment. The consequences of these oversights have ranged from costly to catastrophic.

It is important to establish teamwork and proper communication among the operations, safety, engineering, and maintenance personnel at the start of the job so that cach can fulfill his role in the total effort. A list of key players and where or how they can be reached during the overhaul period should be made available at the start of the job.

Planning

If this is a planned overhaul, as opposed to a forced outage, so much the better. Take full advantage of the planning period to make a visual inspection of the machine before the shutdown. Pay particular attention to the condition of the foundation, anchor bolts, piping, instruments, and look for leaks. It is a good idea to keep an "evergreen" list of required maintenance items in the equipment folder. Encourage personnel who frequently go on the machinery deck to make written note of any problems.

Take a final check of vibration, performance, alignment, and mechanical health data just prior to the shutdown. A small shirt-pocket size tape recorder or palm-size computer with voice recognition software is particularly useful to record notes; it leaves the hands free to manipulate instruments, etc. The data can then be transferred to spreadsheets back in the office. If you are working a forced outage, the most recent set of data will have to do. Compare the most recent information to previous readings and develop a list of anticipated problems. Translate all of this information into a detailed job plan per Figure 2-7, or as available from CMMS software programs.

In our experience, machines are normally shut down for overhaul due to fouling (restricted performance); excessive vibration (ingestion of a liquid slug, a loose piece of hardware, the failure of a mechanical component or misalignment); misoperation (surge, lube oil supply failure, etc.); or when the whole process unit is shut down for a T/A. In general, we do not open machines that are running satisfactorily just for inspection. At every convenient opportunity one should inspect externally accessible components, such as couplings, and also check items such as rotor float and shaft alignment, and all tripping devices and general instrumentation.

In the case of steam turbines, the overspeed trip bolt and the steam trip and throttle valve have proven to be the least reliable—and yet most important—safety devices in the train. A check of these two components is mandatory during major shutdowns, and checks should be made at every other opportunity. We believe these are the most important checks performed during a shutdown. In addition, one should "exercise" the trip and throttle valve weekly by moving the stem in and out manually several turns on the hand wheel to preclude the buildup of deposits that would prevent the machine from tripping during a shutdown condition.

	JOB PLAN: Overhaul/Inspection and Modifications		MWO#:	63051		
			PKG#:	3051-02		
	EQUIPMENT NO.: LC-01-B		PAGE 3	OF 8		
SEQ	DESCRIPTION		CRAFT	NO.	HRS	MH
38	Turn upper half of casing over with the aid of 80-ton		MW	3	2	
	crane, load onto truck, take to cleaning site and offload.					
39	Make sure all diaphragms and guide vanes are properly		MW	2	2	
	marked for position in compressor case.					
40	Remove all anti-rotation locks on shaft and impeller		MW	2	2	
	labyrinths and diaphragms and guide vanes.					
41	Remove upper half labyrinths and make ready for storage		MW	4	2	
	or dispose of properly.					
42	Order DROTT/"Go Devil" crane and operator to pull upper		DROTT	1	4	
	half diaphragms at cleaning site.		EQ OPER	1	4	
43	Remove upper half diaphragms with DROTT; clean, inspect		MW	4	4	
	and make ready for storage.					
44	Remove upper halves of seal housing end walls on each end		MW	4	2	
	of compressor.					
45	Pull lower halves of end walls up against casing in their		MW	4	1	
	original position on each end.					
46	Re-assemble radial bearings and thrust bearing to make		MW	4	2	
	compressor ready for rotor wheel clearances.					
47	Push rotor against active side of thrust and take all		MW	2	3	
	axial and radial clearances on diaphragms, shaft and im-					
	peller labyrinths, etc.					
48	Check and record clearances.		MW	1	1	
49	Dis-assemble radial bearings and thrust bearing to make		MW	4	2	
	ready for rotor lift.					
50	Using nylon slings or soft rope chokers, lift rotor out		MW	4	1	
	of casing, keeping it level to lower half of case.					
51	Lower rotor downstairs in crane bay, put in rotor stand.		MW	3	1	
52	Order truck and trailer and driver to take rotor to		FLATBED	1	1	
	cleaning area.		DRIVER	1	1	
53	Load onto truck, take to cleaning site and offload.		MW	2	1	
	Clean, inspect and make ready for storage; record data.		MW	2	2	
55	Order flatbed and driver to take cleaned and preserved		FLATBED	1	2	
	rotor and diaphragms to final plant destination.		DRIVER	1	2	
			MW	2	1	

Figure 2-7. Sample sheet of turbomachinery turnaround job plan.

Consider an example of the penalty associated with the failure of a trip circuit at one plant. A 10,000 horsepower, steam turbine-driven compressor train failed to trip during a condition which had caused the compressor to fill with liquid while at full speed. The resulting loads led to catastrophic failure of both the compressor and coupling and allowed the steam turbine to overspeed to destruction. The repair bill for parts and labor came to well over $1,000,000. The cause of the wreck was eventually traced to the buildup of deposits on a brass piston in the hydraulic shutdown system which was in a hot dead-ended oil circuit. The heat caused the oil to decompose over a long period of time, in turn causing the piston to stick. As a result, the plant now checks trip circuits more frequently. They have also installed redundant electronic backup trip devices on large equipment trains.

Spare Parts

The time to check spare parts is not in the middle of the night following an emergency phone call from the operating department manager. Most large companies have some degree of computer control on the warehousing and reordering of spare parts. But how many times have you been lied to by a computer? There is no substitute for a hands-on check of parts by a knowledgeable individual. Part numbers must be checked because the item on the shelf is not necessarily the one you expected to find.

Many major plants allocate special boxes to major machinery spare parts storage. The boxes have individual compartments for labyrinths, seals, bearings, etc. A list on the lid details all parts inside, their location in the box, the manufacturer's part number and the company's stock number. Once filled, the box is sealed and stored in the usual manner. During an overhaul, the box is taken to the field and some or all of the parts consumed. The box is then returned to the warehouse with a list of consumed parts to be replaced. A computerized call-file system should be used to keep tabs on rotors that are out of the plant for repairs and delivery of other critical spare parts.

When checking spare parts, it is important to recognize that not only must the part be the right size, it also must be in good condition. Handling and improper storage, as well as deterioration with time, are a few of the hazards associated with a warehousing operation. A nicked O-ring or a carbon seal face out of flatness could require a second shutdown to correct the problem. The use of an optical flat, a set of micrometers, and a knowledgeable pair of eyes can be invaluable in detecting a defective part.

Also, remember that just because the part came from the factory, doesn't necessarily mean it is the right one for your machine. While equipment manufacturers have various quality control procedures, they too rely on human beings, and errors do occur. In addition, some parts have a finite shelf life (case split line sealant is an example) and must be fresh when the time comes to use them.

This is also the time to check on the availability of special (custom fabricated) tools. These should be kept in a separate box, inventoried at regular intervals, and generally treated as a valuable spare part or essential resource. Delaying an overhaul for several hours to fabricate a special seal nut wrench is time and money wasted. Alignment brackets and coupling "solo" plates fall into this category.

The Spare Rotor

By far the most critical single spare part is the spare rotor. Most companies purchase the spare rotor at the time the machine is purchased and require a four-hour mechanical test to ensure integrity prior to acceptance of the machine. It would be prudent to check the spare rotor after every transport event. This means a runout or rotor bow check upon receipt from the manufacturer, as well as a check of the preservative used for long-term storage. A runout check is also performed at the time the rotor is check-balanced and prepared for installation. Be sure to obtain a rotor runout diagram and balance report at that time.

Rotors of all sizes are often stored vertically in a remote temperature controlled storage building as shown in Figure 2-8. If a user opts for horizontal storage, the rotors must be placed on substantial stands and should be turned 180° two to four times a year. These stands must employ rollers rather than lead or Teflon® material at the support points. In cases where sheet Teflon® is placed between the storage cradle and the rotor there is some risk of filling the microscopic pores of a shaft journal which could prevent the formation of an adequate oil film on startup and could cause bearing failure.

Rotors must, of course, be handled with great care. Nylon slings should be used to prevent damage and all lifts should be made under the watchful eye of a competent individual. Never hesitate to call a halt to a lifting operation if the possibility of damage exists. You are being paid to look out for the company's interests and a rotor worth $200,000 to $1,000,000 or more is well worth a lot of care and concern. The rotor must be slung so that it is horizontal and its center of gravity is located under the hook, and it must be moved very slowly. Consider your vibration monitoring probes when removing a rotor from storage in preparation for installation. Record the rotor's serial number and verify that it is not positioned in a way that will interfere with a thrust position eddy current probe. Some

Figure 2-8. Spare rotors must be removed from storage and cleaned and inspected prior to the turnaround.

users also report success with degaussing and/or micropeening techniques to minimize electrical runout in the areas viewed by the radial eddy current probes. Others report some success rolling the rotor on a balance stand with the areas under the probes directly on the balance rollers.

Diagrams

A critical dimension diagram (Figure 2-9) and associated tabular records (Figures 2-10 through 2-12) have proven invaluable in the middle of the night during a complicated overhaul. A critical dimension diagram is a tabulation or sketch recording critical data such as bearing and labyrinth clearances, rotor float, seal clearances, coupling advance, coupling bolt torque, etc. The document must clearly show maximum and

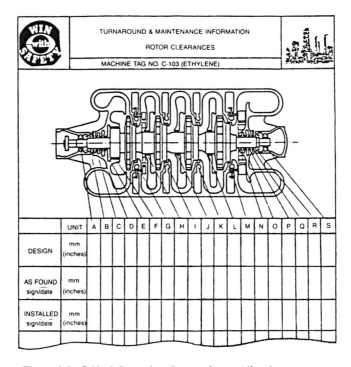

Figure 2-9. Critical dimension diagram for centrifugal compressor.

minimum values, as well as spaces for "as found" and "as left" conditions. Any warning notes such as internal bolts, left-hand threads, or other critical steps should be clearly flagged on this sheet. Clearances should be properly labeled as to diametral or radial, metric or English units, to avoid confusion.

An alignment diagram, as shown in the chapter on machinery alignment, complete with estimated thermal growth and desired readings, is mandatory. This should be available from previous alignment work. If it is not, and if reverse dial indicator alignment techniques are not well known and practiced at your plant, we would strongly recommend implementation of such a program. The techniques and procedures have been the subject of many papers.

The records and documents described thus far have proven to be time saving and hence, money saving, and are well worth the effort. Another useful item involves preplanning the allowable limits on the desired shaft position. It is impractical to expect the field crew to place a compressor or turbine in the exact position as shown on the alignment graph. If allowable limits are known in advance (not necessarily by the field crew, but by

GENERAL

MACHINE NO.____K-6251_____   DATES OF OVERHAUL_____
SERIAL NO.____2-6-2264_____   (INCLUSIVE)
INSTALLED ROTOR SERIAL NO._____
REMOVED ROTOR SERIAL NO._____
ENGINEER IN CHARGE_____

MISCELLANEOUS

WEIGHTS: TOP HALF OF CASE _11.700_ LBS. ROTOR____1390___ LBS.
ROTOR: LENGTH (O.A.)_____78___IN. LARGEST DIA.__25 1/2__IN.
ROTOR: 1ST CRITICAL___4000__RPM 2ND CRITICAL_____RPM
ACCEPTABLE CLEANING MATERIALS: **BLASTING: WALNUT HULLS**____
___**CLEANING: INHIBITED COLD CLEANER**_____

CASE BOLT TORQUE:	DIAMETER	ELONGATION	TORQUE (FT.LBS.)
MAIN	2 1/4 IN.	0.025 IN.	8000
ENDS	1 5/8 IN.	0.018 IN.	3500

CASE SEALANT COMPOUND: __RTV-60_____
DATE OF MANUFACTURE_____AMOUNT REQUIRED _FOUR 1/4 LB_

COUPLING

	SPECIFICATION	CHECK	
		INITIALS	AMOUNT
DRIVER HUB: ADVANCE	0.144 IN.		
BOLT TORQUE	27 FT LBS		
BLU CHECK	85%		
DRIVEN HUB: ADVANCE	0.168 IN.		
BOLT TORQUE	27 FT LBS		
BLU CHECK	85%		
SHAFT END SPACING	18 IN ± 1/64		

MAXIMUM HYDRAULIC PRESSURE _25,000 PSIG EXP./4000 PSIG PUSH_

Figure 2-10. Critical data tabulation.

the engineer in charge of the overhaul), a decision or compromise can be made in a rational manner depending on need for the machine and time available to achieve acceptable alignment. Under no circumstances should alignment be compromised beyond a few thousandths off the desired position nor excessive pipe strain be permitted on the machine. The search for absolute perfection will, however, generally be rewarded with time consuming frustration and an ultimate compromise in any case.

Miscellaneous Items

Any good shutdown/overhaul plan should include an inspection of auxiliary components. During the overhaul period is the time to clean lube

JOURNAL BEARINGS

	SPECIFICATION	CHECK	
		INITIALS	AMOUNT
H.P. END: DIAMETER	4.125 IN.		
CRUSH	0.000 TO -0.001		
CLEARANCE	0.005 TO 0.0075		
L.P. END: DIAMETER	4.125 IN.		
CRUSH	0.000 TO -0.001		
CLEARANCE	0.005 TO 0.0075		

NOTES: 1. USE MANDRIL TO CHECK CLEARANCE
2. CLEARANCES ARE DIAMETRAL

THRUST BEARING

	SPECIFICATION	CHECK		
		BEFORE	AFTER	INITIALS
AXIAL CLEARANCE	0.015 TO 0.022			
COLLAR RUN OUT	0.0005 MAX			
COLLAR FIT	0.000 TO -0.001			
OIL CONTROL RING CLEARANCE	0.005 TO 0.009			
TOTAL ROTOR FLOAT	0.480 IN.			
% TO H.P. END	60%			
% TO L.P. END	40%			

ACTIVE DIRECTION: **OUTBOARD**

THRUST NUT THREAD: **LEFT HAND**

OIL SEALS

	SPECIFICATION		CHECK	
	DIAMETER	CLEARANCE	AMOUNT	INITIALS
LABYRINTHS				
LP INNER STEP	5.250 IN.	0.008 TO 0.0105		
LP OUTER STEP	5.000 IN.	0.008 TO 0.0105		
HP	5.000 IN.	0.008 TO 0.0105		
SEALS				
HP INNER RING	4.500 IN.	0.003 TO 0.004		
HP OUTER RING	4.500 IN.	0.015 TO 0.017		
LP INNER RING	4.500 IN.	0.003 TO 0.004		
LP OUTER RING	4.500 IN.	0.015 TO 0.017		

NOTE: CLEARANCES ARE DIAMETRAL

Figure 2-11. Critical dimension tabulation.

oil coolers, replace filters, overhaul lube oil pumps, etc. But beware of introducing dirt into the system. Many a clean lube set and newly overhauled machine have been damaged by a few seconds of careless maintenance activity.

The instrumentation associated with the machinery train should also be checked and calibrated. Again, a list and adequate record keeping prac-

DIAPHRAM LABYRINTH CLEARANCE

		SPECIFICATION		CHECK	
		DIAMETER	CLEARANCE	AMOUNT	INITIAL
STAGE 1	LARGE	17.125 IN.	0.022 TO 0.027		
STAGE 2	LARGE	15.575 IN.	0.022 TO 0.027		
STAGE 2	SMALL	7.500 IN.	0.012 TO 0.016		
STAGE 3	LARGE	14.200 IN.	0.022 TO 0.027		
STAGE 3	SMALL	7.000 IN.	0.012 TO 0.016		
STAGE 5	LARGE	14.200 IN.	0.022 TO 0.027		
STAGE 5	SMALL	7.500 IN.	0.012 TO 0.016		
STAGE 6	LARGE	12.950 IN.	0.022 TO 0.027		
STAGE 6	SMALL	7.500 IN.	0.012 TO 0.016		
STAGE 7	LARGE	12.950 IN.	0.022 TO 0.027		
STAGE 7	SMALL	7.500 IN.	0.012 TO 0.016		

				BEFORE	AFTER	INITIAL
BALANCE PISTON		14.375 IN.	0.012 TO 0.016			

				AMOUNT	INITIAL
FINAL MID SPAN RUN OUT		0.001 TIR MAX			

NOTE: CLEARANCES ARE DIAMETRAL

SPECIAL NOTES OR PRECAUTIONS: _____

Figure 2-12. Clearance tabulation.

tices are a must. The list should include all set points and complete information on any rebuilt instruments placed back in service. The engineer in charge of the overhaul will normally delegate this task to the *instrument group* after collaborating on the list with this group and the operating department to pinpoint any troublesome items. Key shutdown instruments such as low oil pressure and high discharge temperature should, of course, receive as accurate a test as practical.

The Factory Serviceperson

Most machinery manufacturers' manuals recommend the use of a factory serviceperson. It is most important to know whom you are getting and what his qualifications are. One maintenance manager had the unsettling experience of shaking the hand of a serviceman from a major supplier of gas turbines who then mentioned that he normally worked on steam turbines and this would be his first gas turbine. The point is, after several years in the field, assuming continuity of plant personnel, the user many times will know more about the machine than its manufacturer. Factory field servicemen lead a rough life: 16–20 hour or more shifts are common, as well as frequently being away from home. Attrition is high. You will find some very good and some very bad ones. Keep a list of those to invite back, as well as those you would rather not use again.

The Overhaul

During the course of the overhaul, it is very important to keep track of the job on an hour-to-hour, shift-to-shift basis. Companies using a shift log or diary for this purpose have found it to be an invaluable communications tool. The critical dimension diagram and the alignment diagram should likewise be kept available for ready reference, as should the turnaround book or T/A package mentioned earlier.

The use of a good quality camera and a capable photographer to document details of the overhaul is strongly recommended. A good T/A package will include a pictorial sequence of assembly and disassembly steps, as we will show later. After all, if you do a good job, it will be five to eight years or longer before anyone sees the inside of the machine again. Be sure to use a digital camera, and keep in mind the merits of video taping to provide training films for the maintenance department.

Before the actual shutdown of the machine is the time to take a final set of hot alignment data, if such a program is currently in use at your plant. There are, of course, several accepted methods for checking hot alignment. One is the use of eddy current probes, either inside the coupling guard or on the machine cases. A second is the optical method using a transit and targets on the machine train. A third is the use of a telescoping measuring rod with reference points on each machine and benchmarks on the foundation. A fourth and most up-to-date method may involve laser optics.

If you don't currently check running hot alignment (as opposed to the old method requiring a shutdown/alignment check, which has proven to be both inaccurate and unreliable), we would strongly recommend evaluating the various systems to see which one best fits your needs. As with reverse dial indicator alignment, a good hot alignment method can be a real money saver. Several items should be checked after the machine has stopped turning, but before the actual disassembly begins:

1. The coupling: If it is a gear coupling, is it free to move on the gear teeth? Have you considered upgrading to a contoured diaphragm coupling?
2. Look for broken coupling bolts. Broken bolts can indicate several problems, the most likely being incorrect bolt torque on installation, the wrong bolt material, or mismachined coupling flanges.
3. Get a sample of coupling grease, if a grease-packed coupling is used.
4. When removing the coupling, remember to turn the nuts and prevent the bolt head from turning so as to avoid wearing the body-fitted bolts. In a double-keyed coupling, be sure to check that the keys are marked as to their location.
5. Be sure to keep the coupling bolts and nuts together as individual assemblies. Do not plan to reuse the nuts more than twice. If any doubt exists in this area, a new set of match-weighed nut and bolt assemblies is cheap insurance.
6. Check and record rotor float within the thrust bearings, and note also the spacing between shaft ends.
7. Check the total rotor float with the thrust bearing removed, and note the rotor position relative to the machine case. Check nozzle stand-off in the case of a steam turbine, or position between diaphragms in the case of a compressor.
8. When removing the thrust bearing, be sure to measure and tag any thrust shims used for thickness and location (inboard or outboard).

Opening the Machine

Before actually opening a major piece of machinery, take time to review the critical steps in the operation. Attempting to remove an upper half casing without first removing internal (nonexposed) bolting or lifting the casing without using guide pins can result in a much longer and more expensive overhaul. Be especially careful when opening lube oil lines. The loss of a flow control orifice or the introduction of dirt into the system can cause serious problems during the machine startup.

As the machine comes apart, take lots of pictures, make written notes, and/or use a handheld computer or tape recorder to document what you see. It's amazing how much detail will be lost and how difficult it is to accurately reconstruct events hours or days—let alone years—after they have occurred.

One major petrochemical company operates four identical 20,000 horsepower steam turbines which, due to a series of blade problems, had to be opened a total of 31 times in an eight-year period. They recognize the importance of rotor charts to keep track of rotor movements and modifications, as shown in Figure 2-13. When the first blade in the first rotor failed, it was not apparent that they were in for such a lengthy problem. The rotor movement chart was laboriously constructed from memory when they were halfway into the program and had added a sixth rotor to the system (four installed and two spares).

If you plan to remove compressor diaphragms, be sure to match mark them as to their position in the case. Inadvertent mixing of inlet guide vanes could alter machine performance! Be careful to stone down any match marks which are placed in a machined area, such as the casing split line. When the top half of a horizontally split compressor is removed, it is a good idea to position the rotor with its thrust bearing as it was before shutdown and check to see if the impellers are centered with the diffuser flow passages.

Inspection

As the machine is being opened, pay particular attention to visible deposits. On machined sealing surfaces you may find telltale tracks of a leak or wire drawing. Such leaks may indicate a need to check the flatness and fit of the surfaces with lead wire or Plasti-Gage®, or simply better

| DATE | MACHINE NUMBER | | | | SPARE |
	300	350	500	550	
5/70 (NEW)	A-1	B-1	C-1	D-1	E-1
7/70				C-1	C-2
9/70		B-2			C-2
10/70			C-2		C-2
10/70				E-2	D-3
10/70			D-3		C-3
10/70				C-3	E-3
11/70				E-3	C-4
1/71		B-3			C-4
4/71				C-4	E-4
5/71		E-4			B-4
6/71			B-4		D-5
6/71			D-5		B-5
7/71	B-5			C-5	A-7
8/71		E-5			A-7
9/71	B-6				A-7
6/72	A-7		C-5	F-3	B-10, D-9
2/74		D-9	B-10		C-9, E-12
9/75	C-9				A-12, F-12
12/75	C-9/6				A-12, E-12
6/76	A-12		E-12		B-12, C-12
10/78		B-12		C-12	D-12, F-12

NOTE: LETTER CORRESPONDS TO MFG. ROTOR SERIAL NO.
NUMBER REFERS TO SPECIFIC DESIGN MODIFICATION

Figure 2-13. Rotor history chart.

attention to bolt torquing requirements. Fouling inside the flow passages of the machine will likely not be distributed uniformly from one end to another. In a compressor, the gas will get hotter with each successive stage. With some gases this will bake the deposits in the latter stages; with other gases, heavy, wet deposits will form in the first stages of the machine. Get a sample of the deposits to determine, first, what they are in order to see if they can be eliminated from the process. Failing that, test to see if they can be dissolved in some suitable solvent, for either on-line or off-line washing, in order to delay a subsequent machine overhaul. While compressor manufacturers shy away from on-line full-speed washing, knowledgeable users have had very good experiences with both this technique and with off-line washing when the machine is slow rolled while half full of the wash liquid. When choosing a wash fluid be sure it is compatible with all components in the machine, such as O-rings, as well as the process. On-line abrasive cleaning with walnut hulls, etc., has found wide acceptance with gas turbine users, but is not without its problems. Plugged orifices, airbleed passages, and the like are common. The total subject of on-line or off-line cleaning is beyond the scope of this text, but it is well worth considering in specific situations as it is a real time and money saver.

The bearings, journals, and seals should be visually checked for signs of distress. One frequent problem has been that of babbitt fatigue. While the aftermarket has been offering bearings with babbitting less than 0.010 in. thick for a number of years, some machinery manufacturers have resisted change in this area. Nevertheless, industry experience with thin babbitt bearings has been excellent to date.

Labyrinths can also tell a story which needs to be read and analyzed. Deep grooves in the impellers or shaft spacers are indications of a shaft excursion at some time in the operating cycle. Worn or corroded labyrinths indicate loss of efficiency, and, if found over the balance piston, could lead to a thrust bearing failure. As with bearings, new materials, such as Vespel® high-performance graphite-filled polymers able to combat the corrosion problem, are now coming into the after-market. Rubs could indicate misoperation, such as running at or near a rotor critical speed or in surge; a rotor dynamics problem; a thermal bow, or similar difficulty. Observations regarding location, depth, and distribution of the rubs are the keys to a proper analysis.

Cleaning

When cleaning fouled components—rotors and diaphragms, etc.— make sure the work is done in a remote location. Sand or nut hulls used for this purpose will usually find a way of invading the wrong parts of the machine, such as bearings and seals. The rotor should be carefully checked at this time for debris lodged in the gas passages. We know of instances where a rag or piece of metal was jammed in an inaccessible place in an impeller. The use of a small dental mirror and a thorough inspection by hand can reveal much of this debris.

It is fairly common practice to inspect a rotor using magnetic particle or dye penetrant techniques. This is a strongly recommended step; it can turn up defects which could otherwise prove to be highly damaging during a subsequent running period. In one such instance we uncovered an undesirable manufacturing technique which has been practiced for many years. The magnetic particle and subsequent dye penetrant inspections showed several cracks around the eye of the fifth stage impeller in a multistage barrel compressor rotor installed in relatively clean hydrogen service. Up to this point the overhaul had been a routine matter, but now took on far more serious implications. It seems that this particular compressor manufacturer had been in the habit of overspeed testing impellers and then trimming the eye labyrinth area to size, thus weakening the most critical structural area of the impeller. The explanation given (to compensate for bore stretch during overspeed) is, of course, unacceptable. Most manufacturers will now readily guarantee maximum allowable expansion in the diameter of the eye of an impeller as a function of the diameter before the overspeed test. This is the only acceptable way to buy compressor impellers, either as part of a new machine or as a replacement part.

Reassembly

Once the machine has been opened and all parts cleaned and inspected, the reassembly procedure can begin. There are many critical phases involved with this operation, one of the most important being care in handling the rotor. Large heavy rotors (over approximately 2,500 lbs) require special handling and, in some cases, special guide fixtures should be fabricated to avoid damaging components. This is particularly necessary with gas turbines which have many exposed, fragile parts. A solid rotor cradle is also a very necessary item. Do not jeopardize your most valuable spare part by failing to protect it during the course of an overhaul or during transit to or from the storage warehouse.

When fitting housings and other components with multiple O-rings in blind areas, we have found that it is usually beneficial to first remove the O-rings and fit the housing by hand to check the alignment of the assembly. Blind dowels or concealed shims can be located in this manner with pencil marks. The O-ring fits should be touched lightly with Grade 600 wet or dry emery paper to remove any burrs, and then checked carefully by hand. Lubricate the O-rings with a suitable grease or oil. A cut O-ring, worth very little in itself, can bear heavily on the success of an overhaul.

Bearing clearance is one of the most important checks during reassembly. We have found that after several years of operation, the pads of a tilting pad journal bearing will wear small depressions in the support ring or housing which can open the clearance beyond specifications. Also, replacement pads may not be within tolerance. The only proper way to check bearing clearance with this type of bearing is by using a mandrel the size of the journal and a flat plate. The clearance in a sleeve-type journal bearing can be checked with Plasti-Gage®. Be sure to torque the bearing cap bolts correctly or you may get a false reading.

When assembling bearings be sure the anti-rotation dowels are in place and look to be sure the oil dam (if used in that particular bearing) is in the correct direction of rotation. Some of these steps will sound obvious, but each one results from a problem experienced in the field. It is also useful to check the alignment of oil supply holes in the housing with oil feed grooves in the bearing. For want of a $^3/_8$-in, groove in the housing of a replacement bearing, one user lost a high speed shaft and impeller assembly in a plant air compressor package.

Before the upper half of the casing of a horizontally split machine is bolted in place, a final rotor mid-span bow check is recommended. This is particularly useful if you, as the responsible engineer on an overhaul, have not been able to personally witness all rotor movements during the course of the job.

Coupling hub fit is another area requiring consideration. The assumption that the taper is correct provides a false sense of security. By lightly bluing the shaft and transferring the bluing to the coupling bore, the fit can be properly checked. It is prudent to require at least 85 percent contact. If the contact pattern is not acceptable, the question of whether to lap or not to lap needs to be addressed. We will not lap using the coupling half for obvious reasons, but will lap using a ring and plug gauge set. The advance of the coupling on the taper must be correct and should be witnessed and recorded by a knowledgeable individual. Coupling bolts must be torqued to the coupling manufacturer's specifications as the clamping force, not the bolt body, is generally the means of transmitting the torque.

As the machine goes back together, fill in the information in the critical dimension diagram. Labyrinth and bearing clearances, total rotor float, thrust clearance, coupling advance, bolt torque, etc., should all be measured and logged. Shaft alignment and cold baseline data for comparison with hot growth data taken after startup should also be logged on the appropriate sheet. Remember to check the shaft end gap, as not all rotors are created equal and the wrong dimension could damage your coupling. When leaning into an open machine, it is well to remember to remove all loose objects from shirt pockets!

There are some other checks which may or may not have been incorporated in the critical dimension diagram, most notable of them being whether the rotor is free to turn and whether oil is flowing to and from the proper places. This latter item can be viewed just prior to bolting bearing caps or covers in place, assuming the oil lines have been reconnected. On some machines with internal oil tubing, it is possible to have oil flow showing in the main oil drain sight flow indicator while no oil is reaching the bearings or seals!

Documenting What You've Done

Following the overhaul, the startup will need to be monitored. If you don't have a fixed-base monitoring system, use a portable real time analyzer and a modern recorder to obtain baseline vibration data for comparison with previous operating information. Hot alignment readings can usually be taken several hours after startup. Machine performance will normally be checked after the process has stabilized which, on some machines, can be as long as several days after startup. All of this information provides a very useful check on the success of the overhaul and should be taken at the outset of a run and not delayed until a "convenient" time several weeks from startup.

As soon as the machine is operating satisfactorily, do the paper work, i.e., update your computer log. Many engineers shy away from this duty and use the excuse of day-to-day business pressures to delay or even forget this very necessary chore. While the events are still fresh in your mind, sit down and finish the job. In documenting an equipment overhaul, consider using the following format:

1. *Basic Machine Data*—A brief description of the machine, including manufacturer, model number, number of stages and other physical parameters, serial number, date purchased, date of last overhaul and reason for current overhaul.
2. *Performance, Vibration, and Mechanical Health Data*—A comparison of pre- and post-overhaul levels. Performance and vibration data for the train, including process flow, pressure and temperature, machine case, and eddy current probe vibration levels, as well as oil supply pressure, temperature, and oil return temperature. The performance data should be sufficient to accurately assess the machine's condition. Calibrated instruments are required.

3. *Spare Parts*—A complete list of spare parts for the machine, as well as a list of parts actually consumed. Include machine manufacturer's part number, as well as company warehouse stock number.
4. *Critical Dimension Diagram*—Complete with factory specifications, as-found dimensions (logged during disassembly), and as-overhauled dimensions. This information must include items such as total rotor float, thrust clearance, rotor position within the total float, labyrinth clearance, radial bearing clearance, nozzle stand-off, coupling bluing check, and coupling advance.
5. *Rotor run-out diagram and balance report.*
6. *Shaft Alignment Diagram*—A shaft alignment diagram showing desired readings based on anticipated thermal growth data, "as found" readings (prior to overhaul), "as left" readings after overhaul, and actual measured thermal growth data.
7. *Photographs of the overhaul.*
8. *A discussion of the overhaul.* Refer to appropriate photographs throughout.
9. *Recommendations:*
   - For future overhauls
   - For reconditioning worn but reusable parts
   - For on-line cleaning, if applicable
   - For redesigned parts, if applicable
10. *Shift logs and backup data as required.*

In writing a report, decide what went right and what went wrong. Fully identify the causes in each case so that your successor can benefit from your experiences. Send a list of spare parts used in the overhaul to the warehouse controller. While you hope you won't need parts in a hurry, don't bet on it! Decide if you plan to invite the factory serviceperson back for a subsequent overhaul. In either case, put his name on your report so no confusion exists on this point. Go back to the machine manual and make notes in the margin on any errors that may have appeared in the printed material.

Nonstandard Parts

Once a new machine has operated for a year, it is well to remember that the guarantee has probably elapsed. In addition, bear in mind that the original equipment manufacturer's parts were generally a design compromise which took into account a competitive marketplace and existing, available designs in the manufacturer's shop. Any parts that fail to stand up should not necessarily be replaced by standard parts. There are many excellent aftermarket manufacturers of components and many specialized tools such as multiplane milling machines and overspeed spin pits for individual components. Aerospace technology and materials are beginning to filter down to the aftermarket also. None of the above should be construed as an indictment of the equipment manufacturer, but when his spare part pricing, policies, and failure to solve design problems mount to a point where it becomes necessary to put properly engineered aftermarket components into a machine, do not hesitate to do what is best for your company.

Chapter 3
Machinery Foundations and Grouting*

What's an Epoxy?

According to the Handbook of Epoxy Resins,[1] the term *epoxy* refers to a chemical group consisting of an oxygen atom bonded with two carbon atoms already united in some other way. The simplest epoxy is a three-membered ring. There is no universal agreement on the nomenclature of the three-membered ring. There is division even on the term epoxy itself—the Europeans generally prefer the term *epoxide*, which is doubtless more correct than the American *epoxy*.

In addition to providing a history of the development of epoxy resins, the handbook states that the resins are prepared commercially thus:

1. By dehydrohalogenation of the chlorohydrin prepared by the reaction of epichlorohydrin with a suitable di- or polyhydroxyl material or other active-hydrogen-containing molecule.
2. By the reaction of olefins with oxygen-containing compounds such as peroxides or peracids.
3. By the dehydrohalogenation of chlorohydrins prepared by routes other than by route 1.

Dozens of distinct types of resins are commercially available, and the term *epoxy resin* is generic. It now applies to a wide family of materials. Both solid and liquid resins are available.

---

*E. M. Renfro, P. E. Adhesive Services Company, Houston, Texas.

There are other liquid resins such as phenolics, polyesters, acrylics, etc., which cure in similar fashion, but the epoxy resins possess a rather unique combination of properties. The liquid resins and their curing agents form low-viscosity, easy-to-modify systems. They can cure at room temperatures, without the addition of external heat and they cure without releasing by-products. They have low shrinkage compared to other systems. They have unusually high bond strengths, excellent chemical resistance, high abrasion resistance, and good electrical insulation properties. The basic properties can be modified by blending resin types, by selection of curing agents (hardeners), the addition of modifiers, and by adding fillers.

Perhaps the most valuable single property of the epoxy resins is their ability to cure, thus converting from liquids to tough, hard solids. This is accomplished by the addition of a curing agent. Some agents promote curing by catalytic action, while others participate directly in the reaction and become part of the resin chain. Depending upon the particular agent, curing may be accomplished at room temperature with heat produced by exothermic reaction, or may require application of external heat. The epoxies will react with over 50 different chemical groupings, but the basic curing agents employed in the epoxy resin technology are Lewis bases, inorganic bases, primary and secondary amines, and amides.

An entire spectrum of properties can be obtained in a cured epoxy resin system by careful selection of resins, careful selection of curing agents, varying the ratio of resin to curing agent and by including additives or fillers. The resins and curing agents, themselves, may even be blends. As an illustration of the spectrum of obtainable properties, a cured epoxy system may be as soft as a rubber ball or so hard that it will shatter when dropped. Epoxies can be formulated to be either sticky or tack free. They can be formulated to either melt or char when heated; to release tremendous amounts of heat when curing or they may require heat for curing; to bond tenaciously to sandblasted steel, even under cryogenic conditions, or have relatively little bond; or to be either tough or friable.

Epoxy Grouts

Grout is a broad term covering all of those materials used in a wide variety of applications which include clinking for cracks, fissures, or cavities; a mortar for tile and other masonry; a support for column footings; a sealant for built-in vessels; or a mortar for setting heavy machinery. This text, however, is concerned with those epoxy-based materials used in setting heavy machinery and in repairing concrete foundations. Specifications for Portland cement grouting and epoxy grouting of rotating equipment, as well as a checklist for baseplate grouting, can be found in the appendices at the end of this chapter.

The need for a machinery grout is created by a combination of circumstances occurring in the construction of foundations. Many of these circumstances are unfavorable to concrete, thereby complicating its use. This condition is brought about primarily because it is impossible to pour a concrete foundation to within the tolerances usually required for precision leveling and alignment of dynamic equipment. Even if such exact placement were possible it would be further complicated by the fact that concrete shrinks while curing.

Furthermore, the laitance or weak surface created when simple concrete is cast or troweled would not provide sound support for machinery requiring precision alignment. It has therefore become a standard practice in construction of foundations to pour the concrete to a level slightly above the desired grade, and after curing, chip away the surface to remove the laitance. The machinery is then positioned on the foundation, leveled and aligned to within proper tolerances with the aid of jack screws, wedges, shims, etc., and the gap grouted in solidly to establish integrity between the machine base and the concrete foundation below.

When improperly installed machinery breaks loose, the static forces to which the foundation is subjected do not act alone. Vibratory forces of high magnitude will also exist. Given enough time, this will usually cause cracks in the foundation that allow lubricating oil to penetrate deep into the foundation and proceed to degrade the concrete. It therefore becomes necessary to repair the cracked foundation, remove or repair oil-soaked concrete and regrout in order to re-establish the integrity of the system. Epoxy grouting materials have long been used for these repairs.

The specific use for which grout is intended should be taken into consideration when evaluating the properties of a prospective grout. It is equally important to ascertain the conditions under which a manufacturer obtained his test data.

Some properties contribute to the long service life or performance of a grout while others facilitate the ease of installation or grout placement. In evaluating a prospective grout, performance characteristics should take preference over ease of placement characteristics. These properties are of key importance:

- Nonfoaming—Without a doubt, the single most important characteristic of a grout from a performance standpoint is its ability to stabilize and disperse any air introduced with the aggregate or entrained during normal, nonviolent mixing. Otherwise, a weak, foamy surface would develop soon after pouring, and be unable to maintain alignment. Surface foam can always be eliminated by selecting the proper aggregate and maintaining viscosity of the mixed grout with proper aggregate ratio. This ratio cannot be fixed for all temperature conditions because the viscosities of the liquid ingredients change with temperature as do other hydrocarbons. The aggregate ratio will increase as the temperature of the ingredients becomes higher. Incorporating air release agents and surface defoamers in the grout formulations does improve the *appearance* of the exposed foundation shoulders, but does not prevent entrapment of air bubbles under the equipment base. Even with a time lapse between grout mixing and grout placement, air cannot be properly released because of the difference in rise rates of various size air bubbles, particularly in "soupy" mixes.

- Dimensional Stability—Three causes of dimensional change in grouts are shrinkage while curing, thermal expansion or contraction from temperature changes, and stress deformation or creep. Shrinkage in epoxy grout systems can occur if the formulation contains nonreactive volatile solvents that can, with time, gradually evaporate from the grout. This material loss usually results in shrinkage or cracking. Shrinkage is also theoretically possible in cases where improper ratio of resin to curing agent exists as a result of dispensing error or as a result of poor or incomplete mixing. Shrinkage is virtually nonexistent in properly formulated and properly mixed epoxy grout.

- Grout Expansion—Thermal expansion coefficients of grouts should be compared with the rate of thermal expansion of concrete and steel since it will be sandwiched between the two materials. Concrete and steel have about the same linear coefficient of thermal expansion. Unfilled epoxy resin systems expand or contract at about ten times the rate of concrete and steel. The high rate of expansion of unfilled epoxy does not cause problems when the epoxy is of a nonbrittle formulation and is present only in thin films, as in pressure grouting. When aggregate is added to the liquid epoxy/curing agent mixture to form a mortar, the linear coefficient of thermal expansion can be reduced to the range of $1.2–1.4 \times 10^{-5}$ in./in.°F, or about twice the rate for concrete or steel. When reviewing properties of a grout, compressive strengths should be considered along with the modulus of elasticity (the slope of the stress-strain curve). Generally, the more rigid the material, the steeper the slope and the higher the modulus of elasticity. Rubber, for example, is elastic according to the lay definition, but relatively nonelastic according to the technical definition.

- Strength—There are several methods of measuring strength of a grout. It can be measured under compression, tension, impact, and under flexure. Bond strength, shear strength and cleavage are measurements of adhesion rather than strength. Usually when strength of a grout is mentioned, it is the ultimate compressive strength that is implied. The term yield strength should be reserved for tensile tests of metals which work-harden before reaching the ultimate strength. Grouting materials do not work-harden, and there is but one peak in the stress-strain curve. More important than the ultimate strength, however, is the proportional limit, because beyond that level of stress, the material is permanently distorted and will not return to its original dimension after the load is removed. Data from compression tests can be used for design calculations because static loads are usually known and dynamic loads can be reasonably estimated. Grout is seldom placed under tension, except at rail ends, etc., during start-up. The tensile strength of the grout is important, because if it is known at the operating temperature, the maximum distance between expansion joints can be calculated. In addition to the tensile strength, tensile modulus of elasticity, operating temperature range, and linear coefficient of thermal expansion must be known.

This should illustrate that epoxy grouts are sophisticated products. There are literally thousands of possible resin/curing agent combinations. Developing, manufacturing, and marketing of epoxy grouts is *not* the business for small time formulators with bath tub and boat paddle type equipment. Prospective epoxy grout suppliers should be screened on the basis of their technology and capabilities. If the reader retains nothing more than this one fact, he will have learned within a short period what others have learned through great anguish over a long period and at considerable expense.

Proper Grout Mixing Is Important[2]

Epoxy grouts must be properly mixed if adequate strength is to be maintained at operating temperatures. The strength of epoxy grouts is the result of dense cross-linkage between resin and hardener molecules. Dense cross-linkage cannot occur in either resin-rich or resin-poor areas. Poorly mixed grout, which may appear to be strong at room temperature, can soften and creep under load at temperatures in the operating range.

Epoxy grouts are three-component products. They have an epoxy resin, a hardener, and a graded aggregate. The resin and hardener serve as an adhesive in the mortar while the aggregate serves as a filler to reduce costs. The addition of an aggregate will lower the coefficient of thermal expansion of the mortar to more closely approach that of concrete and steel.

Aggregates also serve as heat sinks to absorb the heat released by curing, and thereby, allow thicker pours.

Both resin and hardener molecules are surface-active, which means that either is capable of clinging to a surface. That is why it is so critical that the resin and hardener be premixed for a minimum of three minutes before adding aggregate. Use of a paint mixer for premixing these adhesive components is preferred over the stick-and-bucket method because it provides more thorough mixing and will not usually whip air into the mix.

The aggregate used in preparing an epoxy grout mortar is a key factor in minimizing the loss of load bearing area caused by the rising of entrapped air after grout placement. Aggregate quality is also a key in minimizing the potential for run-away curing, edge lifting of the grout on foundation corners, loss of bond to the machinery base and stress cracking of the grout.

Most aggregates have about 25–30 percent voids regardless of particle sizes or gradation. The liquid components of an epoxy grout have a density of about 9 lbs per gallon while the aggregate exhibits a bulk density of about 14–16 lbs per gallon. The particle density is much higher. Because of this difference in densities, the aggregate falls to the bottom of the mix and is not immediately wetted. When the liquid and aggregate are blended together, air that was present in the aggregate as well as air introduced into the mortar during mixing has a tendency to rise. The rate at which air bubbles rise is governed by both the size of the bubble as well as the viscosity of the mortar. At any given viscosity, the rise rate increases as the size of the bubble increases; therefore, it is important to keep the size of the bubbles as small as possible. The size of the bubbles is determined by the space between aggregate particles.

The linear coefficient of thermal expansion of unfilled epoxy grout is about ten times greater than that of concrete or steel or $6–8 \times 10^{-5}$ in./in. °F. When aggregate is added to form a mortar, the linear coefficient of thermal expansion is reduced, and the more aggregate added, the closer it approaches the coefficient of concrete and steel. It is important that the thermal expansion coefficient of epoxy mortar approach that of concrete and steel in order to minimize edge lifting on foundation corners and to minimize stress cracking of the grout when temperatures fall below the curing temperature. The ratio of aggregate to epoxy adhesive in the mortar should be as high as possible without exceeding the point at which the mortar becomes permeable. As stated earlier, most commercial epoxy grout mortars have a thermal expansion coefficient of about $1.2–1.4 \times 10^{-5}$ in./in. °F.

Most epoxy adhesives cure by exothermic reaction, i.e., they release heat on curing. If an epoxy grout cures too fast, high curing temperatures are reached and locked-in stresses may be created after heat dissipation.

Aggregate serves as a heat sink. Consequently, it is usually desirable to have as high an aggregate loading as possible. Because the hydroxide ion accelerates the curing of epoxy resins and because water contains hydroxide ions, it is important that the aggregates used in preparing the mortar be kiln dried. As little as one ounce of water per cubic foot of mortar will dramatically increase curing rates. This small amount of moisture is not detectable by sight or touch. Kiln drying is a common practice with bagged aggregates. Even low cost blasting sands are kiln dried.

The viscosity of the mortar is determined by the viscosity of the liquid (which is determined by temperature), the shape and the amount of aggregate as well as the amount of surface area present in the aggregate. The greater the surface area the greater the viscosity of the mortar. While high viscosity in an epoxy mortar is helpful in reducing the rise rate of air bubbles it also reduces the fluidity of the mortar. A powder aggregate would certainly eliminate air rising problems, but unfortunately, a paste consistency would be reached long before an adequate quantity of aggregate is added to significantly reduce the linear coefficient of the mortar expansion.

A high aggregate loading can be accomplished in mortar without eliminating its fluidity and without creating a permeable mortar by careful grading of near-spherical aggregates. Theoretically, the selection of each particle size should be the largest that will fit in the space between particles of the next larger size. The amount of each grade present should be that which fills these spaces without significantly increasing total volume of the aggregate. The variation in particle size should not be so great as to cause classification of the aggregate in the mortar before curing; otherwise, a gradient in coefficents of thermal expansion may be created between the top and bottom of the grout. The diameter of the largest particles should be no more than $1/10$ to $1/15$ the thickness of the grout under the load bearing surface of the machinery. The largest particle size in most commercial epoxy grouts is about $1/8$ in. Epoxy grout manufacturers usually recommend a minimum grout thickness of $1\frac{1}{2}$ in.

Because the adhesive components are organic materials, the viscosities change with temperature. More aggregate is sometimes required when preparing mortar in hot climates than is required when mixing at conventional room temperatures. The proper consistency or viscosity of the mortar is observed when the divot falls free and does not cling to a clean mortar hoe when a gentle chop is made in the mix.

When utilizing a concrete mixer or a mortar mixer for preparing the grout, it is important that mixing after aggregate addition be carried out only long enough to coat all aggregate particles uniformly. Otherwise, a froth may be generated from air whipped into the mix. Ideal mixer speeds are usually about 20 rpm.

Job Planning

If the equipment is being installed in original construction, grouting should be scheduled for a time compatible with critical path sequences. If the equipment to be grouted is in service, it may be advantageous to schedule regrouting during a normal downtime or during a turnaround period. In either case, work planning should be carried out in detail well in advance of the actual time the work is to be done. Proper planning reduces job site problems.

The equipment manufacturer should be informed well in advance in order to alert his service personnel if their presence is required to supervise leveling and alignment prior to grouting. The grout manufacturer should also be alerted if field supervision of grouting is expected. Early communication with these parties will allow them to make necessary arrangements with minimum inconvenience. Last minute notification seldom accomplishes these objectives.

A clear understanding of what is expected from a contractor will minimize extra charges which usually arise after the work is complete. Contract details should include provisions identifying parties responsible for furnishing utilities, materials, etc. The extent of work should also be accurately defined. For example, responsibility for disposal of waste, dressing and painting the foundation, backing-off on the jack screws, and torquing the anchor bolts should be considered.

It would also be prudent to prepare simple, itemized field checklists to be used by personnel involved in equipment installation and grouting. Typical sample checklists can be found in the appendices at the end of this chapter.

Table 3-1 is a materials check list for epoxy grouting. Orders for materials not available locally should be placed with lead time reserved for order processing, packaging, shipping, etc. A good rule-of-thumb is to place orders sufficiently in advance to allow three times the normal time required for unencumbered transit, if it can be anticipated that materials are available from stock.

All grouting materials should be stored indoors in a dry area and preferably at room temperature. Containers stored outside may in the summertime reach temperatures as high as 140°F, particularly if the containers and bags are in direct sunlight. The speed of most chemical reactions is doubled with each 10°C (18°F) rise in temperature. Consequently, it is quite probable that epoxy grout that has been stored outside in the summertime will have an excessively hot cure. When this occurs, the grout cures in a thermally expanded state, and after cooling, creates locked-in stresses. Excessive cracking will result as these stresses are relieved.

Most epoxy formulations do not cure well without accelerators at temperatures below 60°F, and not at all at temperatures below 35°F. Consequently, grouting materials stored at cold ambient temperatures require several days to cure. When this occurs, it is possible that equipment alignment conditions will change before the mortar has set, resulting in a poor installation. Furthermore, when the mortar is cold, it is viscous and very difficult to place.

Conventional Grouting

Concrete Characteristics

Foundation design and machinery installation require more expertise and precision than are usually practiced. Perhaps due to a shortage of skilled manpower, the construction industry has given less attention to technical details. Since there is generally some knowledge—"Everybody knows a little bit about concrete; and aren't foundations just big blocks of concrete?"—grouting is often taken for granted. Consequently, a high percentage of compressors are installed improperly. Many foundations must be renovated or the equipment regrouted long before the life of the equipment is exhausted. Here are a few common problems that can be avoided by putting a little effort into proper design and installation.

Communication links between the equipment manufacturer, grout manufacturer, design engineers, and construction and maintenance personnel are poor. Equipment manufacturers sometimes provide minimum foundation mass and unbalanced forces data but they do not design foundations. Data provided by grout manufacturers are often misinterpreted. Design engineers seldom are provided feedback data on performance of their design once the project is completed. Maintenance personnel rarely have the opportunity to provide input during the planning stage.

The consequences of improper installation are severe. Machinery installation costs often exceed $1,000 per horsepower and the loss of revenue due to idle machinery has advanced at a pace even higher than the rise in fuel costs. Large reciprocating compressor crankshafts are prone to break if the machine is poorly supported on its foundation. Crankshafts are not "hardware store" items. With some equipment manufacturers now relying on foreign sources for their larger crankshafts, logistics of spare parts supply are getting more complex. All the more reason, then, to protect the machinery by doing an adequate grouting job.

Concrete is the most widely used construction material in the world. Because it is so common it is often taken for granted, and therefore it has

Table 3-1
Materials Checklist for Epoxy Grouting

Item	Conventional Grouting & Regrouting	Pressure-Injection Regrouting by the Shoulder Removal Method	Pressure-Injection Regrouting. Thru-the-case Method	Foundation Repairs
Air compressor, manifold and hoses	X	X	X	X
Auxiliary lighting	X	X	X	X
Bolt studs, couplings and dies (for repairing broken anchor bolts)	X	X	–	–
Carpenter tools	X	X	–	X
Copper tubing, tubing cutters compression fittings, grease fittings, automatic injection equipment & grease guns equipment and grease guns	–	X	X	X
Exhaust blower or fan	X	X	X	X
Fire extinguishers	X	X	X	X
Flash lights (for gas atmospheres)	X	X	X	X
Gas detection equipment	X	X	X	X
Grouting materials (epoxy grout, aggregate, accelerators & cleaning solvent)	X	X	X	X
Housekeeping equipment (broom, mop, rags pails, shovel & industrial vacuum cleaner)	X	X	X	X
Lumber for forms, plywood, Celotex (for casting expansion joints, drop-in anchors, chamfer strips, nails, electricians putty (DuxSeal), body putty & wax	X	X	–	–
Mechanics tools	X	X	X	X

Item	1	2	3	4
Mortar mixing equipment (mortar mixer, wheel barrow, hoes & shovels)	–	–	x	x
No. 6 Rebar for pinning grout corners	–	–	x	–
No. 11 Rebar for pinning block to mat	x	–	–	–
Paint mixer (for mixing resin & curing agent)	x	x	x	x
Pneumatic drill, bits, taps & tap wrench	–	x	–	–
Pneumatic right angle grinders, stones, & paint (for dressing & painting the foundation)	x	–	x	x
Pneumatic rivet busters, spare parts and chisels	x	–	x	x
Pneumatic rock drill, extensions and bits	x	x	x	x
Post-tensioning assemblies, epoxy putty anchors)	x	–	–	–
Protective covering (equipment & floor)	x	x	x	x
Safety equipment (dust mask, face shields, hard hats, safety glasses, ear protection, rubber gloves & first aid kit)	x	x	x	x
Sand blaster with dead man control, blasting grit and ventilated hood	–	–	x	x
Shelter (for heating grouting area in cold weather or for containing dust)	x	x	x	x
Silicone rubber (for expansion joints and sealing the engine base)	x	x	x	x
Space heating equipment & fuel	x	x	x	x
Torque wrench and sockets	x	x	x	x
Tubular insulation for isolating anchor bolts	–	–	x	x
Urethane foam (for isolating oil pan)	–	–	x	x

also become one of the most abused materials. For good foundation design, these factors must be considered:

- Proper chemistry
- Proper water/cement ratio
- A quality aggregate
- Low amount of entrained air
- Proper placement
- An acceptable temperature range for curing
- Moist curing conditions

A detailed analysis of each of these considerations would be beyond the scope of this text; however, the listing serves to illustrate the fact that concrete is a complex material. For our purposes, a brief description of the mechanism of concrete curing will suffice.

Concrete is composed of a graded aggregate, held together by a hardened paste of hydraulic cement and water. The thoroughly mixed ingredients, when properly proportioned, make a plastic mass which can be cast or molded to shape, and upon hydration of the cement, becomes rock-like in strength and hardness and has utility for many purposes, including machinery foundations. Fresh cement paste is a plastic network of cement particles in water. Once the paste has set, its volume remains approximately constant. At any stage of hydration the hardening paste consists of hydrates of the various ingredients in the cement which are referred to collectively as the "gel." It also contains crystals of calcium hydroxide, unhydrated cement, impurities, and water-filled spaces called capillary pores. The gel water is held firmly and cannot move into the capillaries, so it is not available for hydration of any unhydrated cement. Hydration can take place only in water within the capillaries. If the capillary pores are interconnected after the cement paste has cured, the concrete will be permeable. The absence of interconnected capillaries is due to a combination of suitable water to cement ratio and sufficiently long moist curing time. At least seven uninterrupted days of moist curing time are required for machinery foundations. Even test cylinders of concrete taken at the jobsite from the pours are often allowed to cure under water for twenty-eight days before testing.

Concrete which has not been allowed to cure properly, even though ingredients are properly mixed in the correct ratio, may be weak and friable or it may be only slightly under ultimate strength, depending upon the humidity and ambient temperature present when curing. Improperly cured concrete will also be permeable and therefore less resistant to degradation from lubricating oils or other materials that may be present.

An illustration of hairline cracks caused by shrinkage of concrete during curing can be seen in Figure 3-1 and Figure 3-2. Figure 3-1 is a photograph of the cambered surface of an airport runway which as been grooved with a diamond saw to facilitate draining of rain water in an attempt to reduce hydroplaning of aircraft in wet weather. In this photograph a 50 percent solution of epoxy grout liquid (without aggregate) in acetone was poured on the surface of the runway. Note the degree of penetration into the concrete between furrows as the solution drains away. In the photograph of Figure 3-2 the highly volatile solvent has all but evaporated from the surface, exposing the wetted crack openings like a fingerprint. Before wetting with the solution, cracks were invisible to the naked eye. This condition exists in most concrete machinery foundations and is caused by water loss from the capillary pores in the concrete while curing. This water loss causes shrinkage which would not be experienced if the concrete had been immersed in water for 28 days like the samples from each pour that are usually sent to the laboratory for testing. While such shrinkage cracks do not constitute structural failure in machinery foundations, they do provide a path for the penetration of lubricating oils into the foundation. One interesting fact was that cored concrete samples from this runway typically had 6,000 psi compressive strength.

It is good construction practice to seal the surface of a foundation with a good quality epoxy paint as soon as the forms are removed. This sealing

Figure 3-1. A photograph of a cambered and grooved surface of an airport runway. Note the degree of penetration between furrows as a low viscosity solution of epoxy adhesive in acetone is poured on and drained away from the surface (courtesy of Adhesive Services Company).

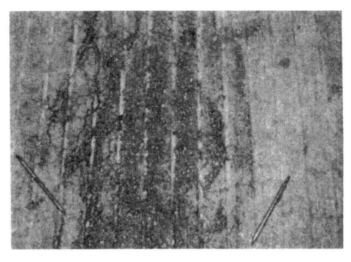

Figure 3-2. Hairline curing cracks become visible as the solvent in Figure 3-1 evaporates from the surface (courtesy of Adhesive Services Company).

of the foundation accomplishes two objectives. First, it seals in water and encourages more complete curing of the concrete, and second, it prevents penetration of lubricating oils into the foundation after start-up. This sealing is particularly important in areas such as around the oil pan trough which are usually flooded with oil. Paint will not usually stick to concrete unless the surface has been sandblasted to remove the laitance or unless a penetrating primer has been applied before painting. Some specialty coating manufacturers provide special primers for epoxy coatings when used on concrete. Most of these special primers contain either acetone or ketone solvents which are low in viscosity and water soluble. When utilizing these primers, care must be taken to prevent build-up of flammable vapors and breathing or contact with eyes or skin. Read the warning labels on the containers.

Methods of Installing Machinery[1]

The four common methods of installing compressors in the order of increasing foundation load requirements are shown in Table 3-2. Static load ranges shown in the first column are relatively low compared with the strength of the supporting concrete. What complicates the situation is the combination of additional anchor bolt load, dynamic load and dramatically lowered epoxy grout strength due to rising temperatures.

Table 3-2
Typical Loadings for the Various Methods of
Installing Compressors

Type of installation	Typical static load from equipment weight on load bearing surface[1]	Typical total load on load bearing surfaces[2]	Minimum compressive strength of epoxy grout at operating temperature necessary to prevent creep
Skid mounting (equipment not shown)	5 — 10 psi	50 — 100 psi	300 psi
Base embedment mounting	20 — 40 psi	200 — 400 psi	1200 psi
Rail mounting	50 — 100 psi	500 — 1000 psi	3000 psi
Sole plate mounting	100 — 200 psi	1000 — 2000 psi	6000 psi

[1] Data are calculated from weight of equipment and do not include additional load due to anchor bolts.
[2] Total load estimate is the sum of static load + dynamic load + anchor bolt load.

Skid mounting is an equipment packaging concept whereby partial erection of the compressor and its related equipment are carried out under shop conditions where quality control can be closely monitored. This concept is ideal for equipment destined for offshore or remote locations where accessibility and accommodations are limited or where skilled manpower is not available. Packaging works well on portable units in the lower horsepower range.

Job-site skid installation is progressively more difficult with increasing compressor size because of the number of structural members required. Most packagers do not provide access holes to permit grouting of internal structural members. Those internal "I" beams anchored to the equipment above are critical. Consequently, with typical factory design, grout placement must be accomplished from the edges of the skid. Placement of grout prepared to the proper consistency is difficult and often the critical members are left unsupported. When this occurs, a suspension bridge effect is created, allowing excessive vibration to occur when the equipment is operating. The obvious solution to this grouting problem is to cut access holes in the field. This should be done only with the manufacturer's approval, since otherwise the warranty may be voided. After grouting, all access holes should be covered.

As mentioned earlier, most compressors leak oil. Because skids are fabricated by strip welding rather than seal welding, oil gradually seeps into the skid cavities. To reduce this fire hazard it is common to provide openings between cavities for oil drainage. With the usual inconsistencies in grout level, complete oil drainage is not possible. Oil degradation of cement grouts and concrete has long been recognized. With this in mind, skids which are to be permanently installed should be installed with epoxy grout. Bond strength of epoxy grout helps to anchor internal structural members that have no anchor bolts in the concrete.

The embedment method of installing machinery is by far the oldest method. For short crankshaft gas engine compressors in the middle horsepower range, this method is preferred because it provides a "key" to resist lateral movement. On long crankshaft equipment in the higher horsepower range, thermal expansion of the foundation can cause crankshaft distortion problems. Foundation expansion is uneven due to heat losses around the outer periphery of the foundation and results in center "humping." The effects of humping can be avoided by installing the equipment on rails or sole plates. The air space between the foundation and equipment provides room for thermal growth without distorting the equipment frame. The air space also allows some heat dissipation through convection.

Exercise caution when installing equipment on sole plates—grout properties are taxed to the absolute maximum when sole plates are designed for static loads in the 200 psi range and then installed under equipment with high operating or oil sump temperatures. This is particularly true during the first few hours of operation until the grout passes through its period of secondary curing. Refer to the typical physical properties of epoxy grouts as shown in Table 3-3. Rails should be as short as possible and all rails and sole plate corners should be rounded to a 2-in. radius to minimize stress risers in the grout.

In recent years there has been a concerted effort to replace steel chocks with epoxy chocks. This involves the use of liquid epoxy grout which is poured in place, and after curing, forms a nonmetallic chock. One of the advantages of this method of installing machinery is that it is not necessary to have a machined surface on the engine base in contact with the chock. This method of engine installation has been utilized for many years

Table 3-3
Typical Physical Properties of Epoxy Grouts

Compressive strengths After curing 7 days at 77° F. and testing at 77° F..................	8,000 — 10,000 psi.
After curing 7 days at 77° F., heating for 8 hrs. at 160° F. and testing at 160° F.......................	1,000 — 3,000 psi.
After curing 7 days at 77° F., heating for 7 days at 160° F. and testing at 160° F.......................	4,000 — 7,000 psi.
After curing 7 days at 77° F., heating for 7 days at 160° F., cooling to 77° F. and testing at 77° F...........	12,000 — 18,000 psi.
Tensile strengths After curing at 77° F for 7 days......	1,600 — 2,000 psi.
After curing 7 days at 77° F., heating to 160° F. for 7 days, cooling to 77° F. and testing at 77° F...........	2,300 — 3,000 psi.
Linear coefficient of thermal expansion Unfilled grout cured for 7 days at 77° F. and tested at 77° F........	$6 - 8 \times 10^{-5}$ in./in.°F.
Filled grout cured for 7 days at 77° F. and tested at 77° F........	$1.2 - 2.0 \times 10^{-5}$ in./in.°F.
Bond strength Cured at 77° F. for 7 days at tested at 77° F.......................	200 — 400 psi.
Modulus of elasticity Cured at 77° F. for 7 days and tested at 77° F..................	320,000 psi.

in the marine industry on diesel engines. The forces imparted by Diesel engines driving propulsion systems are quite different from the forces imparted by integral gas engine compressors. For example, in the Diesel engine propulsion system, the forces are primarily those involving torque as imparted by the crankshaft at the output end of the engine. In integral gas engine compressors, cyclic lateral forces, created primarily by the compressor stages, are involved. On some compressors, the lateral forces are so great that the engine base is fretted by steel chocks. It stands to reason that epoxy chocks would be much less abrasion resistant than steel chocks.

While there are numerous reports of "satisfactory installations" involving integral gas engine compressors on epoxy chocks, the fact is that this technique has not been utilized long enough to ascertain life expectancy. The authors are not aware of any installations where the anchor bolts have been retorqued after several months of operation or where follow-up data have been taken from bench marks or other datum points such as tooling balls. In other words, the creep characteristics of epoxy chocks have not at this time been evaluated to the satisfaction of the authors. Further, some manufacturers do not provide test temperatures for the physical properties reported in their technical literature. Remember, the physical properties of epoxy grouts, unlike cement-based grouts, are reduced drastically with rising temperatures. Because of the lack of good data and experience, this method of installation should be classed as experimental and utilized only at the equipment owner's risk.

Anchor Bolts: Overview

The stretching of an anchor bolt between the bottom of the sleeve and the bottom of the nut (Figure 3-3) is desirable to create a spring effect that will absorb impact without fatiguing when the bolt is tightened to proper torque. Isolating the bolt from the epoxy grout prevents bonding that can cause temporary stretching over a short section, resulting in loose bolts soon after start-up as the bond fatigues. Isolating the bolts also prevents short radius flexing of the bolt if lateral movement develops. Anchor bolts are designed for hold-down purposes and not as pins to restrict lateral movement[4].

Original Anchor Bolt Installations

It is a standard practice to install anchor bolts in a foundation at the same time the reinforcing steel cage is fabricated and installed. Typically,

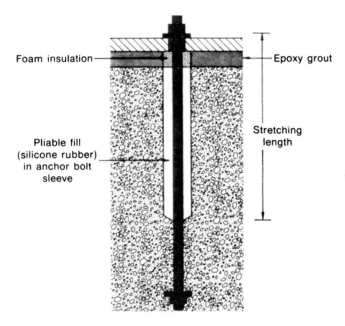

Foam insulation

Epoxy grout

Pliable fill
(silicone rubber)
in anchor bolt
sleeve

Stretching
length

Figure 3-3. A typical anchor bolt installation which allows freedom for equipment growth from thermal expansion.

the anchor bolts are located with the aid of a template created from engineering drawings. It should be a common practice to isolate the upper portion of the bolt with a sleeve. The purposes of the sleeve are twofold. First, it allows stretch of the bolt during torque application. Second, it provides a degree of freedom for the anchor bolt, which compensates for minor positioning errors. The proper terminology for these sleeves is "anchor bolt sleeves." These sleeves are often, but incorrectly, referred to as "grout sleeves."

As mentioned earlier, grout should never be placed in anchor bolt sleeves because bonding to the anchor bolt by the grout, particularly epoxy grout, prevents proper stretching and defeats the main purpose of the sleeves. The stretching of an anchor bolt between the bottom of the sleeve and the bottom of the nut is desirable to create a spring effect that will absorb impact without fatiguing when the bolt is tightened to proper torque. Bolt load should be calculated to prevent separation between the bottom surface of the nut and the machine boss when the bolts are subjected to operating forces, and in cases involving cyclic loading, to protect the bolt from fatigue effects of alternating tensile and compressive stresses.

Figure 3-3 is a sketch illustrating proper anchor bolt installation. Molded polyethylene sleeves are manufactured for the popular bolt sizes.

They are designed so the ends of the sleeve fit tightly around the bolt in order to center the sleeve, prevent concrete from entering the sleeve when the concrete foundation is poured, and, at the same time, prevent water, applied to the foundation for moist concrete curing, or rainwater from entering the sleeve.

After the concrete has cured, the surfaces to be in contact with grout are chipped away to expose the coarse aggregate. Immediately before positioning the equipment on the foundation, the upper end of the sleeve is cut off even with the top of the foundation and removed. Care must be taken to assure that water will not enter the sleeves and be allowed to freeze and crack the foundation, particularly on outdoor installations. After the equipment has been positioned on the foundation, leveled, and aligned, the grout sleeves are filled with a pliable material such as a castable polysulfide-epoxy joint sealant or closed-cell polyurethane sleeve.

Filling the sleeve with a pliable material allows for movement and stretch, and at the same time prevents accumulation of lubricating oil in the sleeve after equipment startup. Lubricating oil, in time, will degrade concrete.

Figure 3-4 is a photograph illustrating the cracking of a foundation at an anchor bolt, with the plane of the crack perpendicular to the crankshaft. This crack was caused by grout being placed in the anchor bolt sleeve during original construction, thereby restricting movement of the bolt.

Figure 3-5 shows the foundation after regrouting. The exposed portion of the anchor bolt was isolated with a tubular closed-cell polyurethane sleeve prior to repouring the epoxy grout. An expansion joint was installed to prevent new cracks from forming. After the grout has cured and the forms have been removed, the expansion joints and the outer periphery of the machine base where the grout contacts the boss are sealed with oil-resistant silicone rubber. The silicone provides a barrier against infiltration of oil and other liquids into the foundation.

Anchor Bolt Replacement

When anchor bolt failure is such that complete replacement is necessary, it can be accomplished using techniques consistent with the sketch shown in Figure 3-6. This sketch is of a typical replacement anchor bolt in an ideal installation. Complete replacement of an anchor bolt is possible without lifting or regrouting the machine. This is accomplished by drilling large-diameter vertical holes, adjacent to the anchor bolt to be replaced and tangent to the boss of the machine. Once the cores have been removed, access is gained to concrete surrounding the anchor bolt. After the surrounding concrete is chipped away, a two-piece and sleeved anchor

Figure 3-4. A photograph showing foundation cracks at an anchor bolt. This crack is in a plane perpendicular to the crankshaft.

bolt is installed. After the replacement anchor bolt has been installed, epoxy grout is poured to replace the concrete chipped from around the original bolt and to replace the concrete removed by the coring.

This procedure utilizes an air-powered diamond coring machine, as illustrated in Figure 3-7. Because the machine is air powered, it can be used in hazardous environments without creating a danger from sparks of open electric motors. Further, because a lot of power can be delivered by small air motors, the size of the coring machine is relatively small. With proper gear reduction, a hole as large as 16 inches in diameter can be drilled with this machine. Figure 3-8 shows 12-inch-diameter cores that have been removed with this machine. In the course of obtaining these cores, it was necessary to core through a No. 11 (1.375″-diameter) rebar, a cross section of which can be seen in this illustration.

Figure 3-5. Foundation after regrouting. Note the expansion joint at the anchor bolt and that the outer periphery of the machine base has been sealed with a fillet of silicone rubber.

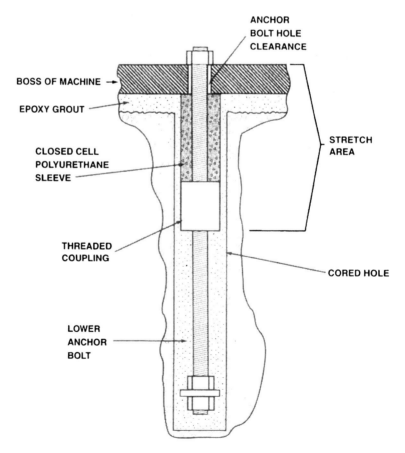

Figure 3-6. Replacement anchor bolt.

Figure 3-7. Air-powered diamond coring machine used in replacing anchor bolts without regrouting the machinery.

Figure 3-8. Twelve-inch-diameter cores removed in the course of complete replacement of an anchor bolt. Note the cross section of a No. 11 rebar in the core.

Figure 3-9 depicts a dual anchor bolt installation where both anchor bolts have been replaced and grouting is in progress. This picture was taken after the first pour of epoxy grout. Note that sleeving has not yet been installed on the upper stud above the coupling nut. Before the second pour was made, a split closed-cell polyurethane sleeve was installed to isolate the upper stud and coupling.

Figure 3-9. Replacement of dual anchor bolts after the first pour of epoxy grout. The isolation sleeving has not yet been installed.

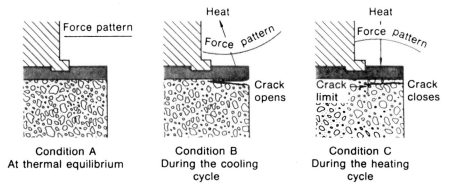

Figure 3-10. Stresses at foundation corners caused by cyclic temperature.

Outdoor Installations

Because epoxy grout and concrete absorb and dissipate heat rather slowly, cyclic temperatures cause uneven thermal expansion or contraction. This unequal expansion produces unequal stresses. The weak link is the tensile strength of the concrete, and cracking occurs in the corners where stress risers exist (Figure 3-10). In Condition A, the system is in thermal equilibrium and no stresses exist. During the cooling cycle shown in Condition B, the grout surface contracts first. This causes a thermal gradient within the grout which produces stress that promotes edge lifting. As the temperature conditions are reversed during the heating cycle, as shown in Condition C, cracks have a tendency to close. As the cycle is repeated, cracking progresses until it reaches a point under the edge of the equipment where compressive loading exists. Because tension is required for cracking, cracking cannot continue into an area which is under compression. Thicker grout increases the tendency for cracking until the cross-sectional dimensions of grout and shoulder are about equal. After grout thickness equals or exceeds the shoulder width, tendency for cracking is greatly reduced due to the inflexibility of this configuration. While cracking of this nature does not cause immediate operating problems, it does provide a path for oil to penetrate into the foundation. Over an extended period of time support will be diminished as oil degradation of the concrete proceeds.

Cracks of this nature can be virtually eliminated by utilizing one of the design techniques illustrated in Figure 3-11. Solution A is based on the transfer of stress away from the corner. This technique also changes the stress from tension to shear. Solution B transfers the stress away from the corner to a shear area on the back side of the key. Solution C changes the usual cross section dimensions of the shoulder, making a relatively inflexible configuration. Other designs, such as feather edging the grout, can be used, but all are based on eliminating stress risers at the foundation corners.

Expansion Joints

The small differences in thermal expansion rates between concrete or steel and an aggregate-filled epoxy grout become increasingly important

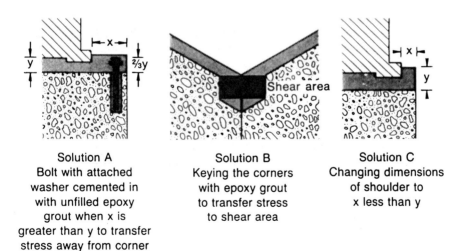

Solution A	Solution B	Solution C
Bolt with attached washer cemented in with unfilled epoxy grout when x is greater than y to transfer stress away from corner	Keying the corners with epoxy grout to transfer stress to shear area	Changing dimensions of shoulder to x less than y

Figure 3-11. Designs to eliminate stress risers in foundation corners.

as the length of the grouted equipment increases. Cracking can be expected near regions of anchor bolts or at rail or sole plate ends, unless care is taken in the design to eliminate stress risers. This is particularly true during equipment startup or shutdown where a temperature gradient might be created or when brittle grouting materials are employed. For example, during startup of rail-mounted equipment, the rails begin to grow first as a result of thermal expansion because they are in contact with the equipment base and conduct heat well. In order to prevent rail growth, opposing forces must be created equal to the compressive strength of the steel in the rail. Such forces would be well over 50 times the tensile strength of the grout. If the grout has enough elasticity to allow rail growth without extensive cracking, it is probably too soft to maintain support without creep. The grout should have adequate compressive strength to maintain alignment without creep.

The obvious solution is to install expansion joints. Expansion joints, as shown in Figure 3-5, can be cast into the foundation when the epoxy grout is poured. After the foundation has been dressed, the surface of the expansion joint and the outer periphery of the machine base is sealed with silicone rubber.

Postponement of Regrouting Is Risky

Real or perceived economic conditions in industry encourage postponement of routine maintenance of operating equipment. As a result, machinery foundations fail at an increasing rate during these periods. The most serious type of failure is foundation cracking in a plane parallel to the crankshaft. These cracks may be caused by inadequate design or by operating conditions that exert excessive forces on the foundation. Unless these foundation cracks are repaired at the time of regrouting, grout life will be greatly reduced (usually to about 10 percent of its normal life).

Lateral dynamic forces are generated by compressor pistons and by some power pistons. Theoretically, if a machine were perfectly balanced, there would be no forces exerted on the foundation other than dead weight. Under such a condition, there would be no need for anchor bolts. In reality, a perfectly balanced reciprocating machine has never been built. No experienced engineer would ever consider operating reciprocating equipment without anchor bolts.

After establishing the fact that unbalanced forces do exist on well-designed and -maintained equipment, consider what happens when maintenance is postponed. Take the ignition system, for example. Everyone knows what to expect from an automobile with the engine idling after one or two spark plugs have been disconnected. Imagine the same circumstances with a large industrial gas engine compressor running at 100 percent capacity. Next, suppose there are lubricating oil leaks that puddle on the foundation shoulder. If any movement exists between the machine and grout, oil will penetrate voids caused by the movement, and hydraulically fracture any remaining bond between the machine base and grout. As movement between the machine and grout increases, forces exerted on the foundation increase at an exponential rate, because of change in direction and impact.

At 330 rpm there are 475,200 cycles per day. Over 20 years the foundation sees the stresses of 3.4 billion cycles. Most reciprocating equipment is expected to last more than 20 years.

The tensile strength of concrete is only about 10 percent of its compressive strength. Because of this weakness in tension, reinforcing steel is embedded in concrete to carry the tensile loads. The placement of reinforcing steel should be chosen with consideration as to the source and direction of the external forces applied to the foundation. According to this reasoning, the preponderance of reinforcing steel in a reciprocating engine/compressor foundation should be placed in the upper portion of the block in a direction perpendicular to the crankshaft. Weighting the placement of steel in this location would reduce the tendency for cleavage-type failures that sometimes begin at the top of the foundation in the notch below the oil pan and extend through the block to the mat below.

The notch provided in the top of a foundation for the oil pan creates a perfect location for stress risers. A moment is created by lateral dynamic forces multiplied by the distance between the machine base and the transverse reinforcing steel in the foundation below. The possibility of a foundation cracking at this location increases as the depth of the notch increases. The further the distance between the horizontal forces and transverse reinforcing steel, the greater the moment.

Figure 3-12 illustrates a method of repairing such cracks by drilling horizontal holes spaced from one end of the foundation to the other end. This series of holes is placed at an elevation of just below the oil pan trough. A high-tensile alloy steel bolt is inserted into each hole and anchored at the bottom of the hole. Next, a small-diameter copper injection tube is placed in the annular space around the bolt; the end of the hole is then sealed and the nut tightened to an appropriate torque to draw the two segments of the block back together. An unfilled or liquid epoxy is injected into the annular space around the bolt. Air in the annular space around the bolt is pressed into the porous concrete as pressure builds. After the annular space has been filled, injection continues, and the crack is filled and sealed from the inside out.

This repair method places the concrete in compression, which would otherwise be in tension. The compressive condition must be overcome

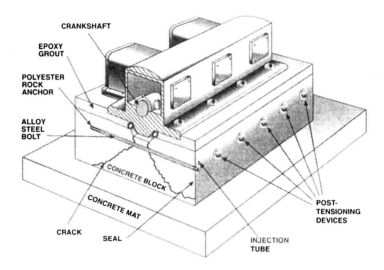

Figure 3-12. Method of repairing compressor foundations that are cracked parallel to the crankshaft.

before a crack could possibly reoccur. As a result, the repaired foundation is much stronger than the original foundation. This technique is often used when the concrete in the foundation is of poor quality.

Preparation of Concrete Surfaces Prior to Grouting

It has been estimated that 90 percent of the heavy equipment installed today on original installations was installed utilizing faulty grouting techniques. Because vibration and alignment problems with heavy machinery are solved (or should be solved) in the direction from the ground up, it is logical that grouting errors should be discussed beginning with surface preparation of the concrete.

Many early grout failures can be attributed to poor surface preparation of the concrete prior to grouting. Because the grouting problems associated with poor surface preparation are so widespread, it is obvious that few understand the difference between good and poor surface preparation.

The only *good method* of preparing a concrete surface prior to grouting is to chip away the surface with a chipping gun to expose coarse aggregate. This means at least a minimum of $^1/_2''$ to $1''$ of the surface must be removed. Poor methods of surface preparation include raking the surface of concrete prior to curing, intermittent pecking of the surface with a chipping gun, sandblasting the surface after the concrete has cured, and roughening the surface with a bushing tool (a spiked potato masher). Distinguishing between good and poor concrete surface preparation requires an understanding of bleeding of fresh concrete pours and mechanisms involving hydration of cement. Bleeding of freshly placed concrete is a form of separation where water in the mix tends to rise to the surface. In the course of bleeding, some of the solid ingredients classify near the surface. Classifying of concrete ingredients is a form of sedimentation.

If the bleeding rate is faster than the evaporation rate, the rising water brings to the surface a considerable amount of the fine cement particles, along with any residual silt or clay that may have been present in the aggregate.

In the course of concrete mixing, some of the hard and adherent clay and silt coatings will be ground loose from the surface of the aggregate. These loosened particles migrate to the surface of the concrete while the concrete is vibrated to gain proper compaction. The migration is enhanced by bleeding. This process promotes the formation of heavy laitance at the surface and results in a porous, weak, and nondurable concrete surface.

If the bleeding rate is slower than the evaporation rate, the water loss at the surface prevents proper hydration of the cement near the surface. Improper hydration of the cement at the surface also results in a weak and nondurable concrete surface. Further, water loss while the cement paste is in its plastic state causes a volume change commonly known as plastic shrinkage. While 1 percent plastic shrinkage is considered normal, excessive water loss through evaporation leads to surface cracking. Figure 3-13 illustrates proper chipping of a concrete surface prior to grouting. Note

Figure 3-13. Properly prepared concrete surface ready for grouting.

the fact that coarse aggregate has been exposed. Also note that the coarse aggregate is fractured in the process of chipping. Fracturing of coarse aggregate while chipping confirms the good bond of the cement to the aggregate. This observation is a good indication of quality concrete.

In summary, regardless of the bleeding or the evaporation rate, the concrete surface will be weak. The internal tensile strength of concrete can be estimated to be about 8 to 10 percent of its compressive strength. For example, the tensile strength of 4,000 lb concrete is usually 320 to 400 psi. The tensile strength of concrete at the surface is only about 50 psi. Because of this weakness, the surface of the concrete must be removed prior to grouting if good bonding is expected.

Repairing Failures Between Block and Mat

Until recent years, foundation failures on reciprocating equipment between the concrete block and concrete mat were a rare occurrence. At present, this type of failure is becoming more common. This increase in failures can be attributed to poorer construction practices and postponement of equipment maintenance. Before the concrete block is poured, the mat must be chipped to expose course aggregate. This is the only good method of removing the laitance from the surface of the mat and providing an anchor pattern between the block and mat. This requires chipping away at least $\frac{1}{2}''$ to 1" from the surface of the mat. Sandblasting, raking the concrete surface prior to curing, or roughening the surface with a bushing tool as a means of surface preparation is unacceptable. These methods do not remove all the laitance, nor do they expose course aggregate in the concrete.

Lateral dynamic forces are generated by most reciprocating equipment, and in particular with gas-engine compressors. Consider what happens when maintenance is postponed. One would certainly expect distress from an automobile engine with each cylinder operating at a different pressure because of blow-by from defective piston rings. Imagine the same circumstances with a large industrial gas-engine compressor running at full capacity. Next, suppose there are portions of defective grout on the foundation shoulder. If any movement exists between the machine and grout, ever-present spilled oil will penetrate voids caused by the movement, and hydraulically fracture any remaining bond between the machine base and grout. As movement between the machine and grout increases, forces exerted on the foundation increase at an exponential rate; they change their direction and impact billions of times over the life of the machine. Operating under these conditions ultimately results in foundation cracking, separation between the block and mat, or both types of failure.

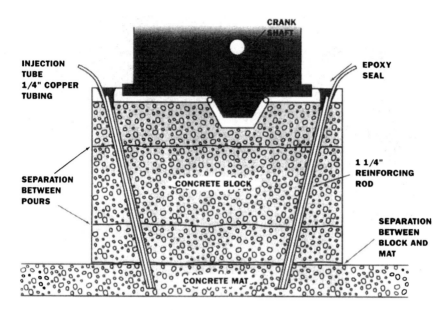

Figure 3-14. Method of repairing compressor foundations where the block has separated from the mat.

Figure 3-14 illustrates a method of repairing separation between the block and mat. Vertical, or near-vertical, holes are drilled through the foundation and into the mat. These holes are usually placed in the foundation around the outer periphery of the equipment. Next, rebar is placed in the holes along with an injection tube, and the entrance of the hole is sealed with an epoxy material. After the seal cures, the annular space around the rebar is pressure-filled with an epoxy liquid, and any cracks that the holes cross are then pressure-grouted from the inside out as pressure builds. The curing of the injected epoxy completes the foundation repair.

Grouting Skid-Mounted Equipment

A skid is a steel structure, used as a shipping platform, that is subsequently installed on a concrete pad or foundation at the job site. This installation concept, most often called "packaging," allows the manufacturer to factory assemble a unit under shop conditions. Packages are frequently complete with accessories, instruments and controls. The cost of packaging is usually much less than would be required for field assembly, particularly where the job site is in a remote part of the world.

Figure 3-15. A typical skid-mounted integral gas engine compressor complete with accessories, controls, and instrumentation.

When the installations are temporary, and relocation of the equipment at a later date is anticipated, cement grouts are generally used. Because cement grouts do not bond well to steel surfaces, lifting the skid at a later date is relatively easy. On the other hand, when the installation is to be permanent, epoxy grouts are generally utilized. The advantage of an epoxy grout lies in the fact that it bonds extremely well to both concrete and steel. Epoxy grouts also provide an oil barrier to protect the underlying concrete foundation. Concrete exposed to lubricating oils over a long period of time can become severely degraded and lose all its structural properties.

A typical skid-mounted integral gas engine compressor is shown in Figure 3-15. When proper techniques are carried out during the original installation, the grout should contact the entire lower surfaces of all longitudinal and transverse "I" beams. Complete contact is necessary in order to prevent vibration when the unit is placed in operation.

Figure 3-16 shows a foundation pad where a skid has been removed leaving the cement grout intact. This photograph illustrates proper cement grouting. Note the impression left in the grout by the lower flange of the longitudinal and transverse "I" beams. Virtually 100 percent grout contact was obtained on these load-bearing surfaces.

Figure 3-17 is an example of poor grout placement. Note the lack of support in the center where most of the machinery weight is concentrated. Long, unsupported spans are an invitation to resonant vibration problems and to progressive sagging of the beams with age. Progressive sagging

Figure 3-16. Proper skid grouting.

Figure 3-17. Poor grout placement on a similar installation.

eventually causes continual misalignment problems. Further, the anchor bolts on the compressor side of the crankcase are attached to one of the internal longitudinal beams. When the equipment is at rest, there may be perfect alignment; however, when the equipment is running, the beam may be flexing much the same as a suspension bridge. If this is true, fatigue of the crankshaft and bearing damage may result.

The obvious solution to this defect is to grout-in the unsupported sections. Since cement grout will not bond well to itself or concrete, any regrouting should be carried out with an epoxy grout because of the inherent bonding properties of epoxies. Some epoxies will even bond to oily surfaces.

Grouting of Oil-Degraded Concrete

In establishing guidelines for the use of epoxy materials on oil-saturated concrete, the expected results should be compared with the properties of good concrete because these were the criteria invoked when the installation was originally designed.

The compressive strength of good concrete will vary from 2,500 to 7,000 psi depending upon its cement content, curing conditions, etc. The internal tensile strength should be about 8 to 10 percent of the compressive strength or 200 to 700 psi. The tensile strength at the surface of formed concrete may be as low as 75 psi and the surface of a steel-troweled floor may be as low as 50–100 psi due to laitance on the surface. Consequently, good surface preparation must be carried out before a satisfactory bond of epoxy to concrete can be obtained.

Experience has definitely shown that the best method of preparing a concrete surface for bonding is through mechanical scarification to remove surface laitance. This scarification can be accomplished by chipping away the surface, sandblasting or grinding in this order of preference. At one time acid washing was widely respected as a means of surface preparation, but this practice has not proved reliable. When contaminants, such as oil or grease, are present, special consideration should be given to surface preparation and epoxy thickness.

Although concrete can absorb oil, the process is, fortunately, relatively slow. Once oil has been absorbed a gradual degradation in both tensile and compressive strengths will follow and given enough time the compressive strength of the concrete may be reduced to the point where it can be crumbled between the fingers. Preventive measures, such as sealing the concrete with an epoxy sealer to provide a barrier, can avoid this problem. This is usually done at the time of original construction. Remedial measures can also be used once the problem has occurred. Most of these remedial techniques involve surface preparation, patching, or transfer of loading.

The importance of epoxy grout thickness is better understood when it is recognized that in solid materials, forces resulting from compressive loading are dispersed throughout the solid in a cone-shaped pattern with the apex at the point of loading. In tensile point loading the force pattern is such that, on failure, a hemispherically shaped crater remains. Consequently, the weaker the concrete, the thicker the epoxy covering should be in order that loading can be sufficiently distributed before force is transferred to the concrete. For example, a severely oil degraded foundation may be capped with a thick layer of epoxy grout in much the same manner as a dentist caps a weak tooth. If you can contain a weak material, you can maintain its strength.

There have been many repair jobs with epoxy grout on foundations of reciprocating machinery where oil degradation was so severe that it was impossible to remove all oil-soaked concrete before regrouting. In such cases, regrouting can sometimes be done with the equipment in place. Such repairs are accomplished by chipping away the oil grout from the foundation shoulder and as far as one-half of the load bearing area under the equipment. It is important that enough grout remain to support the equipment while repairs are being conducted. The advantage of removing some of the old grout under the equipment is to provide a structurally sound area after repairs, equivalent to that supporting area which would be available had the equipment originally been installed on rails or sole plates. Once this is done, the equipment can be pressure-grouted as discussed later in this chapter. Enough concrete is removed to round off the shoulders to a cross-sectional radius of $1\frac{1}{2}$ to 2 ft. Then vertical holes can be drilled into the exposed concrete with a pneumatic rock drill. Usually these holes are placed two ft (about 60 cm) apart and are drilled to a depth to provide penetration through the remaining oil-soaked concrete and at least two ft into the undamaged concrete below. In addition, holes can be drilled in the remaining part of the foundation shoulders at such angles so as to cross below the oil pan at an elevation of approximately two to three ft below the pan or trough. Afterward, additional horizontal reinforcing steel can be installed and wired to the vertical members which were earlier cemented into the good concrete with an epoxy adhesive. The purpose of the new reinforcing steel is to transfer as much load as possible to an area where the concrete was unaffected by oil degradation.

Pressure-Injection Regrouting

Pressure-grouting is a repair process whereby equipment can be reaffixed on the foundation without lifting the equipment, without completely chipping away the old grout, and without repositioning and complete regrouting. Pressure grouting should not, however, be considered a panacea. Nevertheless, when properly used it can be a valuable tool.

Shoulder Removal Method

Pressure-injection regrouting techniques offer equipment operators important advantages of reduced downtime, lower labor costs, and less

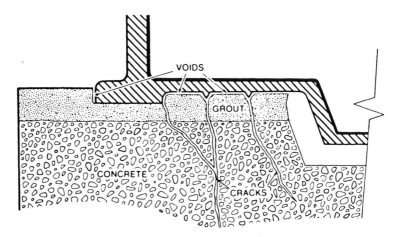

Figure 3-18. Illustrating the damage done to a compressor foundation before making repairs by pressure-injection regrouting, shoulder removal method (courtesy Adhesive Services Company).

revenue lost from idle equipment. These techniques make possible satisfactory and long life regrouts with machine downtime at a minimum.

Figure 3-18 shows typical damage before making repairs. Figure 3-19 is the first step in conducting repairs where the old grout is chipped away along with any damaged or oil soaked concrete. It is desirable to remove enough old grout from beneath the machine so that a load bearing area equivalent to a rail or sole plate mounting can be provided by the epoxy grout once it has been poured. If the foundation itself is cracked, it should be repaired before proceeding further. Otherwise, the effectiveness of the regrout will not be maintained over a long period of time.

After the old grout has been chipped away, holes are drilled into the remaining grout for the installation of injection tubes, and are usually spaced 18 in., or approximately 45 cm, apart (see Figure 3-20). The copper tubing installed in these holes should be sealed with epoxy putty or electrician's putty as illustrated in Figure 3-21.

Before installing forms, all anchor bolts should be isolated to provide at least a $1/4$ in. barrier. This minimizes the possibility of later stress cracking of the grout shoulder and also allows stretching of the anchor bolt from the bottom of the nut to the bottom of the anchor bolt sleeve when torquing the anchor bolts. Forms should be designed to provide a grout level of at least one in. above the machine base (see Figure 3-22). This raised shoulder acts as an effective horizontal restraint for the machine and thereby reduces lateral movement. Forms should be near liquid tight to contain the epoxy grout mortar. Any holes in the forms can be plugged

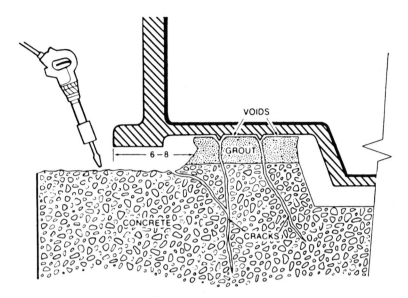

Figure 3-19. The first step of repair is to remove the shoulder and about ½ of the load-bearing area using a pneumatic jumbo rivet buster (courtesy Adhesive Services Company).

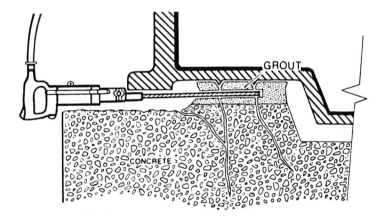

Figure 3-20. Drilling of injection holes into the old grout using a pneumatic drill (courtesy Adhesive Services Company).

with electrician's putty before pouring the mortar. Forms should be waxed with a quality paste wax before installation in order to facilitate easy removal.

Once the forms have been poured and the grout cured for approximately 24 hours, pressure-injection regrouting is carried out to provide a liquid

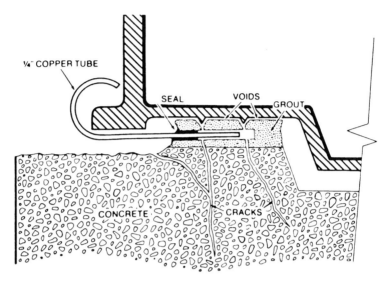

Figure 3-21. Installing copper tubes for pressure-injection (courtesy Adhesive Services Company).

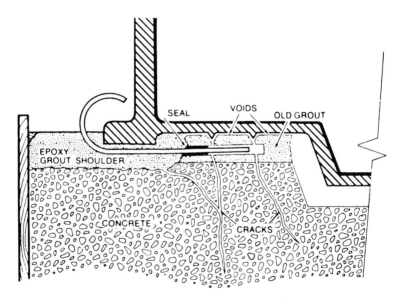

Figure 3-22. Installing forms and repouring the shoulder with an epoxy grout (courtesy Adhesive Services Company).

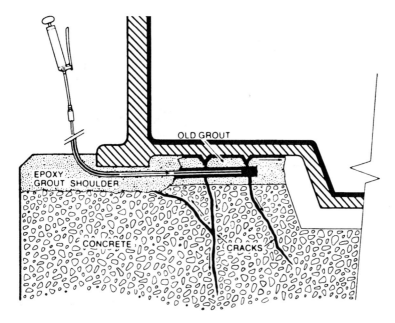

Figure 3-23. Removing forms and pressure-injecting an epoxy adhesive into the cracks and voids under the machine base. Excess epoxy is drained into the trough under the oil pan (courtesy Adhesive Services Company).

shim of epoxy between the remaining old grout and the equipment base as illustrated in Figure 3-23. Once this shim of grout has cured, the forms are removed and the foundation dressed and painted according to Figure 3-24.

Through-the-Case Method

Occasionally, in spite of the best intentions, proper techniques are not used and grout failure results. The cause might be air trapped under the equipment when grouting, or foam under the equipment because of improper grout preparation, or loss of adhesion caused by improper surface preparation. Whatever the cause, there is movement that must be stopped. If the grout and foundation are in good structural condition, pressure-grouting through the case may be a practical solution to the problem. Refer to Figure 3-25 for a good illustration of this grouting method.

With this procedure, the equipment is shut down and the oil removed and cleaned from the crankcase. Holes are drilled through the case and tapped to accommodate grease fittings. Usually, holes are drilled on about

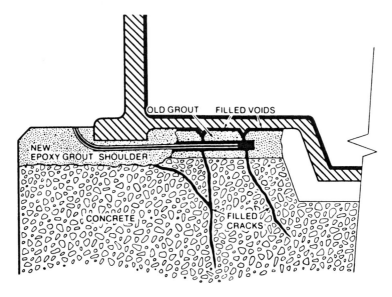

Figure 3-24. The foundation is dressed and painted, thereby completing the repairs (courtesy Adhesive Services Company).

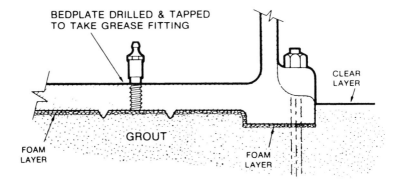

Figure 3-25. Method of rectifying grout installation where surface foam was present[3].

two-ft centers. Alignment is then checked and corrected as necessary. A grease fitting is installed in one of the holes near the center and pressure grouting is begun. Dial indicator gauges should be used to confirm that pressure grouting is being accomplished without lifting the equipment. Pressure grouting should proceed in both directions from the center. As soon as clear epoxy escapes from the adjacent hole, a grease fitting is installed and injection is begun at the next location. This step-wise procedure is continued until clear epoxy is forced from all sides of the equipment. After curing, alignment is rechecked and the equipment returned to service.

Pressure Grouting Sole Plates[3]

Occasionally when installing sole plates, a contractor will fail to clean them properly before grouting. Later this will cause loss of adhesion and result in excessive movement. Pressure grouting of sole plates can be accomplished with a relatively high degree of success if proper techniques are used. Refer to Figure 3-26 for an illustration of this procedure.

A small pilot hole is drilled through the sole plate at a 45° angle beginning about one-third of the distance from the end of the sole plate. The hole is then counterbored. The pilot hole is reamed out and tapped to accommodate $\frac{1}{8}$ in. pipe. Fittings are installed and epoxy can be injected while the equipment is running until the epoxy begins to escape around the outer periphery of the sole plate as illustrated. Usually, oil is flushed out from beneath the sole plate along with the epoxy. Flushing should be continued until clear epoxy appears. It is not uncommon that flow will channel to the extent that epoxy will escape from only 30 to 40 percent of the sole plate circumference during the first injection. The epoxy will

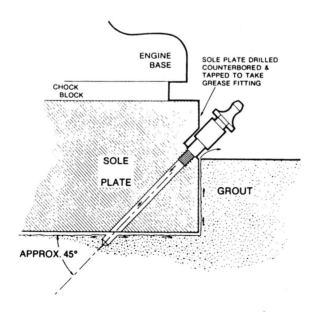

Figure 3-26. Pressure grouting of sole plates[3].

begin to gel in about 15 minutes at the operating temperature of 160° to 180°F. A second injection is carried out after sufficient gelling has been accomplished to restrict flow to ungrouted areas. Two or three injections at about 15-minute intervals are usually required to effect 100 percent coverage under the sole plate. Epoxy will bond through a thin lubricating oil film at about 150 to 200 psi.

One person should be assigned for every two to three sole plates that are loose. To the extent possible, pressure grouting of all sole plates should proceed at the same time. Once the grouting is complete, the equipment should be shut down for at least 6 hours to allow the epoxy to cure. Alignment should be checked and chocks machined or shimmed, if necessary, before the equipment is returned to service.

This technique has also been used without shutting the equipment down. However, it is somewhat less reliable under these circumstances. Nevertheless, if for some unknown reason a sole plate comes loose after it has been grouted, the process can be repeated. It is a simple matter to remove the fittings, rebore and retap the hole, and thereby use the same injection site.

Pressure grouting sole plates seldom changes alignment. We attribute this to the fact that excess epoxy is squeezed from beneath the sole plate by the weight of the equipment. For purposes of illustration, assume the equipment is being aligned with the aid of jack screws having hexagonal heads and ten threads per in. One revolution of the screw would raise or lower the equipment $^{100}/_{1000}$; moving the screw one face would create a change of $^{1}/_{6}$ this amount or $^{16}/_{1000}$. However, one face change on a jack screw is scarcely detectable when measuring web deflections. Nevertheless, the film thickness of epoxy under the sole plate should be far less than $^{16}/_{1000}$. Thus, alignment should not be changed when equipment is pressure-grouted.

Prefilled Equipment Baseplates: How to Get a Superior Equipment Installation for Less Money*

Why Be Concerned

Proper field installation of rotating equipment has a tremendous impact on the life-cycle cost of machinery. According to statistical reliability analysis, as much as 65 percent of the life-cycle costs are determined during the design, procurement, and installation phases of new machin-

---

*Contributed by Todd R. Monroe, P.E., and Kermit L. Palmer, Stay-Tru® Services, Inc., Houston, Texas.

ery applications[5]. While design and procurement are important aspects for any application, the installation of the equipment plays a very significant role. A superb design, poorly installed, gives poor results. A moderate design, properly installed, gives good results[5].

A proper installation involves many facets, such as good foundation design, no pipe strain, and proper alignment, just to name a few. All of these issues revolve around the idea of reducing dynamic vibration in the machinery system. Great design effort and cost are expended in the construction of a machinery foundation, as can be seen in Figure 3-27. The machinery foundation, and the relationship of $F = ma$, is extremely important to the reliability of rotating equipment. Forces and mass have a direct correlation on the magnitude of vibration in rotating equipment systems. The forces acting on the system, such as off-design operating conditions, unbalance, misalignment, and looseness, can be transient and hard to quantify. An easier and more conservative way to minimize motion in the system is to utilize a large foundational mass. Through years of empirical evidence, the rule of thumb has been developed that the foundation mass should be three to five times the mass of the centrifugal equipment system.

Figure 3-27. Construction of machinery foundation.

How well the machinery system is joined to the foundation system is the key link to a proper installation and to reduced vibration. The baseplate, or skid, of the machinery system must become a monolithic member of the foundation system. Machinery vibration should ideally be transmitted through the baseplate to the foundation and down through the subsoil. "Mother earth" can provide very effective damping, i.e., modification of vibration frequency and attenuation of its amplitude. Failure to do so results in the machinery resonating on the baseplate, as shown in Figure 3-28. Proper machinery installation results in significant increase in mean time between failures (MTBF), longer life for mechanical seal and bearings, and a reduction in life-cycle cost[6].

The issue is to determine the most cost-effective method for joining the equipment baseplate to the foundation. Various grouting materials and methods have been developed over the years, but the quest always boils down to cost: life-cycle costs versus first cost. It's the classic conflict between the opposing goals of reliability professionals and project personnel. Machinery engineers want to use an expensive baseplate design with epoxy grout; project engineers want to use a less expensive baseplate design with cementitious grout.

A new grouting method, the Stay-Tru® Pregrouted Baseplate System, and a new installation technique, the Stay-Tru® Field Installation System,

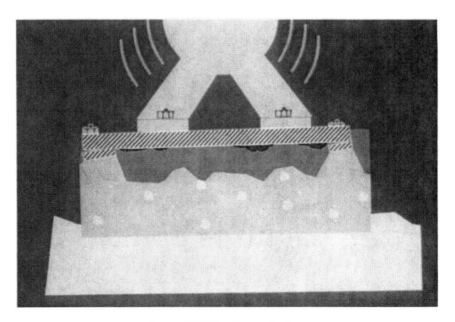

Figure 3-28. Machinery vibration.

bridge the gap, and utilizing these two systems satisfies the requirements of both job functions.

Conventional Grouting Methods

The traditional approach to joining the baseplate to the foundation has been to build a liquid-tight wooden form around the perimeter of the foundation, and fill the void between the baseplate and the foundation with either a cementitious or epoxy grout. There are two methods used with this approach, the two-pour method, shown in Figure 3-29, and the one-pour method, shown in Figure 3-30.

The two-pour method is the most widely used, and can utilize either cementitious or epoxy grout. The wooden grout forms for the two-pour method are easier to build because of the open top. The void between the foundation and the bottom flange of the baseplate is filled with grout on the first pour, and allowed to set. A second grout pour is performed to fill the cavity of the baseplate, by using grout holes and vent holes provided in the top of the baseplate.

The one-pour grouting method reduces labor cost, but requires a more elaborate form-building technique. The wooden grout form now requires a top plate that forms a liquid-tight seal against the bottom flange of the

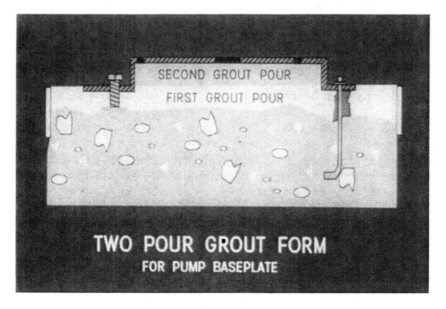

Figure 3-29. Two-pour grout method.

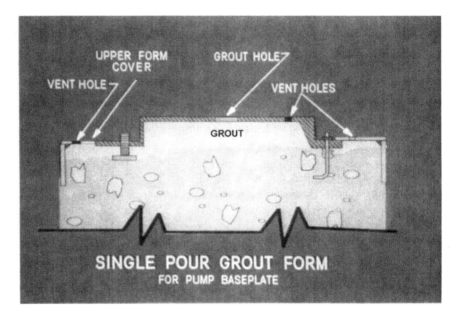

Figure 3-30. One-pour grout method.

baseplate. The form must be vented along the top seal plate, and be sturdy enough to withstand the hydraulic head produced by the grout. All of the grout material is poured through the grout holes in the top of the baseplate. This pour technique requires good flow characteristics from the grout material, and is typically used for epoxy grout applications only.

Field Installation Problems Explained

Grouting a baseplate or skid to a foundation requires careful attention to many details. A successful grout job provides a mounting surface for the equipment that is flat, level, very rigid, and completely bonded to the foundation system. Many times these attributes are not obtained during the first attempt at grouting, and expensive field-correction techniques have to be employed. The most prominent installation problems involve voids and distortion of the mounting surfaces.

Voids and Bonding Issues

As shown in Figure 3-31, the presence of voids at the interface between the grout material and the bottom of the baseplate negates the very

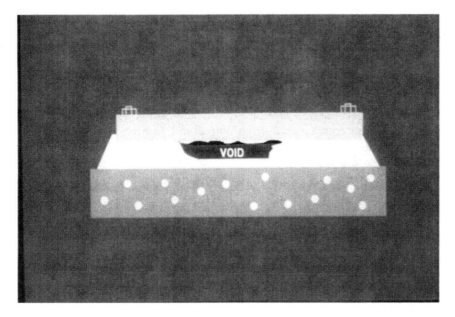

Figure 3-31. Grout void under baseplate.

purpose of grouting. Whether the void is one inch deep, or one-thousandth of an inch deep, the desired monolithic support system has not been achieved. Voids prevent resonance of the foundation system and preclude the dampening of resonance and shaft-generated vibration.

The creation of voids can be attributed to a number of possible causes:

- Insufficient vent holes in baseplate
- Insufficient static head during grout pour
- Nonoptimum grout material properties
- Improper surface preparation of baseplate underside
- Improper surface primer

Insufficient vent holes or static head are execution issues that can be addressed through proper installation techniques. Insufficient attention usually leaves large voids. The most overlooked causes of voids are related to bonding issues. These types of voids are difficult to repair because of the small crevices to be filled.

The first issue of bonding has to do with the material properties of the grout. Cementitious grout systems have little or no bonding capabilities. Epoxy grout systems have very good bonding properties, typically an average of 2,000 psi tensile adherence to steel, but surface preparation and primer selection greatly affect the bond strength. The underside of the baseplate must be cleaned, and the surface must be free of oil, grease, moisture, and other contaminants. All of these contaminants greatly reduce the tensile bond strength of the epoxy grout system.

The type of primer used on the underside of the baseplate also affects the bond between the epoxy grout and the baseplate. Ideally, the best bonding surface would be a sandblasted surface with no primer. Since this is not feasible for conventional grouting methods, a primer must be used, and the selection of the primer must be based on its tensile bond strength to steel. The epoxy grout system bonds to the primer, but the primer must bond to the steel baseplate to eliminate the formation of voids. The best primers are epoxy based, and have minimum tensile bond strength of 1,000 psi. Other types of primers, such as inorganic zinc, have been used, but the results vary greatly with how well the inorganic zinc has been applied.

Figure 3-32 shows the underside of a baseplate sprayed with inorganic zinc primer. The primer has little or no strength, and can be easily removed with the tip of a trowel. The inorganic zinc was applied too thickly, and the top layer of the primer is little more than a powdery matrix. The ideal dry film thickness for inorganic zinc is three mils, and is very hard to

Figure 3-32. Soft inorganic zinc primer.

achieve in practice. The dry film thickness for this example is 9 to 13 mils, as shown in Figure 3-33.

The consequences of applying epoxy grout to such a primer are shown in Figure 3-34, a core sample taken from a baseplate that was free of voids for the first few days. As time progressed, a void appeared, and over the course of a week the epoxy grout became completely "disbonded" from the baseplate. The core sample shows that the inorganic zinc primer bonded to the steel baseplate, and the epoxy grout bonded to the inorganic zinc primer, but the primer delaminated. It sheared apart because it was applied too thickly and created a void across the entire top of the baseplate.

Distortion of Mounting Surfaces

Another field installation problem with costly implications is distortion of the baseplate's machined surfaces. This distortion can be either induced prior to grouting due to poor field leveling techniques, or generated by the grout itself.

Figure 3-33. Dry film thickness indicator.

Figure 3-34. Baseplate core sample with zinc primer.

Baseplate designs have become less rigid over time. Attention has been focused on the pump end of the baseplate to provide enough structural support to contend with nozzle load requirements. The motor end of the baseplate is generally not as rigid, as shown in Figure 3-35. The process of shipping, lifting, storing, and setting the baseplate can have a negative impact on the motor mounting surfaces. Although these surfaces may have initially been flat, there often is work to be done when the baseplate reaches the field.

Using the system of jack bolts and anchor bolts of Figure 3-36, the mounting surfaces can be reshaped during the leveling process, but the

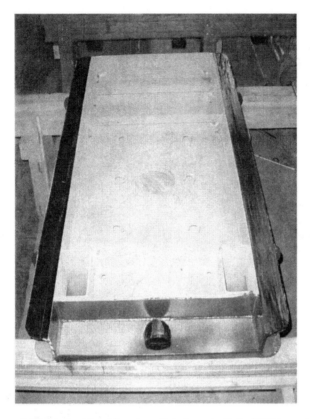

Figure 3-35. Underside of American Petroleum Institute (API) baseplate.

concepts of flatness and level have become confused. Flatness cannot be measured with a precision level, and unfortunately this has become the practice of the day. A precision level measures slope in inches per foot, and flatness is not a slope, it is a displacement. In the field, flatness should be measured with either a ground straightedge or bar and a feeler gauge, as shown in Figure 3-37, not with a level. Once the mounting surfaces are determined to be flat, then the baseplate can be properly leveled. This confusion has caused many baseplates to be installed with the mounting surfaces out of tolerances for both flatness and level.

The other issue of mounting surface distortion comes from the grout itself. All epoxy grout systems have a slight shrinkage factor. While this shrinkage is very small, typically 0.0002″/in, the tolerances for flatness and level of the mounting surfaces are also very small. The chemical reaction that occurs when an epoxy grout resin and hardener are mixed together results in a volume change that is referred to as shrinkage.

Figure 3-36. Anchor bolt and jack bolt system.

Figure 3-37. Flatness and coplanar check.

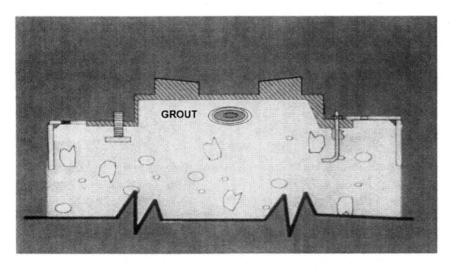

Figure 3-38. Grout cure and mounting surface distortion.

Chemical cross-linking and volume change occur as the material cools after the exothermic reaction. Epoxy grout systems cure from the inside out, as shown in Figure 3-38. The areas closest to the baseplate vs. grout interface experience the highest volume change.

Baseplates with sturdy cross-braces are not affected by the slight volume change of the grout. For less rigid designs, the bond strength of the epoxy grout can be stronger than the baseplate itself. Referring back to Figure 3-38, after the grout has cured the motor mounting surfaces become distorted and are no longer coplanar. Tolerances for alignment and motor soft foot become very difficult to achieve in this scenario. This "pull-down" phenomenon has been proven by finite element analysis (FEA) modeling and empirical lab tests jointly performed by a major grout manufacturer and an industrial grout user.

Hidden Budget Busters

Correcting the problems of voids and mounting surface distortion in the field is a very costly venture. Repairing voids takes a lot of time, patience, and skill to avoid further damage to the baseplate system. Field machining the mounting surfaces of a baseplate also involves commodities that are in short supply: time and money.

The real problem with correcting baseplate field installation problems is that the issues of "repair" are not accounted for in the construction budget. Every field correction is a step backward in both time and money. For a fixed-cost project, the contractor must absorb the cost. In a cost-plus project, the client is faced with the cost. Either way, the parties will have a meeting, which is just another drain on available time and money.

Pregrouted Baseplates

The best way to solve a problem is to concentrate on the cause, rather than developing solutions addressing the effects. The answer for resolving field installation problems is not to develop better void repair procedures or field machining techniques, it is to eliminate the causes of voids and mounting surface distortion.

A new baseplate grouting system has been developed to address the causes of field installation problems. The term *pregrouted baseplate* sounds simple enough, but addressing the causes of installation problems involves far more than flipping a baseplate over and filling it up with grout. In that scenario, the issues of surface preparation, bonding, and mounting surface distortion still have not been addressed. A proper pregrouted baseplate provides complete bonding to the baseplate underside, contains zero voids, and provides mounting surfaces that are flat, coplanar, and colinear within the required tolerances. To assure that these requirements are met, a good pregrout system will include the following.

Proper Surface Preparation

Baseplates that have been specified with an epoxy primer on the underside should be solvent washed, lightly sanded to remove the grossly finish, and solvent washed again. For inorganic zinc and other primer systems, the bond strength to the metal should be determined. There are several methods for determining this, but as a rule of thumb, if the primer can be removed with a putty knife, the primer should be removed. Sandblasting to an SP-6 finish is the preferred method for primer removal. After sandblasting, the surface should be solvent washed, and grouted within 8 hours.

Void-free Grout Installation

By its very nature, pregrouting a baseplate greatly reduces the problems of entrained air creating voids. However, because grout materials are highly viscous, proper placement of the grout is still important to prevent developing air pockets. The baseplate must also be well supported to prevent severe distortion of the mounting surfaces due to the weight of the grout.

A side benefit to using a pregrouted baseplate system is the ability to successfully use cementitious grouts as the fill material. With conventional installation methods, cementitious grout is very difficult to place and has no bond strength to the metal baseplate. With the pregrout system, an epoxy-based concrete adhesive can be applied to the metal prior to the placement of the grout, as shown in Figure 3-39. This technique provides bond strength equal to the tensile strength of the cementitious grout, which is around 700 psi.

For epoxy grout systems, flow ability is no longer an issue, and highly loaded systems can now be employed. Adding pea gravel to the epoxy grout system increases the yield, increases the strength, and reduces the shrinkage factor. Figure 3-40 shows an application using a high-fill epoxy grout system.

Figure 3-39. Epoxy bond adhesive for cementitious grout.

Figure 3-40. High-fill epoxy grout system.

Postcuring of the Grout

As mentioned earlier, epoxy grout systems undergo a slight volume change during the curing process. For conventional installation methods, this physical property creates distortion. While the effects are greatly reduced with the pregrouted system, it is still necessary to allow the epoxy grout to fully cure before any inspection or correction to the mounting surfaces is performed. Figure 3-41 shows a time vs. cure chart that can be used for epoxy grout systems.

For cementitious grout systems, the material should be kept wet and covered for at least 3 days to help facilitate the curing process. While cementitious grout systems are nonshrink and don't induce distortion to the mounting surfaces, the postcuring process helps to achieve full compressive strength. To further enhance the curing process, after 24 hours the grout surface can be sealed with an epoxy resin to prevent contamination and water evaporation (Figure 3-42).

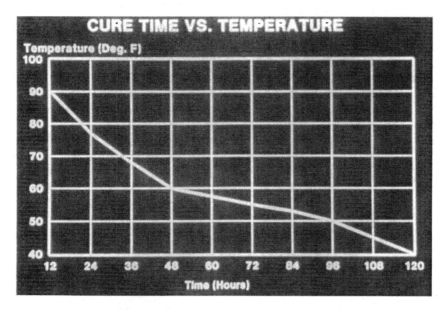

Figure 3-41. Epoxy grout cure time vs. temperature.

Mounting Surfaces

Once the pregrout baseplate has been fully cured, a complete inspection of the mounting surfaces should be performed. If surface grinding of the mounting surfaces is necessary, then a postmachining inspection must also be performed. Careful inspection for flatness, coplanar, and relative level (colinear) surfaces should be well documented for the construction or equipment files. The methods and tolerances for inspection should conform to the following:

Flatness. A precision ground parallel bar is placed on each mounting surface. The gap between the precision ground bar and the mounting surface is measured with a feeler gauge. The critical areas for flatness are within a 2″ to 3″ radius of the equipment hold down bolts. Inside of this area, the measured gap must be less than 0.001″. Outside the critical area, the measured gap must be less than 0.002″. If the baseplate flatness falls outside of these tolerances, the baseplate needs to be surface ground.

Coplanar. A precision ground parallel bar is used to span across the pump and motor mounting pads in five different positions, three lateral and two

Figure 3-42. Epoxy sealer for cementitious grout.

diagonal. At each location, the gap between the precision ground bar and the mounting surfaces is measured with a feeler gage. If the gap at any location along the ground bar is found to be more than 0.002″, the mounting pads are deemed non-coplanar, and the baseplate will need to be surface ground.

Relative Level (Colinear). It is important to understand the difference between relative level and absolute level. Absolute level is the relationship of the machined surfaces to the earth. The procedure for absolute level is done in the field, and is not a part of this inspection. Relative level is an evaluation of the ability to achieve absolute level before the baseplate gets to the field.

The procedure for this evaluation is based on a rough level condition. A Starrett 98 or similar precision level is placed on each machine surface and the rough level measurement, and direction recorded for each machine surface. The rough level measurements of each surface are then compared to each other to determine the relative level. The difference between the rough level measurements is the relative level. The tolerance for relative level is 0.010″/ft.

Field Installation Methods for Pregrouted Baseplates

The use of a proper pregrouted baseplate system eliminates the problem areas associated with field installations. The baseplate has been filled with grout that has properly bonded and is void free. All the mounting surfaces have been inspected, corrected, and documented to provide flat, coplanar, and colinear surfaces. The next step is to join the prefilled baseplate to the foundation system. This can be done using either conventional grouting methods or a new grouting method that is discussed later.

Field Leveling

Knowing that the mounting surfaces already meet flatness and coplanar tolerances makes field leveling of the baseplate very easy. Because the prefilled baseplate is very rigid, it moves as a system during the leveling process. The best method is to use a precision level for each mounting surface. This gives you a clear picture of the position of the baseplate to absolute level. The level must also fit completely inside the footprint of the mounting surface to read properly. If the level is larger than the mounting surface, use a smaller level or a ground parallel bar to ensure that the ends of the level are in contact with the surface.

With the levels in position, adjust the jack bolt and anchor bolt system to the desired height for the final grout pour, typically $1\frac{1}{2}$ to 2 inches for conventional grout. With the grout height established, the final adjustments for level can be made. The baseplate should be leveled in the longitudinal or axial direction first, as shown in Figure 3-43, and then in the transverse direction, as shown in Figure 3-44.

Conventional Grouting Method

Using the conventional method for installing a pregrouted baseplate is no different from the first pour of a two-pour grout procedure. After the concrete foundation has been chipped and cleaned, and the baseplate has been leveled, grout forms must be constructed to hold the grout (Figure 3-45). To prevent trapping air under the prefilled baseplate, all the grout

Figure 3-43. Field leveling in axial direction.

material must be poured from one side. As the grout moves under the baseplate, it pushes the air out. Because of this, the grout material must have good flow characteristics. To assist the flow, a head box should be constructed and kept full during the grouting process.

Hydraulic Lift of a Pregrouted Baseplate

It is important when using a head box that the pregrouted baseplate is well secured in place. The jack bolt and anchor bolt system must be tight, and the anchor bolt nut should be locked down to the equivalent of 30 to 45 ft-lbs.

The bottom of a pregrouted baseplate provides lots of flat surface area. The specific gravity of most epoxy grout systems is in the range of 1.9 to 2.1. Large surface areas and very dense fluids create an ideal environment for buoyancy. Table 3-1 shows the inches of grout head necessary to begin lifting a pregrouted American National Standards Institute (ANSI) baseplate. During the course of a conventional grouting procedure, it is very common to exceed the inches of head necessary to lift a pre-filled baseplate. For this reason, it is very important to assure that the baseplate in

Figure 3-44. Field leveling in transverse direction.

locked down. As a point of interest, the whole range of American Petroleum Institute (API) baseplates listed in Appendix M of API 610 can be lifted with 9 inches of grout head.

Baseplate Stress Versus Anchor Bolt Torque

With the necessity of using the jack bolt and anchor bolt system to lock the pregrouted baseplate in position, it is important to determine if this practice introduces stresses to the baseplate. It is also important to remember that any induced stresses are not permanent stresses, provided they remain below the yield strength of the baseplate. The anchor bolts will be loosened, and the jack bolts removed, after the grout has cured.

Figure 3-45. Pregrout installation using conventional method.

An FEA analysis was performed on a pregrouted ANSI baseplate and a pregrouted API baseplate. The baseplates that were analyzed had six anchor bolt and jack bolt locations, used $^3/_4''$ bolts, and was based on 45 ft-lbs and 100 ft-lbs of torque to the anchor bolts. The 100 ft-lbs of torque was considered to be extremely excessive for leveling and locking down a baseplate, but was analyzed as a worst-case scenario.

The peak local stress loads for 45 ft-lbs was 14,000 psi, and 28,000 psi for 100 ft-lbs. Most baseplates are fabricated from ASTM A36 steel, which has a yield stress of 36,000 psi. As Figure 3-46 shows, the stresses are very localized and decay very rapidly. The result of the FEA analysis shows that the effect of locking down the pregrouted baseplate does not induce any detrimental stresses.

New Field Grouting Method for Pregrouted Baseplates

Conventional grouting methods for nonfilled baseplates, by their very nature, are labor and time intensive. Utilizing a pregrouted baseplate with

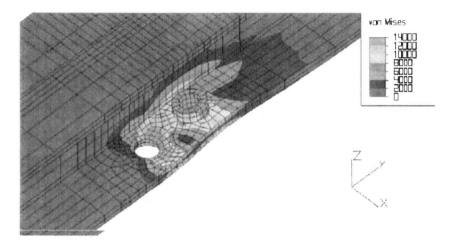

Figure 3-46. Stresses due to 45 ft-lbs anchor bolt preload.

conventional grouting methods helps to minimize some of the cost, but the last pour still requires a full grout crew, skilled carpentry work, and good logistics. To further minimize the costs associated with baseplate installations, a new field grouting method has been developed for pregrouted baseplates. This new method utilizes a low-viscosity, high-strength epoxy grout system that greatly reduces foundation preparation, grout form construction, crew size, and the amount of epoxy grout used for the final pour.

While there may be other low-viscosity, high-strength epoxy grout systems available on the market, the discussion and techniques that follow are based on the flow and pour characteristics of Escoweld® 7560. This type of low viscosity grout system can be poured from $^1/_2''$ to $2''$ depths, has the viscosity of thin pancake batter, and is packaged and mixed in a liquid container. As shown in Figure 3-47, this material can be mixed and poured with a two-man crew.

Concrete Foundation Preparation

One of the leading conflicts on epoxy grout installations is the issue of surface preparation of the concrete foundation. Removing the cement lattice on the surface of the concrete is very important for proper bonding, but this issue can be carried to far (Figure 3-48). Traditional grouting methods require plenty of room to properly place the grout, and this requires chipping all the way to the shoulder of the foundation. Utilizing

Figure 3-47. Mixing of low viscosity epoxy grout.

a low-viscosity epoxy grout system greatly reduces the amount of concrete chipping required to achieve a proper installation.

The new installation method allows for the chipped area to be limited to the footprint of the baseplate (Figure 3-49). A bushing hammer can be used to remove the concrete lattice, and the required depth of the final grout pour is reduced to $\frac{3}{4}''$ to $1''$.

New Grout-forming Technique

With the smooth concrete shoulder of the foundation still intact, a very simple "2 × 4" grout form can be used (Figure 3-50). One side of the simple grout form is waxed, and the entire grout form is sealed and held in place with caulk (refer back to Figure 3-49). While the caulk is setting up, a simple head box can be constructed out of dux seal. Due to the flow characteristics of the low-viscosity epoxy grout, this head box does not need to be very large or very tall.

The low viscosity epoxy grout is mixed with a hand drill, and all the grout is poured through the head box to prevent trapping an air pocket under the baseplate.

Figure 3-48. Chipping of concrete foundation.

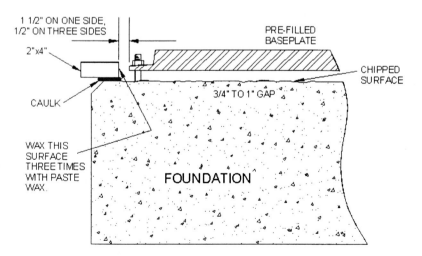

Figure 3-49. New grout installation technique.

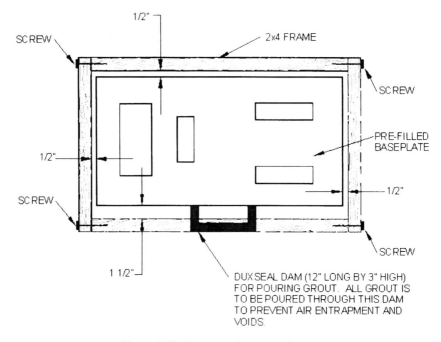

Figure 3-50. New grout-forming technique.

This new installation method has been used for both ANSI- and API-style baseplates with great success. With this technique, field experience has shown that a pregrouted baseplate can be routinely leveled, formed, and poured with a two-man crew in 3 to 4 hours.

Field Installation Cost Comparison

The benefits of using a pregrouted baseplate with the new installation method can be clearly seen when field installation costs are compared. This comparison looks at realistic labor costs, and does not take any credit for the elimination of repair costs associated with field installation problems, such as void repair and field machining.

Years of experience with grouting procedures and related systems point to an average-size grout crew for conventional installations as eight men. As of 2004, an actual man-hour labor cost of $45/hr can be easily defended when benefits and overhead are included.

A cost comparison can be developed, based on the installation of a typical API baseplate using epoxy grout, for the conventional two-pour

Table 3-4
Lifting Forces for ANSI Baseplates

Grout Head Pressure Required to Lift a Pregrouted Baseplate

ANSI Type Baseplates

Base Size	Length (in)	Width (in)	Height (in)	Volume (in³)	Base Weight (lbs)	Epoxy Grout Weight (lbs)	Equalizing Pressure (psi)	Grout Head (in)
139	39	15	4.00	2,340	93	169	0.45	6.22
148	48	18	4.00	3,456	138	250	0.45	6.22
153	53	21	4.00	4,452	178	322	0.45	6.22
245	45	15	4.00	2,700	108	195	0.45	6.22
252	52	18	4.00	3,744	150	271	0.45	6.22
258	58	21	4.00	4,872	195	352	0.45	6.22
264	64	22	4.00	5,632	225	407	0.45	6.22
268	68	26	4.25	7,514	283	544	0.47	6.47
280	80	26	4.25	8,840	332	639	0.47	6.47
368	68	26	4.25	7,514	283	544	0.47	6.47
380	80	26	4.25	8,840	332	639	0.47	6.47
398	98	26	4.25	10,829	407	783	0.47	6.47

Density of Grout 125 lbs/ft³
Specific Gravity 2.00

procedure and a pregrouted baseplate using the new installation method. The following conditions apply:

Baseplate dimensions:	$72'' \times 36'' \times 6''$
Foundation dimensions:	$76'' \times 40'' \times 2''$ (grout depth)
Labor cost:	$45/hr
Epoxy grout cost:	$111/cubic ft

A baseplate with the listed dimensions can be pregrouted for $2,969. This cost would include surface preparation, epoxy grout, surface grinding, and a guaranteed inspection.

Table 3-2 shows a realistic accounting of time and labor for the installation of a typical API baseplate. The total installed cost for a conventional two-pour installation is $6,259. The total installed cost for a pregrouted baseplate, installed with the new installation method, is $4,194. That's a cost savings of almost 50 percent. More importantly, the installation is void-free and the mounting surfaces are in tolerance.

Table 3-5
Cost Comparison for Two-Pour vs. New Method

Installation Labor Cost for Two-Pour Procedure		Installation Labor Cost for Stay-Tru System	
Leveling of Baseplate Millwright: 2 men × 4 hr × $65/h	520	**Leveling of Base Plate** Millwright: 2 men × 1 hr × $65/hr	130
Forming of Baseplate 4 men × 4 hr × $hr =	720	**Forming of Base Plate** 2 men × 2 hr × $hr =	180
First Pour Grout Setup Time 8 men × 1.0 hr × $hr =	360	Grout Setup Time 2 men × 1.0 hr × $hr =	90
Grout Placement 8 men × 2.0 hr × $hr =	720	Grout Placement 2 men × 2.0 hr × $hr =	180
Grout Clean-up 8 men × 1.0 hr × $hr =	360	Grout Cleanup 2 men × 1.0 hr × $hr =	90
Additional Cost Forklift & driver; 1 hr × $45 =	45	**Additional Cost** Wood Forming Materials =	50
Supervisor: 4.0 hr × $hr =	180		
Mortar Mixer =	100		
Wood Forming Materials =	100		
Second Pour Grout Setup Time 8 men × 1.0 hr × $hr =	360		
Grout Placement 8 men × 2.0 hr × $hr =	720		
Grout Cleanup 8 men × 1.0 hr × $hr =	360		
Additional Cost Forklift & driver; 1 hr × $45 =	45		
Supervisor: 4.0 hr × $hr =	180		
Mortar Mixer =	100		
LABOR COST	4,570	**LABOR COST**	670
ADDITIONAL COST	300	**ADDITIONAL COST**	50
GROUT COST	1,389.56	**STAY-TRU COST**	2,969
		GROUT COST (7560)	505.3
TOTAL PER BASE	$6,259.56	TOTAL PER BASE	$4,194.30

Consider Prefilled Baseplates

It is possible to satisfy the concerns of both the project engineer and the machinery engineer regarding rotating equipment installation. The issues of first costs versus life-cycle costs can be reconciled with this new approach to machinery field installations. As an added bonus, the term *repair* can be eliminated from the grouting experience.

References

1. Lee, H. and Neville, K., *Handbook of Epoxy Resins*, McGraw-Hill, New York, 1967, Page 1-1.
2. Adhesive Services Company Sales Literature and Advertising Copy "The Foundation Report" as issued 1983 and 1984.
3. Renfro, E. M., "Five Years with Epoxy Grouts," 19th annual meeting of the Gas Compressor Institute, Liberal, Kansas, Preprint, April 4–5, 1972.
4. Bemiller, Clifford C., "Advances in Setting and Grouting Large Compressor Units," Cooper-Bessemer Company.
5. Barringer, P. and Monroe, T., "How to Justify Machinery Improvements Using Reliability Engineering Principles," *Proceedings of the Sixteenth International Pump Users Symposium*, Turbomachinery Laboratory, Texas A&M University, College Station, Texas, 1999.
6. Myers, R., "Repair Grouting to Combat Pump Vibration," *Chemical Engineering*, August 1998.

Bibliography

Exxon Chemical Company, "Escoweld 7505 Epoxy Grout," OFI 68-251, Houston, Texas.

Exxon Chemical Company, U.S.A., Product Publication Number OFCA-74-1500, 1974, Page 12.

Renfro, E. M., "Foundation repair techniques," *Hydrocarbon Processing*, January, 1975.

Renfro, E. M., "Good foundations reduce machinery maintenance," *Hydrocarbon Processing*, January 1979.

Renfro, E. M., "Preventative design/construction criteria for turbomachinery foundations," C9/83, *Institute of Mechanical Engineers*, London, February 1983.

U.S. Department of Interior, Bureau of Reclamation, *Concrete Manual*, eighth edition, 1975, Page 1.

Waddell, Joseph J., *Concrete Construction Handbook*, McGraw-Hill, New York, 1968, Pages 6–12.

Appendix 3-A

Detailed Checklist for Rotating Equipment: Horizontal Pump Baseplate Checklist

1. CONCRETE FOUNDATION ROUGHED UP TO PROVIDE BOND FOR GROUT. _____

2. CONCRETE FOUNDATION CLEAN AND FREE OF OIL, DUST AND MOISTURE. BLOWN WITH OIL FREE COMPRESSED AIR. _____

3. FOUNDATION BOLT THREADS UNDAMAGED. _____

4. FOUNDATION BOLT THREADS WAXED OR COVERED WITH DUCT SEAL. _____

5. BASEPLATE UNDERSIDE CLEAN AND FREE OF OIL, SCALE, AND DIRT. _____

6. EIGHT POSITIONING SCREWS, TWO PER DRIVER PAD. _____

7. BASEPLATE WELDS CONTINUOUS AND FREE OF CRACKS. _____

8. MOUNTING PADS EXTEND 1 INCH BEYOND EQUIPMENT FEET EACH DIRECTION. _____

9. MOUNTING PADS MACHINED PARALLEL WITHIN 0.002 INCH. _____

10. SHIM PACKS OR WEDGES ON TWO SIDES OF EACH FOUNDATION BOLT. SHIMS ARE STAINLESS STEEL. _____

11. BASEPLATE RAISED TO PROPER HEIGHT PER DRAWING. _____

12. PAD HEIGHTS PERMIT 1/8" MINIMUM SHIM UNDER DRIVER FEET. _____

13. ALL LEVELING DEVICES MAKE SOLID CONTACT WITH CONCRETE AND BASEPLATE. _____

14. ALL MACHINED SURFACES ON BASE LEVEL WITH 0.0005 IN./FOOT IN TWO DIRECTIONS 90° OPPOSED USING "MASTER LEVEL" (0.0005 IN./DIVISION) WITH ANCHOR BOLT NUTS SNUGGED DOWN. _____

15. FOUNDATION AND BASEPLATE PROTECTED FROM DIRT AND MOISTURE CONTAMINATION. _____

WHEN ALL OF ABOVE ACCEPTED, BASEPLATE LEVELING IS ACCEPTED AND BASEPLATE CAN BE GROUTED.

Appendix 3-B

Specification for Portland Cement Grouting of Rotating Equipment

I. SCOPE

1. THIS SPECIFICATION COVERS PORTLAND CEMENT GROUTING OF MECHANICAL EQUIPMENT ON CONCRETE FOUNDATIONS.

II. MATERIALS

1. PORTLAND CEMENT SHALL CONFORM TO ASTM C-150.

2. FILLER SAND AGGREGATE SHALL CONFORM TO ASTM C-33 FINE AGGREGATE.

3. ONLY CLEAN, FRESH WATER SHALL BE USED IN PREPARING THE GROUT.

4. ALL MATERIALS SHALL BE STORED INDOORS AND KEPT DRY, FREE OF MOISTURE IN ITS OWN ORIGINAL SHIPPING CONTAINERS.

5. STORAGE TEMPERATURE SHALL BE MAINTAINED BETWEEN 40°F AND 90°F. GROUTING SHALL NOT BE DONE DURING FREEZING WEATHER.

II. PREPARATION OF FOUNDATION

1. FOUNDATION BOLT THREADS SHALL BE EXAMINED FOR STRIPPED OR DAMAGED THREADS. THESE THREADS SHALL BE RE-CHASED TO CLEAN-UP OR THE FOUNDATION BOLTS REPLACED IF NECESSARY. THE FOUNDATION BOLTS AND THEIR THREADS SHALL BE PROTECTED DURING THE LEVELING AND GROUTING OPERATIONS.

2. THE CONCRETE FOUNDATION SHALL BE DRY AND FREE OF OIL. IF OIL IS PRESENT IT SHALL BE REMOVED WITH SOLVENT PRIOR TO CHIPPING.

3. THE CONCRETE SHALL BE CHIPPED TO EXPOSE ITS AGGREGATE, APPROXIMATELY 1/2 INCH DEEP SO AS TO REMOVE ALL LAITACE AND PROVIDE A ROUGH SURFACE FOR BONDING. LIGHT HAND TOOLS SHALL BE USED IN CHIPPING.

4. AFTER CHIPPING, THE EXPOSED SURFACE SHALL BE BLOWN FREE OF DUST USING OIL-FREE COMPRESSED AIR FROM AN APPROVED SOURCE.

5. AFTER THE FOUNDATION HAS BEEN CHIPPED AND CLEANED,
    IT SHALL BE COVERED SO AS TO PREVENT IT FROM BECOMING
    CONTAMINATED WITH OIL OR DIRT.

6. FOUNDATION BOLTS SHALL BE WAXED OR COVERED WITH DUCT
    SEAL OVER THEIR ENTIRE EXPOSED LENGTH. IF THE BOLTS
    ARE SLEEVED, THE SLEEVES SHALL BE PACKED WITH NON-
    BONDING MATERIAL TO PREVENT THE ANNULAR SPACE AROUND
    THE BOLT FROM BEING FILLED WITH GROUT.

IV. PREPARATION OF BASEPLATE

1. SURFACES OF THE BASEPLATE WHICH WILL COME IN CONTACT
    WITH THE GROUT SHALL BE CLEAN AND FREE OF ALL OIL,
    RUST, SCALE, PAINT, AND DIRT. THIS CLEANING SHALL BE
    DONE SHORTLY BEFORE THE LEVELING OF THE BASEPLATE
    AND PLACEMENT OF THE GROUT.

V. FORMING

1. ALL FORMING MATERIAL COMING INTO CONTACT WITH THE
    GROUT SHALL BE COATED WITH COLORED PASTE WAX -- THE
    COLORING TO CONTRAST WITH THE FORMING MATERIAL COLORS.

2. FORMS SHALL BE MADE LIQUID TIGHT TO PREVENT LEAKING
    OF GROUT MATERIAL. CRACKS AND OPENINGS SHALL BE
    SEALED OFF WITH RAGS, COTTON BATTING, FOAM RUBBER OR
    CAULKING COMPOUND.

3. CARE SHALL BE TAKEN TO PREVENT ANY WAX FROM CONTACT-
    ING THE CONCRETE FOUNDATION OR THE BASEPLATE.

4. LEVELING SCREWS, IF USED, SHALL BE COATED WITH WAX.

VI. WETTING THE FOUNDATION

1. AT LEAST EIGHT (8) HOURS PRIOR TO PLACEMENT OF THE
    GROUT THE CONCRETE SHALL BE WETTED WITH FRESH WATER
    AND KEPT WET UP TO THE ACTUAL PLACEMENT OF THE GROUT.

2. EXCESS WATER IS TO BE REMOVED F OM THE CONCRETE JUST
    PRIOR TO THE PLACEMENT OF THE GROUT.

VII. MIXING THE GROUT

1. THE MIXING EQUIPMENT SHALL BE CLEAN AND FREE OF ALL
    FOREIGN MATERIAL, SCALE, MOISTURE, AND OIL.

2. THE CEMENT AND SAND FILLER SHALL BE HAND BLENDED DRY
    TO FORM A HOMOGENEOUS MIXTURE.

3. FRESH WATER SHALL BE ADDED TO THE DRY MATERIALS AND
    HAND BLENDED TO WET ALL THE DRY MATERIAL. WATER
    QUANTITY SHALL BE THE LESSER OF THE AMOUNT SPECIFIED
    BY THE CEMENT MANUFACTURER OR 50 PERCENT OF THE CEMENT
    BY WEIGHT.

4. THE RATE OF BLENDING SHALL PREVENT AIR ENTRAINMENT
    IN THE GROUT MIXTURE.

VIII. PLACEMENT

1. A SUITABLE HEAD BOX SHALL BE PREPARED TO HYDRAULICALLY
    FORCE THE GROUT INTO THE BASEPLATE CAVITIES.

2. GROUTING SHALL BE CONTINUOUS UNTIL THE PLACEMENT OF
    GROUT IS COMPLETE UNDER ALL SECTIONS OR COMPARTMENTS

OF THE BASEPLATE. IF MORE GROUT IS NEEDED THAN CAN BE PREPARED IN ONE MIXING BOX, A SECOND BOX SHALL BE EMPLOYED. SUBSEQUENT BATCHES OF GROUT SHOULD BE PREPARED SO AS TO BE READY WHEN THE PRECEDING BATCH HAS BEEN PLACED.

3. NO PUSH RODS OR MECHANICAL VIBRATORS SHALL BE USED TO PLACE THE GROUT UNDER THE BASEPLATE.

4. ONE (1) 4" x 4" x 4" TEST CUBE SHALL BE MADE FROM EACH UNIT OF GROUT PLACED. THE SAMPLE(S) SHALL BE TAGGED WITH THE EQUIPMENT NUMBER ON WHICH THE BATCH WAS USED AND WHERE IN THE FOUNDATION THE BATCH WAS PLACED.

IX. FINISHING

1. WHEN THE ENTIRE FORM HAS BEEN FILLED WITH GROUT, EXCESS MATERIAL SHALL BE REMOVED AND THE EXPOSED SURFACES TROWELLED SMOOTH. THE TROWEL SHALL BE WETTED WITH FRESH WATER. CARE SHALL BE TAKEN TO PREVENT BLENDING WATER INTO THE GROUT SURFACE.

2. FORMS SHALL BE LEFT IN PLACE UNTIL THE GROUT HAS SET AND THEN REMOVED.

3. IF WEDGES WERE USED TO LEVEL THE BASEPLATE, THEY SHALL BE REMOVED WHEN THE GROUT HAS SET BUT NOT CURED. AFTER THE GROUT HAS CURED THE WEDGE VOIDS SHALL BE FILLED WITH ESCOWELD 7501 PUTTY.

4. THE REMAINING CEMENT GROUT SURFACES SHALL BE TROWELLED SMOOTH WITH A TROWEL WETTED WITH FRESH WATER. SHARP GROUT EDGES SHALL BE STRUCK TO GIVE A 1/4" CHAMFER.

5. THE GROUT SHALL BE KEPT WETTED WITH FRESH WATER FOR THREE (3) DAYS.

6. THE GROUT SHALL CURE FOR SEVEN (7) DAYS WITH THE AMBIENT TEMPERATURE NOT LESS THAN 50°F PRIOR TO MOUNTING ANY EQUIPMENT ON THE BASEPLATE.

7. THE TOP OF THE BASEPLATE SHALL BE SOUNDED FOR VOIDS. AFTER ALL THE VOIDS HAVE BEEN DEFINED AND THE GROUT CURED, THE BASEPLATE SHALL HAVE TWO HOLES DRILLED THROUGH TO EACH VOID IN OPPOSITE CORNERS OF THE VOID. ONE OF THE PAIR OF HOLES SHALL BE TAPPED AND FITTED WITH A PRESSURE GREASE FITTING. THE OTHER HOLE SHALL BE USED AS A VENT HOLE FOR ITS VOID. THE VOID SHALL BE FILLED WITH UNFILLED ESCOWELD 7505 EPOXY WITH HAND GREASE GUN. CARE MUST BE TAKEN TO PREVENT LIFTING THE BASEPLATE SHOULD A BLOCKAGE OCCUR BETWEEN THE GREASE FITTING AND THE VENT. WHEN THE VOID HAS BEEN FILLED, THE GREASE FITTING SHALL BE REMOVED AND BOTH HOLES DRESSED SMOOTH.

8. ALL EXPOSED GROUT AND CONCRETE SURFACES SHALL BE SEALED WITH ONE COAT OF UNFILLED ESCOWELD 7505 EPOXY HAND APPLIED WITH A PAINT BRUSH. THE EPOXY SEALER SHALL BE TINTED TO CONTRAST WITH THE CONCRETE AND GROUT.

X. CLEAN-UP

1. IMMEDIATELY AFTER GROUTING IS COMPLETED ALL TOOLS AND MIXING EQUIPMENT SHALL BE CLEANED USING FRESH WATER.

Appendix 3-C

Detailed Checklist for Rotating Equipment: Baseplate Grouting

DATE/BY

1. BASEPLATE SURFACES FREE OF OIL, SCALE, AND DIRT.

2. CONCRETE CLEAN AND FREE OF OIL, DUST, AND MOISTURE.

3. FOUNDATION BOLT SLEEVES PACKED WITH NON-BONDING MATERIAL.

4. FOUNDATION BOLTS WAXED OR COVERED WITH DUCT SEAL.

5. VENT HOLES UNOBSTRUCTED.

6. FORMS TO PROPER HEIGHT.

7. FORMS PROPERLY BRACED.

8. FORMS IN CONTACT WITH GROUT WAXED.

9. GROUTING MATERIALS IN UNOPENED CONTAINERS; DRY AND STORED AT APPROXIMATELY 80°F.

10. SUFFICIENT QUANTITY OF MATERIAL ON HAND AT SITE TO COMPLETE WORK.

11. MIXING BOXES AND TOOLS CLEAN.

THE FOLLOWING SECTIONS APPLY TO EPOXY GROUTING ONLY

12. AMBIENT TEMPERATURE ABOVE 65°F DURING MIXING, POUR AND CURE.

13. NO PARTIAL UNITS OF EPOXY, CONVERTER OR SAND AGGREGATE USED.

14. EPOXY AND CONVERTER HAND BLENDED THREE TO FIVE MINUTES MAXIMUM.

134

DATE/BY

15. FULL UNITS OF AGGREGATE SLOWLY ADDED TO BLENDED LIQUID AND HAND MIXED TO JUST COMPLETELY WET AGGREGATE. _____

16. EPOXY/AGGREGATE MIXTURE NOT EXCESSIVELY MIXED SO AS TO ENTRAIN AIR. _____

17. BATCH PLACED WITHIN ITS POT LIFE. AMBIENT TEMP AT POUR OF _____°F. POUR TIME – START OF LIQUID BLENDING TO END OF POUR _____ MINUTES. _____

18. NO VIBRATOR OR RODDING USED TO PLACE GROUT. _____

19. POUR RATE SLOW ENOUGH TO PERMIT AIR TO ESCAPE. _____

20. GROUT FILLS GROUT HOLES AND VENTS. _____

21. FORMS REMOVED AFTER GROUT CURES. _____

THE FOLLOWING APPLIES TO CEMENT GROUT ONLY

22. FOUNDATION WETTED WITH FRESH WATER AT LEAST EIGHT HOURS PRIOR TO PLACEMENT OF GROUT AND KEPT WET WITH FRESH WATER UNTIL POUR BEGINS. EXCESS WATER TO BE REMOVED JUST PRIOR TO PLACEMENT OF GROUT. _____

23. WATER TO CEMENT RATIO NOT MORE THAN 50% BY WEIGHT. WATER TO BE MEASURED AMOUNT PER VENDOR INSTRUCTIONS. _____

24. SAND-CEMENT MIXED DRY TO HOMOGENEOUS BLEND. _____

25. WATER ADDED TO SAND-CEMENT MIXTURE. _____

26. GROUT MIXED TO WET SAND AND CEMENT. NO AIR ENTRAINED. _____

27. POUR SLOW ENOUGH TO PERMIT AIR TO ESCAPE. _____

28. NO VIBRATOR OR RODDING USED TO PLACE GROUT. _____

29. GROUT FILLS GROUT HOLES AND VENTS. _____

30. FORMS REMOVED AFTER GROUT SETS. _____

31. SET GROUT TRIMMED AND TROWELLED SMOOTH. _____

32. SEVEN DAY CURING TIME WITH AMBIENT TEMPERATURE NOT LESS THAN 50°F. _____

33. GROUTED SURFACES KEPT WET WITH FRESH WATER FOR THREE DAYS AFTER PLACEMENT. _____

THE FOLLOWING APPLIES TO BOTH GROUTS

34. WEDGES REMOVED AND VOIDS FILLED. IF SHIMS USED. DO NOT REMOVE. _____

35. BASEPLATE SOUNDED FOR VOIDS. VOIDS FILLED WITH UNFILLED ESCOWELD 7505 INJECTED THROUGH GREASE FITTING. _____

WHEN ALL OF ABOVE ARE ACCEPTED, GROUTING IS ACCEPTED AND EQUIPMENT CAN BE PLACED ON BASEPLATE.

Appendix 3-D

Specifications for Epoxy Grouting of Rotating Equipment

I. SCOPE

THIS SPECIFICATION COVERS EPOXY GROUTING OF MECHANICAL EQUIPMENT ON CONCRETE FOUNDATIONS.

II. MATERIALS

1. EPOXY SHALL BE ESCOWELD 7505 TWO PART LIQUID (EPOXY RESIN AND CONVERTER).

2. FILLER SHALL BE ESCOWELD 7530 SAND AGGREGATE.

3. ALL MATERIALS SHALL BE STORED INDOORS AND KEPT DRY, FREE OF MOISTURE IN ITS ORIGINAL SHIPPING CONTAINERS.

4. STORAGE TEMPERATURES SHALL BE MAINTAINED BETWEEN 40°F AND 90°F. GROUTING MATERIAL SHALL BE KEPT AT 65°F MINIMUM FOR 48 HOURS PRIOR TO MIXING AND PLACEMENT.

III. PREPARATION OF FOUNDATION

1. FOUNDATION BOLT THREADS SHALL BE EXAMINED FOR STRIPPED OR DAMAGED THREADS. THESE THREADS SHALL BE RECHASED TO CLEAN UP OR THE FOUNDATION BOLT REPLACED IF NECESSARY. THE FOUNDATION BOLTS AND THEIR THREADS SHALL BE PROTECTED DURING THE LEVELING AND GROUTING OPERATIONS.

2. THE CONCRETE FOUNDATION SHALL BE DRY AND FREE OF OIL. IF OIL IS PRESENT IT SHALL BE REMOVED WITH SOLVENT PRIOR TO CHIPPING.

3. THE CONCRETE SHALL BE CHIPPED TO EXPOSE ITS AGGREGATE APPROXIMATELY 1/2 INCH DEEP SO AS TO REMOVE ALL LAITANCE AND PROVIDE A ROUGH SURFACE FOR BONDING. LIGHT HAND TOOLS SHALL BE USED IN CHIPPING.

4. AFTER CHIPPING, THE EXPOSED SURFACES SHALL BE BLOWN FREE OF DUST USING OIL FREE COMPRESSED AIR FROM AN APPROVED SOURCE.

5. AFTER THE FOUNDATION HAS BEEN CHIPPED AND CLEANED, IT SHALL BE COVERED SO AS TO PREVENT IT FROM BECOMING WET.

6. FOUNDATION BOLTS SHALL BE WAXED OR COVERED WITH DUCT SEAL OVER THEIR ENTIRE EXPOSED LENGTH. IF THE BOLTS ARE SLEEVED, THE SLEEVES SHALL BE PACKED WITH NON-BONDING MATERIAL TO PREVENT THE ANNULAR SPACE AROUND THE BOLT FROM BEING FILLED WITH GROUT.

IV. PREPARATION OF BASEPLATE

1. SURFACES OF THE BASEPLATE WHICH WILL COME IN CONTACT WITH THE EPOXY GROUT SHALL BE CLEAN AND FREE OF ALL OIL, RUST, SCALE, PAINT AND DIRT. THIS CLEANING SHALL BE DONE IMMEDIATELY PRIOR TO THE LEVELING OF THE BASEPLATE AND PLACEMENT OF THE EPOXY GROUT.

2. IF THE GROUTING IS TO BE DELAYED BY MORE THAN EIGHT (8) HOURS AFTER THE BASEPLATE IS CLEANED, THE CLEANED BASEPLATE SHALL BE PAINTED WITH ONE COAT OF UNFILLED ESCOWELD 7505 TO GIVE A DRY FILM THICKNESS OF THREE (3) MILS. THIS COAT SHALL BE FULLY DRIED PRIOR TO PLACEMENT OF GROUT.

3. IF THE EPOXY COATED BASEPLATE IS NOT GROUTED WITHIN THIRTY (30) DAYS AFTER BEING PAINTED, THE EPOXYED SURFACE SHALL BE ROUGHED UP WITH A WIRE BRUSH TO REMOVE THE BLOOM OR SHINE. ALL DUST PRODUCED BY BRUSHING SHALL BE WIPED OFF USING A WATER DAMPENED RAG. THESE SURFACES SHALL BE AIR DRIED PRIOR TO PLACEMENT OF GROUT.

V. FORMING

1. ALL FORMING MATERIAL COMING INTO CONTACT WITH THE GROUT SHALL BE COATED WITH COLORED PASTE WAX. THE COLORING TO CONTRAST WITH THE FORMING MATERIALS COLORS.

2. FORMS SHALL BE MADE LIQUID TIGHT TO PREVENT LEAKING OF GROUT MATERIAL. CRACKS AND OPENINGS SHALL BE SEALED OFF WITH RAGS, COTTON BATTING, FOAM RUBBER OR CAULKING COMPOUND.

3. CARE SHOULD BE TAKEN TO PREVENT ANY WAX FROM CONTACTING THE CONCRETE FOUNDATION OR THE BASEPLATE.

4. ANY LEVELING WEDGES THAT ARE TO BE REMOVED AFTER THE GROUT HAS CURED SHALL BE COATED WITH WAX. WEDGES AND/OR SHIMS WHICH ARE TO REMAIN EMBEDDED IN THE GROUT SHALL NOT BE COATED.

5. LEVELING SCREWS, IF USED, SHALL BE COATED WITH WAX.

VI. MIXING

1. AS EPOXY COMPOUNDS HAVE LIMITED POT LIFE AFTER MIXING, BOTH THE ELAPSED TIME FROM BEGINNING OF MIXING TO THE COMPLETION OF THE POUR AND THE AMBIENT TEMPERATURES AT THE BEGINNING OF MIXING AND AT THE COMPLETION OF POUR SHALL BE RECORDED AND GIVEN TO THE TURBOMACHINERY ENGINEER WHO WILL RECORD THE DATA IN THE PERMANENT EQUIPMENT RECORDS.

2.   AMBIENT TEMPERATURE SHALL BE AT LEAST 65°F DURING MIXING, PLACEMENT AND CURING OF THE EPOXY GROUT.

3.   MIXING EQUIPMENT SHALL BE CLEAN AND FREE OF ALL FOREIGN MATERIAL, SCALE, MOISTURE AND OIL.

4.   ALL PERSONNEL HANDLING OR WORKING WITH THE GROUTING MATERIALS SHALL FOLLOW SAFETY INSTRUCTIONS PRINTED ON THE CAN LABELS.

5.   ONLY FULL UNITS OF EPOXY RESIN, CONVERTER AND AGGREGATE SHALL BE USED IN PREPARING THE GROUT.

6.   THE EPOXY RESIN (PART A - BLUE LABEL) AND THE CONVERTER (PART B - RED LABEL) SHALL BE HAND BLENDED FOR AT LEAST THREE (3) MINUTES BUT NO MORE THAN FIVE (5) MINUTES PRIOR TO ADDING THE AGGREGATE FILLER.

7.   IMMEDIATELY AFTER THE LIQUID BLENDING HAS BEEN COMPLETED, THE AGGREGATE SHALL BE SLOWLY ADDED AND GENTLY HAND BLENDED SO AS TO FULLY WET THE AGGREGATE. THE RATE OF BLENDING SHALL PREVENT AIR ENTRAINMENT IN THE GROUT MIXTURE.

VII.   PLACEMENT

1.   A SUITABLE HEAD BOX SHALL BE PREPARED TO HYDRAULICALLY FORCE THE GROUT INTO THE BASEPLATE CAVITIES.

2.   GROUTING SHALL BE CONTINUOUS UNTIL THE PLACEMENT OF GROUT IS COMPLETE UNDER ALL SECTIONS OR COMPARTMENTS OF THE BASEPLATE.   IF MORE THAN ONE UNIT OF GROUTING MATERIALS IS REQUIRED FOR A GIVEN PIECE OF EQUIPMENT, TWO MIXING BOXES SHALL BE EMPLOYED.   SUBSEQUENT BATCHES OF GROUT SHOULD BE PREPARED SO AS TO BE READY WHEN THE PRECEDING BATCH HAS BEEN PLACED.

3.   NO PUSH RODS OR MECHANICAL VIBRATORS SHALL BE USED TO PLACE THE GROUT UNDER THE BASEPLATE.

4.   ONE (1) 4" x 4" x 4" TEST CUBE SHALL BE MADE FROM EACH UNIT OF GROUT PLACED.   THE SAMPLE(S) SHALL BE TAGGED WITH THE EQUIPMENT NUMBER ON WHICH THE BATCH WAS USED AND WHERE IN THE FOUNDATION THE BATCH WAS PLACED.

VIII.   FINISHING

1.   WHEN THE ENTIRE FORM HAS BEEN FILLED WITH GROUT, EXCESS MATERIAL SHALL BE REMOVED AND THE EXPOSED SURFACES TROWELLED SMOOTH.   THE TROWEL SHALL BE WETTED WITH A SMALL AMOUNT OF ESCOWELD 9615 CLEAN-UP SOLVENT.   CARE SHALL BE TAKEN TO PREVENT BLENDING THE SOLVENT INTO THE GROUT SURFACE.

2.   ANY AIR BUBBLES RISING TO THE SURFACE OF THE GROUT SHALL BE REMOVED BY LIGHTLY SPRAYING THE BUBBLED SURFACE WITH ESCOWELD 9615 CLEAN-UP SOLVENT USING A HAND SPRAYER.

3.   FORMS SHALL BE LEFT IN PLACE FOR TWENTY-FOUR (24) HOURS TO PERMIT THE GROUT TO CURE.   THE SURFACE OF THE GROUT SHOULD BE FIRM AND NOT TACKY TO THE TOUCH.

4. IF WEDGES WERE USED TO LEVEL THE BASEPLATE, THEY SHALL BE REMOVED AT THIS TIME AND THEIR VOIDS IN THE GROUT FILLED WITH ESCOWELD 7501 PUTTY. SHIMS SHALL BE LEFT IN PLACE.

5. THE TOP OF THE BASEPLATE SHALL BE SOUNDED FOR VOIDS. AFTER ALL THE VOIDS HAVE BEEN DEFINED THE BASEPLATE SHALL HAVE TWO HOLES DRILLED THROUGH TO EACH VOID IN OPPOSITE CORNERS OF THE VOIDS. ONE OF THE PAIR OF HOLES SHALL BE TAPPED AND FITTED WITH A PRESSURE GREASE FITTING. THE OTHER HOLD SHALL BE USED AS A VENT HOLE FOR ITS VOID. THE VOID SHALL BE FILLED WITH UNFILLED ESCOWELD 7505 WITH A HAND GREASE GUN. CARE MUST BE TAKEN TO PREVENT LIFTING THE BASEPLATE SHOULD A BLOCK-AGE OCCUR BETWEEN THE GREASE FITTING AND THE VENT. WHEN THE VOID HAS BEEN FILLED THE GREASE FITTING SHALL BE REMOVED AND BOTH HOLES DRESSED SMOOTH.

6. ALL EDGES OF THE EPOXY GROUT SHALL BE DRESSED SMOOTH BY GRINDING TO A 1/4 INCH CHAMFER.

IX. CLEAN-UP

1. IMMEDIATELY AFTER GROUTING IS COMPLETED ALL TOOLS AND MIXING EQUIPMENT SHALL BE CLEANED USING EITHER WATER OR ESCOWELD 9615 CLEAN-UP SOLVENT.

2. ALL UNUSED MIXED EPOXY MATERIALS SHALL BE DISPOSED OF IN ACCORDANCE WITH INSTRUCTIONS ON THE EPOXY CONTAINERS.

Appendix 3-E

Specification and Installation of Pregrouted Pump Baseplates

This appendix, or standard procedure, outlines the requirements for specifying and installing pregrouted machinery baseplates. However, this standard does not cover the installation requirements for machinery mounted on sole plates. A typical application for this standard procedure would be the installation or retrofit of an ANSI or API pump.

I. Purpose

The purpose of the standard procedure is to provide specific requirements for prefilling any machinery baseplate, or, in particular, a pump baseplate, with epoxy grout, and machining the mounting surfaces of the baseplate after the grout has cured. Additionally, the standard procedure outlines the requirements for installing the pregrout baseplate in the field, utilizing a special grouting technique for the final grout pour. This special technique makes use of a low-viscosity epoxy grout for the final pour. The technique greatly reduces the field costs associated with traditional installation methods.

By utilizing this standard procedure, baseplate-mounted machinery can be installed with zero voids, eliminating the possibility of expensive field machining, and reducing field installation costs by 40 to 50 percent.

II. Specification of Pregrouted Baseplates

1. The underside of the baseplate to be pregrouted must be sand-blasted to white metal to remove all existing paint, primer, or scale.
2. Any tapped bolt holes that penetrate through the top of the baseplate, such as the coupling guard holding down bolts, must be filled with the appropriate-sized bolts and coated with never-seize to create the necessary space for bolt installation after grouting of the baseplate.
3. Anchor bolt or jack bolt holes, located inside the grouted space of the baseplate, must have provisions for bolt penetration through the baseplate after grouting.
4. If the baseplate has grout holes and/or vent holes, these holes must be completely sealed prior to grouting.
5. All pregrouted baseplates will be filled with catalyzed epoxy grout or a premium nonshrink cement grout.
6. Once the baseplate has been filled with epoxy grout, the grout must be completely cured before any machining is performed.
7. The machining of the baseplate must be set up to assure that the baseplate is under no stress or deformation.
8. Prior to machining, the baseplate must be adjusted and leveled to assure that no more than 0.020" of metal is removed at the lowest point.
9. The baseplate will have two (2) mounting surfaces for the driver, and two (2) to four (4) mounting surfaces for the driven equipment. The flatness tolerance for all these mounting surfaces will be 0.001" per ft. The finished surface roughness must be no more than an 85P profiled surface.
10. The two (2) mounting surfaces for the driver must be coplanar within 0.002". The two (2) to four (4) mounting surfaces for the driven equipment must also be coplanar within 0.002". The original dimensional relationship (elevation) between the driver mounting surfaces and the driven mounting surfaces must be maintained to within 0.020".
11. Once the machining process has been completed, an "as-machined" tolerance record must be taken, and provided with the pregrouted baseplate.

III. General Field Grouting Requirements

1. The epoxy grout utilized for the final field grout pour is a low-viscosity epoxy grout. This grout has a special aggregate and has

the consistency of thin pancake batter. This allows for a very thin final grout pour, with the optimum vertical thickness being 3/4".

2. All grout material components must be stored in a dry and weatherproof area in original unopened containers. Under no circumstances should grouting components be stored outside subject to rain or under a tarpaulin with no air circulation.

3. For optimum handling characteristics, precondition the resin and hardener to a temperature between 64° and 90°F.

4. The work area, including foundation and machinery, must be protected from direct sunlight and rain. This covering (shading) should be erected 48 hours prior to alignment and grouting, and shall remain until 24 hours after placement of the grout, by which time the grout will have cured and returned to ambient temperature. The shading is also to prevent the foundation from becoming wet. It is important that the concrete remain dry prior to grouting.

5. Grouting shall be scheduled to take place during early morning or afternoon hours depending on the surface temperatures.

6. Just before starting the grouting operation, the temperature of the concrete foundation and machinery shall be taken using a surface thermometer. Ideal surface temperatures shall be between 70° and 90°F.

IV. Foundation Preparation

1. The concrete must be chipped to expose a minimum of 50 percent aggregate so as to remove all laitance and provide a rough surface for bonding. Hand-chipping guns only will be used. No jackhammers will be permitted. If oil or grease are present, affected areas will be chipped out until free of oil or grease.

2. The concrete to be chipped should not extend more than 2" outside the "footprint" of the pregrouted baseplate. Low-viscosity epoxy grout can be poured only up to a 2" depth, and should not extend more than 2" from the edge of the baseplate. By limiting the chipped area of the concrete to just outside the foot print of the baseplate, simple forming techniques can be utilized.

3. After chipping, the exposed surface must be blown free of dust and concrete chips using oil- and water-free compressed air from an approved source. Concrete surface can also be vacuumed.

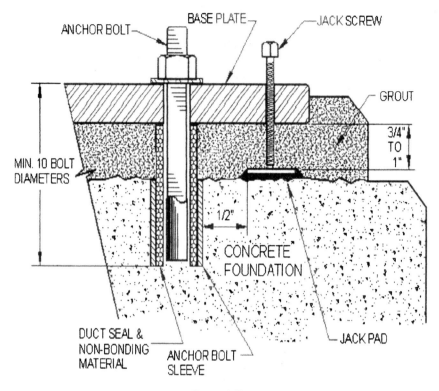

ANCHOR BOLT

BASE PLATE

JACK SCREW

GROUT

3/4" TO 1"

MIN. 10 BOLT DIAMETERS

1/2"

CONCRETE FOUNDATION

DUCT SEAL & NON-BONDING MATERIAL

ANCHOR BOLT SLEEVE

JACK PAD

Figure 3-E-1.

4. After the foundation has been chipped and cleaned, adequate precautions must be taken to ensure that there is no contamination of the concrete surfaces. To prevent debris, loose materials, or parts from falling on the top of the concrete, properly cover the workspace with polyethylene sheet.

5. The foundation bolt threads must be protected during the grouting operations.

6. As regards the bolts (Figure 3-E-1), which will be tensioned after grouting, care must be taken to prevent the bolt surfaces from coming in contact with the epoxy grout. All anchor bolts should have grout sleeves, which must be filled with a nonbonding material to prevent the epoxy grout from filling the grout sleeve. This can be accomplished by protecting the anchor bolt beforehand between the top of the grout sleeve and the underside of the baseplate by wrapping the bolt with foam insulation, Dux-Seal, or other nonbonding material.

V.   Pregrouted Baseplate Preparation

1. Prior to positioning the baseplate over the foundation, the bottom side of the prefilled baseplate must be solvent washed to remove any oil or other contaminants from the surface. After the surface has been cleaned, sand the surface to break the glaze of the epoxy grout.
2. Vertical edges of the baseplate that come in contact with the epoxy grout must be radiused/chamfered to reduced stress concentration in the grout.
3. Vertical jackscrews should be provided at each anchor bolt. The jackscrews will be used to level the pregrouted baseplate. These jackscrews will be removed after the low-viscosity epoxy cures, generally 24 hours after placement at 78°F.
4. Leveling pads should be used under each jackscrew to prevent the baseplate from "walking" while leveling the baseplate. The pads will remain in the grout, and must be made from stainless steel. The pads must have radius edges and rounded corners to reduce stress concentrations in the grout.
5. With the jackscrews and leveling pads in place, level the pregrouted baseplate to 0.002"/foot for API applications and 0.005"/foot for ANSI applications.
6. After the baseplate has been leveled, the jackscrews must be greased or wrapped with Dux-Seal to facilitate their removal once the grout has cured. Wax is not a suitable releasing agent.

VI.   Forming

1. Low-viscosity epoxy grout should only be poured up to a 2" depth, and should not extend more than 2" from the edge of the baseplate. The optimum pour depth is 3/4" to 1". The best wood-forming material for this product is a "2 by 4".
2. Any wood surface coming in contact with epoxy grout shall be coated three times with paste floor wax. (Liquid wax or oil is not acceptable as an alternative.) All forms must be waxed three times *before* the forms are placed on the foundation. Do *not* wax the wood surface that comes in contact with the foundation. This may prevent the silicone sealant from sticking to the form board, and forming a proper seal. Care should be taken to

prevent any wax from falling on the concrete foundation or the baseplate.

3. In most cases, there is very little room between the side of the baseplate and the edge of the foundation. To help position the form boards, it is best to fasten the boards together with wood screws or nails. One side of the form should leave an opening between the board and the baseplate that measures 1″ to 2″. The other three sides should have a separation of $\frac{1}{2}$″ to $\frac{3}{4}$″. The larger side will be used to pour the low-viscosity epoxy grout.

4. Forms shall be made liquid tight. Silicone sealant that does not cure to a hard consistency is best suited for this application. A sealant that remains pliable will facilitate easy removal. The best approach is to apply the sealant directly on the foundation, where the front edge of the form will fall, and then press the form down to create the seal. Check for cracks and openings between the form and the foundation, and apply additional sealant where needed. Allow at least an hour for the sealant to cure before pouring the grout.

5. Because of the small depth of the epoxy grout pour, it will be very difficult to use a chamfer stripe to create a bevel around the outside of the form. The best approach is to use a grinder after the grout has cure to create a bevel.

6. Once the forms boards are in place, a small "head box" can be made using blocks of duct seal. To help create a slight head for the grout, build a duct seal dam on the side of the baseplate with the larger opening. The dam should be about 3″ tall, 12″ long, and form a rectangle by connecting the two short sides direct to the baseplate. The end result will be a 3″ head box that will be used to pour and place all the epoxy grout. The best location for the head box is the midpoint of the baseplate.

VII. Grouting Procedure for Low-Viscosity Epoxy Grout

1. The required number of units of epoxy grout, including calculated surplus, should be laid out close to the grouting location. The $\frac{1}{2}$″ drill and mixer blade should be prepared for the grouting operation.

2. Low-viscosity epoxy grout is a three-component, high-strength, 100 percent solids epoxy grouting compound. The resin, hard-

ener, and aggregate are supplied in a 6-gallon mixing container. One unit produces 0.34 cubic feet of grout.

Inspection of Work Site

Check for:
- Proper shading
- Preparation of concrete, baseplate, jackscrews, leveling pads
- Wood forms properly waxed and sealed
- Foundation bolts wrapped and sealed

Before Mixing

Check for:
- Mixing equipment clean
- Surface temperature of epoxy grout components (<90°F)
- Ambient temperature (<95°F)

While Mixing

Check for:
- Slow drill motor rpm's to avoid entrapping air
- Resin and hardener mixed 3 minutes (use the timer)

Before Pouring

Check for:
- Temperature of concrete foundation (<95°F)
- Temperature of the machinery baseplate (<95°F)

While Pouring

Check for:
- Continuous operation
- Adequate head to fill corners in baseplate

After Pouring

Check for:
- Ambient temperature for the record
- Maintaining head until the grout starts to set

Curing

Check for:

- Work site kept shaded for 24 hours to avoid sharp temperature increase
- Formwork left in place until grout is no longer tacky to the touch

Chapter 4

Process Machinery Piping*

Fundamentals of Piping Design Criteria

It is certainly not within the scope of this text to deal extensively with piping design and installation criteria; however, there are certain fundamentals which can have an impact on machinery reliability. These must be appreciated by the machinery engineer if he is to retain a good overview of the integrated machinery system. Some key installation procedures and verification criteria are, therefore, included for the machinery engineer's benefit.

The design of a piping system consists of the design of pipe, flanges, bolting, gaskets, valves, fittings, and other pressure components such as expansion joints. It also includes *pipe supporting elements* but does not include the actual support structures such as building steel work, stanchions or foundations, etc.

Piping Design Procedure

These steps need to be completed in the design of any piping system:

- Selection of pipe materials
- Calculation of minimum pipe wall thickness for design temperatures and pressures (generally per ANSI B31.3)
- Establishment of acceptable layout between terminal points for the pipe

*We gratefully acknowledge the help received from Wolfgang Schmidt (formerly with Essochem Europe Brussels, Belgium), in compiling this section.

- Establishment of acceptable support configuration for the system
- Flexibility stress analysis for the system to satisfy the design criteria stipulated by ANSI B31.3

This flexibility analysis is intended to verify that piping stresses, local component stresses and forces/moments generated at the terminal points are within the acceptable limits throughout all anticipated phases of normal and abnormal operation of the plant during its life.

Design Considerations

A piping system constitutes an irregular space frame into which strain and attendant stress may be introduced by initial fabrication and erection, and also may exist due to various circumstances during operation. Example: three pumps taking suction from and/or discharging into a common header, as shown in Figure 4-1. One or two of the three pumps removed for shop repair.

Each piping system must be designed with due consideration to these circumstances for the most severe conditions of coincident loading. The following summarizes possible imposed loads that typically need to be considered in a piping design:

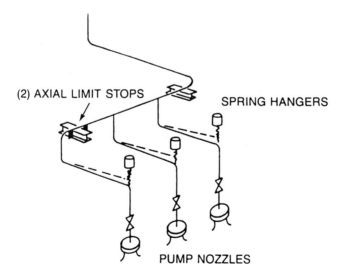

Figure 4-1. Flexibility analysis must consider:
- All pumps operating simultaneously
- Effect of any pump used as nonrunning standby spare, or blocked off for maintenance.

Design Pressure Loads

The pressure at the most severe condition of coincident internal or external pressure and temperature expected during normal operation.

Weight Loads

- Dead weight loads including pipe components, insulation, etc.
- Live weight loads imposed by service or test fluid, snow and ice, etc.

Dynamic Loads

- Design *wind* loads exerted on exposed piping systems
- *Earthquake* loads must be considered for piping systems where earthquake probability is significant
- *Impact* or *surge* loads typically due to water hammer, letdown, or discharge of fluids
- *Excessive vibration* arising from pressure pulsations, resonance caused by machinery excitations or wind loads

Thermal Expansion/Contraction Effects

- Thermal and friction loads due to restraints preventing free thermal expansion
- Loading due to severe temperature gradients or difference in expansion characteristics

Effects of Support, Anchor, and Terminal Movements

- Thermal expansion of equipment
- Settlement of equipment foundations and/or piping supports

The When, Who, What, and How of Removing Spring Hanger Stops Associated with Machinery

Initial Tasks Prior to Machinery Commissioning

- Align machinery without pipe attached
- Adjust pipe for proper fit-up and make connection

- Observe alignment of machinery with pipe being attached. If excessive movement is noted, the pipe is to be disconnected and modified until misalignment is brought within the limits permitted.
- If the pipe is greater than 8 in. NPS, one may need to add sandbags or similar weights to the pipe at the hanger adjacent to machinery to simulate the operating condition of the pipe.
- Pull stops on all system hangers
- Check to determine that no hanger travel indicator moves out of the "$^1/_3$ total travel" cold setting zone. If travel is excessive, refer immediately to the design contractor for modifications of support.
- Adjust the hanger to return travel marker to the "C" position
- Record alignment of machinery
- Reinstall piping system hanger stops

Final Check, Immediately Prior to Machinery Operation

- Disconnect or dismantle piping as necessary
- Flush and/or steam blow
- Repipe and realign
- Weight the hanger adjacent to the machinery
- Pull system pins, check "C" settings and fine tune hangers. If travel is excessive (out of the $^1/_3$ total "C" zone) contact the designated piping engineer for resolution.

Flange Jointing Practices

These steps can be written up in checklist format allowing field personnel to use piping-related guidelines in an efficient manner, as shown in the appendices at the end of this chapter.

The importance of getting flanged joints right the first time cannot be overemphasized if trouble-free performance during startup is desired. In order to obtain an adequate joint the first time we must assure ourselves that the contractor, subcontractor, and the working crews appreciate the importance of quality workmanship needed during each stage of the flange joint building process. This includes materials handling and storage operations, piping prefabrication, erection, and bolting-up procedures. Time spent in covering preventive measures, supervision and crew guidance, and/or training (if needed), and assuring adequate quality control will pay dividends.

Primary Causes of Flange Leakage

Several common causes of flange leakage are hereby outlined to create an awareness of the effects of poor inspection procedure or materials:

Uneven Bolt Stress. Flanges bolted up unevenly cause some bolts to be nearly loose while others are so heavily loaded that they locally crush the gasket. This causes leaks, particularly in high-temperature service where the heavily-loaded bolts tend to relax with subsequent loosening of the joint.

Improper Flange Alignment. Unevenly bolted joints, improper alignment, and especially lack of parallelism between flange faces can cause uneven gasket compression, local crushing, and subsequent leakage. Proper centerline alignment of flanges is also important to assure even compression of the gasket. See Figure 4-2 for general guidance.

Improper Centering of Gasket. A gasket which is installed so that its centerline does not coincide with the flange centerline will be unevenly compressed, thereby increasing the possibility of subsequent leakage. Spiral-wound and double-jacketed high-temperature gaskets are provided with a centering ring or gasket extension to the ID of the bolt circle to facilitate centering of the gasket. Even so, the gasket should be centered with respect to the bolt circle. Certain asbestos replacement gaskets should be cut so that the OD extends to the ID of the bolt circle.

Dirty or Damaged Flange Faces. These are obvious causes for leakage since damage or dirt (including scale) can create a leakage path along the

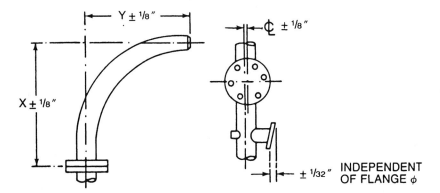

Figure 4-2. Dimensional variations permitted for piping and flanges are independent of pipe size.

flange face. Damage includes scratches, protrusions (e.g., weld spatter) and distortion (warpage) of the flange.

Excessive Forces in the Piping System at Flange Locations. This can occur because of improper piping flexibility design, or by excessive application of force to attain flange alignment. Improper location of temporary or permanent restraints or supports will also cause high flange bending moments and forces.

The Importance of Proper Gasket Selection

The following discussion covers some of the more important factors relating to gasket size and type. Flanges are designed to accommodate specific sizes and types of gaskets (Figure 4-3). When the gasket does not meet the requirements necessary to ensure good seating, or is crushed by the bolt load, leakage will result. Heat exchanges girth flanges are more closely tailored to one specific gasket than are piping flanges per ANSI B16.5. Therefore, somewhat greater latitude is possible with the latter.

Gasket Width

The width of a gasket is considered in the design of a flange. For a given bolt load, a narrow gasket will experience a greater unit load than a wide gasket. It is, therefore, important to determine that the proper width gasket has been used.

- For piping gaskets made of an asbestos-replacing material consult ANSI B16.5
- For double-jacketed, corrugated gaskets consult API 601
- For spiral-wound gaskets consult API 601
- For heat exchanger girth flanges, consult the exchanger drawings

A common reason for gasket leakage is the use of gaskets which are too wide because of the erroneous impression that the full flange face must be covered. This is not true. The above standards should always be followed.

Gasket Thickness

Gasket thickness determines its compressibility and the load required to seat it. The thicker the gasket, the lower the load necessary for seating.

Flat Face

Unconfined Gasket

- Mating faces of both flanges are flat
- Gasket may be ring type, or full face, which covers the entire face both inside and outside the bolts

Raised Face

Unconfined Gasket

- Mating face is flat, but the area inside the bolt holes is raised 1/16" or 1/4"
- Gasket is usually ring type, entirely within bolts
- Flanges may be disassembled easily without springing the flange

Ring Joint

Also Called "API Joint"

- Both flange faces have matching flat-bottomed grooves with sides tapered from the vertical at 23°
- Gasket seats on flat section of flange between bore and ring joint groove
- Garlock spiral wound gaskets can replace solid metal ring gaskets

Male-Female

Semi-Confined Gasket

- Depth of female (recessed) face normally equal to or less than height of male (raised) face, to prevent metal-to-metal contact during gasket compression
- Recessed O.D. normally is not more than 1/16" larger than the O.D. of the male face
- Joint must be pried apart for disassembly

Tongue and Groove

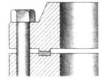

Fully Confined Gasket

- Groove depth is equal to or less than tongue height
- Groove usually not over 1/16" wider than tongue
- Gasket dimensions will match tongue dimensions
- Joint must be pried apart for disassembly

Groove to Flat

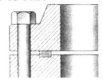

Fully Confined Gasket

- One flange face is flat, the other is recessed
- For applications requiring accurate control of gasket compression
- Only resilient gaskets are recommended—spiral wound, hollow metal O-ring, pressure-actuated, and metal-jacketed gaskets

Figure 4-3. Principal flange configurations.

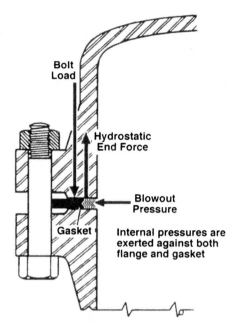

Figure 4-4. Forces acting on a gasket.

All piping flanges are designed to take $^1/_{16}$ in. thick asbestos-replacement gaskets. The $^1/_{16}$ in. thickness assures sufficient compressibility to accommodate slight facing irregularities while having a sufficiently high seating load to prevent blowout. One-sixteenth in. thick gaskets should always be used with ANSI B16.5 flanges unless a specific design check has been made to verify another thickness.

Spiral-wound and double-jacketed gasket thickness should comply with API 601.

Flange Types and Flange Bolt-Up*

Factors Affecting Gasket Performance

A gasket is any deformable material that, when clamped between essentially stationary faces, prevents the passage of media across the gasketed connection (Figure 4-4).

---

* Major portions contributed by Garlock Sealing Technologies, Palmyra, New York 14522.

Compressing the gasket material causes the material to flow into the imperfections of the sealing areas and effect a seal. This bond prevents the escape of the contained media. In order to maintain this seal, sufficient load must be applied to the connection to oppose the hydrostatic end force created by the internal pressure of the system.

Gasket performance depends on a number of factors, including:

1. Gasket metal and filler material: The materials must withstand the effects of:
   a. Temperature: Temperature can adversely affect mechanical and chemical properties of the gasket, as well as physical characteristics such as oxidation and resilience.
   b. Pressure: The media or internal piping pressure can blow out the gasket across the flange face.
   c. Media: The gasket materials must be resistant to corrosive attack from the media.
2. Joint design: The force holding the two flanges together must be sufficient to prevent flange separation caused by hydrostatic end force resulting from the pressure in the entire system.
3. Proper bolt load: If the bolt load is insufficient to deform the gasket, or is so excessive that it crushes the gasket, a leak will occur.
4. Surface finish: If the surface finish is not suitable for the gasket, a seal will not be effected.

Spiral Wound Gaskets Manufactured in Accordance with American Society of Mechanical Engineers (ASME) B16.20

Spiral wound gaskets made with an alternating combination of formed metal wire and soft filler materials form a very effective seal when compressed between two flanges. A "V"-shaped crown centered in the metal strip acts as a spring, giving gaskets greater resiliency under varying conditions.

Filler and wire material can be changed to accommodate different chemical compatibility requirements. Fire safety can be assured by choosing flexible graphite as the filler material. If the load available to compress a gasket is limited, gasket construction and dimensions can be altered to provide an effective seal.

A spiral wound gasket may include a centering ring, an inner ring, or both. The outer centering ring centers the gasket within the flange and acts as a compression limiter, while the inner ring provides additional radial strength. The inner ring also reduces flange erosion and protects the sealing element.

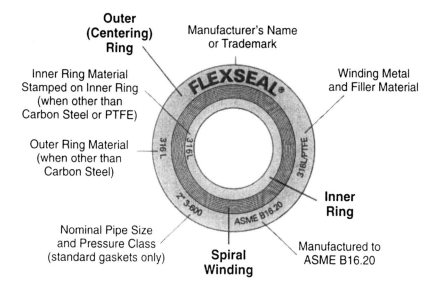

Figure 4-5. Gasket identification markings required by ASME B16.20.

Resiliency and strength make spiral wound gaskets an ideal choice under a variety of conditions and applications. Widely used throughout refineries and chemical processing plants, spiral wound gaskets are also effective for power generation, aerospace, and a variety of valve and specialty applications.

The spiral wound gasket industry is currently adapting to a change in the specification covering spiral wound gaskets. Previously API 601, the new specification is ASME B16.20. These specifications are very similar, and experienced gasket producers follow manufacturing procedures in accordance with the guidelines set forth in the ASME B16.20 specifications. (See Figure 4-5 for markings.)

Torque Tables

Tables 4-1 through 4-4 are representative of tables that were developed to be used with Garlock spiral wound gaskets. They are to be used only as a general guide. Also they should not be considered to contain absolute values due to the large number of uncontrollable variables involved with bolted joints. If there is doubt as to the proper torque value to use, we suggest that the maximum value be used.

(Text continued on page 165)

Table 4-1
Torque Tables for Spiral Wound Gaskets, ASME B16.5

Class 150

Nom. Pipe Size (inches)	Gsk. ID Contact (inches)	Gsk. OD Contact (inches)	Gsk. Area Contact (Sq. in.)	No. of Bolts	Size of Bolts (inches)	Max. Torque per Bolts @ 60 ksi Bolt Stress (ft lb)	Comp. per Bolt @ 60 K (ft lb)	Max. Gsk. Comp. Available (psi)	Min. Gsk. Comp. Recomm. (psi)	Minimum Torque per Bolt (ft lb)	Max. Gsk. Comp. Recomm. Avail. (psi)	Prefer'd Torque (ft lb)
0.5	0.75	1.25	0.79	4	0.50	60	7,560	38,503	10,000	16	30,000	47
0.75	1.00	1.56	1.13	4	0.50	60	7,560	26,712	10,000	22	26,712	60
1	1.25	1.88	1.53	4	0.50	60	7,560	19,713	10,000	30	19,713	60
1.25	1.88	2.38	1.67	4	0.50	60	7,560	18,119	10,000	33	18,119	60
1.5	2.13	2.75	2.39	4	0.50	60	7,560	12,637	10,000	47	12,637	60
2	2.75	3.38	3.01	4	0.63	120	12,120	16,125	10,000	74	16,125	120
2.5	3.25	3.88	3.50	4	0.63	120	12,120	13,861	10,000	87	13,861	120
3	4.00	4.75	5.15	4	0.63	120	12,120	9,406	9,406	120	9,406	120
4	5.00	5.88	7.47	8	0.63	120	12,120	12,974	10,000	92	12,974	120
5	6.13	7.00	9.02	8	0.75	200	18,120	16,071	10,000	124	16,071	200
6	7.19	8.25	12.88	8	0.75	200	18,120	11,253	10,000	178	11,253	200
8	9.19	10.38	18.25	8	0.75	200	18,120	7,945	7,945	200	7,945	200
10	11.31	12.50	22.21	12	0.88	320	25,140	13,584	10,000	236	13,584	320
12	13.38	14.75	30.37	12	0.88	320	25,140	9,933	9,933	320	9,933	320
14	14.63	16.00	33.07	12	1.00	490	33,060	11,995	10,000	408	11,995	490
16	16.63	18.25	44.51	16	1.00	490	33,060	11,884	10,000	412	11,884	490
18	18.69	20.75	63.88	16	1.13	710	43,680	10,940	10,000	649	10,940	710
20	20.69	22.75	70.36	20	1.13	710	43,680	12,415	10,000	572	12,415	710
24	24.75	27.00	91.45	20	1.25	1,000	55,740	12,190	10,000	820	12,190	1,000

Class 300

Nom. Pipe Size (inches)	Gsk. ID Contact (inches)	Gsk. OD Contact (inches)	Gsk. Area Contact (Sq. in.)	No. of Bolts	Size of Bolts (inches)	Max. Torque per Bolts @ 60 ksi Bolt Stress (ft lb)	Comp. per Bolt @ 60 K (ft lb)	Max. Gsk. Comp. Available (psi)	Min. Gsk. Comp. Recomm. (psi)	Minimum Torque per Bolt (ft lb)	Max. Gsk. Comp. Recomm. Avail. (psi)	Prefer'd Torque (ft lb)
0.5	0.75	1.25	0.79	4	0.50	60	7,560	38,522	10,000	16	30,000	47
0.75	1.00	1.56	1.13	4	0.63	120	12,120	43,079	10,000	28	30,000	84
1	1.25	1.88	1.55	4	0.63	120	12,120	31,319	10,000	38	30,000	115
1.25	1.88	2.38	1.67	4	0.63	120	12,120	28,994	10,000	41	28,994	120
1.5	2.13	2.75	2.38	4	0.75	200	18,120	30,517	10,000	66	30,000	197
2	2.75	3.38	3.03	8	0.63	120	12,120	31,983	10,000	38	30,000	113
2.5	3.25	3.88	3.53	8	0.75	200	18,120	41,110	10,000	49	30,000	146
3	4.00	4.75	5.15	8	0.75	200	18,120	28,139	10,000	71	28,139	200
4	5.00	5.88	7.52	8	0.75	200	18,120	19,287	10,000	104	19,287	200
5	6.13	7.00	8.97	8	0.75	200	18,120	16,166	10,000	124	16,166	200
6	7.19	8.25	12.85	12	0.75	200	18,120	16,925	10,000	118	16,925	200
8	9.19	10.38	18.28	12	0.88	320	25,140	16,502	10,000	194	16,502	320
10	11.31	12.50	22.24	16	1.00	490	33,060	23,782	10,000	206	23,782	490
12	13.38	14.75	30.25	16	1.13	710	43,680	23,102	10,000	307	23,102	710
14	14.63	16.00	32.94	20	1.13	710	43,680	26,520	10,000	268	26,520	710
16	16.63	18.25	44.36	20	1.25	1,000	55,740	25,133	10,000	398	25,133	1,000
18	18.69	20.75	63.78	24	1.25	1,000	55,740	20,975	10,000	477	20,975	1,000
20	20.69	22.75	70.25	24	1.25	1,000	55,740	19,044	10,000	525	19,044	1,000
24	24.75	27.00	91.40	24	1.50	1,600	84,300	22,135	10,000	723	22,135	1,600

Tables are based on the use of bolts with a yield strength of 100,000 psi.

Table 4-2
Torque Tables for Spiral Wound Gaskets, ASME B16.5

Class 400

Nom. Pipe Size (inches)	Gsk. ID Contact (inches)	Gsk. OD Contact (inches)	Gsk. Area Contact (Sq. in.)	No. of Bolts	Size of Bolts (inches)	Max. Torque per Bolts @ 60 ksi Bolt Stress (ft lb)	Comp. per Bolt @ 60 K (ft lb)	Max. Gsk. Comp. Available (psi)	Min. Gsk. Comp. Recomm. (psi)	Minimum Torque per Bolt (ft lb)	Max. Gsk. Comp. Recomm. Avail. (psi)	Prefer'd Torque (ft lb)
4	4.75	5.88	9.43	8	0.88	320	25,140	21,329	10,000	150	21,329	320
5	5.81	7.00	11.97	8	0.88	320	25,140	16,807	10,000	190	16,807	320
6	6.88	8.25	16.27	12	0.88	320	25,140	18,540	10,000	173	18,540	320
8	8.88	10.38	22.68	12	1.00	490	33,060	17,493	10,000	280	17,493	490
10	10.81	12.50	30.92	16	1.13	710	43,680	22,600	10,000	314	22,600	710
12	12.88	14.75	40.56	16	1.25	1,000	55,740	21,988	10,000	455	21,988	1,000
14	14.25	16.00	41.56	20	1.25	1,000	55,740	26,826	10,000	373	26,826	1,000
16	16.25	18.25	54.17	20	1.38	1,360	69,300	25,588	10,000	531	25,588	1,360
18	18.50	20.75	69.33	24	1.38	1,360	69,300	23,991	10,000	567	23,991	1,360
20	20.50	22.75	76.39	24	1.50	1,600	84,300	26,485	10,000	604	26,485	1,600
24	24.75	27.00	91.40	24	1.75	3,000	118,800	31,194	10,000	962	30,000	2,885

Class 600

Nom. Pipe Size (inches)	Gsk. ID Contact (inches)	Gsk. OD Contact (inches)	Gsk. Area Contact (Sq. in.)	No. of Bolts	Size of Bolts (inches)	Max. Torque per Bolts @ 60 ksi Bolt Stress (ft lb)	Comp. per Bolt @ 60 K (ft lb)	Max. Gsk. Comp. Available (psi)	Min. Gsk. Comp. Recomm. (psi)	Minimum Torque per Bolt (ft lb)	Max. Gsk. Comp. Recomm. Avail. (psi)	Prefer'd Torque (ft lb)
0.5	0.75	1.25	0.79	4	0.50	60	7,560	38,522	10,000	16	30,000	47
0.75	1.00	1.56	1.13	4	0.63	120	12,120	43,079	10,000	28	30,000	84
1	1.25	1.88	1.55	4	0.63	120	12,120	31,319	10,000	38	30,000	115
1.25	1.88	2.38	1.67	4	0.63	120	12,120	28,994	10,000	41	28,994	120
1.5	2.13	2.75	2.38	4	0.75	200	18,120	30,517	10,000	66	30,000	197
2	2.75	3.38	3.03	8	0.63	120	12,120	31,983	10,000	38	30,000	113
2.5	3.25	3.88	3.53	8	0.75	200	18,120	41,110	10,000	49	30,000	146
3	4.00	4.75	5.15	8	0.75	200	18,120	28,139	10,000	71	28,139	200
4	4.75	5.88	9.43	8	0.88	320	25,140	21,329	10,000	150	21,329	320
5	5.81	7.00	11.97	8	1.00	490	33,060	22,102	10,000	222	22,102	490
6	6.88	8.25	16.27	12	1.00	490	33,060	24,381	10,000	201	24,381	490
8	8.88	10.38	22.68	12	1.13	710	43,680	23,112	10,000	307	23,112	710
10	10.81	12.50	30.92	16	1.25	1,000	55,740	28,840	10,000	347	28,840	1,000
12	12.88	14.75	40.56	20	1.25	1,000	55,740	27,486	10,000	364	27,486	1,000
14	14.25	16.00	41.56	20	1.38	1,360	69,300	33,353	10,000	408	30,000	1,223
16	16.25	18.25	54.17	20	1.50	1,600	84,300	31,127	10,000	514	30,000	1,542
18	18.50	20.75	69.33	20	1.63	2,200	100,800	29,080	10,000	757	29,080	2,200
20	20.50	22.75	76.39	24	1.63	2,200	100,800	31,669	10,000	695	30,000	2,084
24	24.75	27.00	91.40	24	1.88	4,000	138,240	36,298	10,000	1,102	30,000	3,306

Tables are based on the use of bolts with a yield strength of 100,000 psi.

WARNING: Properties/applications shown throughout this brochure are typical. Your specific application should not be undertaken without independent study and evaluation for suitability. For specific application recommendations consult Garlock. Failure to select the proper sealing products could result in property damage and/or serious personal injury.

Performance data published in this brochure has been developed from field testing, customer field reports and/or in-house testing. While the utmost care has been used in compiling this brochure, we assume no responsibility for errors. Specifications subject to change without notice. This edition cancels all previous issues. Subject to change without notice.

Table 4-3
Torque Tables for Spiral Wound Gaskets, ASME B16.5

Class 900

Nom. Pipe Size (inches)	Gsk. ID Contact (inches)	Gsk. OD Contact (inches)	Gsk. Area Contact (Sq. in.)	No. of Bolts	Size of Bolts (inches)	Max. Torque per Bolts @ 60 ksi Bolt Stress (ft lb)	Comp. per Bolt @ 60 K (ft lb)	Max. Gsk. Comp. Available (psi)	Min. Gsk. Comp. Recomm. (psi)	Minimum Torque per Bolt (ft lb)	Prefer'd Torque (ft lb)
0.5	0.75	1.25	0.79	4	0.75	200	18,120	92,284	10,000	22	100
0.75	1.00	1.56	1.13	4	0.75	200	18,120	64,024	10,000	31	100
1	1.25	1.88	1.53	4	0.88	320	25,140	65,555	10,000	49	160
1.25	1.88	2.38	1.67	4	0.88	320	25,140	60,253	10,000	53	160
1.5	1.13	2.75	2.39	4	1.00	490	33,060	55,261	10,000	89	266
2	2.75	3.38	3.01	8	0.88	320	25,140	66,893	10,000	48	160
2.5	3.25	3.88	3.50	8	1.00	490	33,060	75,620	10,000	65	245
3	3.75	4.75	6.68	8	0.88	320	25,140	30,126	10,000	106	319
4	4.75	5.88	9.39	8	1.13	710	43,680	37,222	10,000	191	572
5	5.81	7.00	11.95	8	1.25	1,000	55,740	37,316	10,000	268	804
6	6.88	8.25	16.33	12	1.13	710	43,680	32,090	10,000	221	664
8	8.75	10.13	20.38	12	1.38	1,360	69,300	40,798	10,000	333	1,000
10	10.88	12.25	24.97	16	1.38	1,360	69,300	44,400	10,000	306	919
12	12.75	14.50	37.45	20	1.38	1,360	69,300	37,006	10,000	368	1,103
14	14.00	15.75	40.89	20	1.50	1,600	84,300	41,233	10,000	388	1,164
16	16.25	18.00	47.07	20	1.63	2,200	100,800	42,825	10,000	514	1,541
18	18.25	20.50	68.48	20	1.88	4,000	138,240	40,376	10,000	991	2,972
20	20.50	22.50	67.54	20	2.00	4,400	159,120	47,116	10,000	934	2,802
24	24.75	26.75	80.90	20	2.50	8,800	257,520	63,667	10,000	1,382	4,400

Class 1500

Nom. Pipe Size (inches)	Gsk. ID Contact (inches)	Gsk. OD Contact (inches)	Gsk. Area Contact (Sq. in.)	No. of Bolts	Size of Bolts (inches)	Max. Torque per Bolts @ 60 ksi Bolt Stress (ft lb)	Comp. per Bolt @ 60 K (ft lb)	Max. Gsk. Comp. Available (psi)	Min. Gsk. Comp. Recomm. (psi)	Minimum Torque per Bolt (ft lb)	Prefer'd Torque (ft lb)
0.5	0.75	1.25	0.79	4	0.75	200	18,120	92,284	10,000	22	100
0.75	1.00	1.56	1.13	4	0.75	200	18,120	64,024	10,000	31	100
1	1.25	1.88	1.53	4	0.88	320	25,140	65,555	10,000	49	160
1.25	1.56	2.38	2.51	4	0.88	320	25,140	40,021	10,000	80	240
1.5	1.88	2.75	3.18	4	1.00	490	33,060	41,606	10,000	118	353
2	2.31	3.38	4.75	8	0.88	320	25,140	42,376	10,000	76	227
2.5	2.75	3.88	5.85	8	1.00	490	33,060	45,182	10,000	108	325
3	3.63	4.75	7.40	8	1.13	710	43,680	47,222	10,000	150	451
4	4.63	5.88	10.31	8	1.25	1,000	55,740	43,258	10,000	231	694
5	5.63	7.00	13.63	8	1.50	1,600	84,300	49,464	10,000	323	970
6	6.75	8.25	17.67	12	1.38	1,360	69,300	47,059	10,000	289	867
8	8.50	10.13	23.77	12	1.63	2,200	100,800	50,886	10,000	432	1,297
10	10.50	12.25	31.27	12	1.88	4,000	138,240	53,052	10,000	754	2,262
12	12.75	14.50	37.45	16	2.00	4,400	159,120	67,975	10,000	647	2,200
14	14.25	15.75	35.34	16	2.25	6,360	205,380	92,997	10,000	684	3,180
16	16.00	18.00	53.41	16	2.50	8,800	257,520	77,149	10,000	1,141	4,400
18	18.25	20.50	68.48	16	2.75	11,840	315,540	73,728	10,000	1,606	5,920
20	20.25	22.50	75.55	16	3.00	15,440	379,440	80,363	10,000	1,921	7,720
24	24.25	26.75	100.14	16	3.50	26,000	525,000	83,884	10,000	3,100	13,000

Tables are based on the use of bolts with a yield strength of 100,000 psi.

Table 4-4
Torque Tables for Spiral Wound Gaskets, ASME B16.5

Class 2500

Nom. Pipe Size (inches)	Gsk. ID Contact (inches)	Gsk. OD Contact (inches)	Gsk. Area Contact (Sq. in.)	No. of Bolts	Size of Bolts (inches)	Max. Torque per Bolts @ 60 ksi Bolt Stress (ft lb)	Comp. per Bolt @ 60 K (ft lb)	Max. Gsk. Comp. Available (psi)	Min. Gsk. Comp. Recomm. (psi)	Minimum Torque per Bolt (ft lb)	Prefer'd Torque (ft lb)
0.5	0.75	1.25	0.79	4	0.75	200	18,120	92,284	10,000	22	100
0.75	1.00	1.56	1.13	4	0.75	200	18,120	64,024	10,000	31	100
1	1.25	1.88	1.53	4	0.88	320	25,140	65,555	10,000	49	160
1.25	1.56	2.38	2.51	4	1.00	490	33,060	52,629	10,000	93	279
1.5	1.88	2.75	3.18	4	1.13	710	43,680	54,971	10,000	129	387
2	2.31	3.38	4.75	8	1.00	490	33,060	55,725	10,000	88	264
2.5	2.75	3.88	5.85	8	1.13	710	43,680	59,696	10,000	119	357
3	3.63	4.75	7.40	8	1.25	1,000	55,740	60,260	10,000	166	500
4	4.63	5.88	10.31	8	1.50	1,600	84,300	65,423	10,000	245	800
5	5.63	7.00	13.63	8	1.75	3,000	118,800	69,708	10,000	430	1,500
6	6.75	8.25	17.67	8	2.00	4,400	159,120	72,035	10,000	611	2,200
8	8.50	10.13	23.77	12	2.00	4,400	159,120	80,323	10,000	548	2,200
10	10.63	12.25	29.19	12	2.50	8,800	257,520	105,849	10,000	831	4,400
12	12.50	14.50	42.41	12	2.75	11,840	315,540	89,280	10,000	1,326	5,920

Tables are based on the use of bolts with a yield strength of 100,000 psi.

WARNING: Properties/applications shown throughout this brochure are typical. Your specific application should not be undertaken without independent study and evaluation for suitability. For specific application recommendations consult Garlock. Failure to select the proper sealing products could result in property damage and/or serious personal injury.

Performance data published in this brochure has been developed from field testing, customer field reports and/or in-house testing. While the utmost care has been used in compiling this brochure, we assume no responsibility for errors. Specifications subject to change without notice. This edition cancels all previous issues. Subject to change without notice.

(Text continued from page 157)

All bolt torque values are based on the use of new nuts (ASTM A194, GR 2H) and new bolts (ASTM A193, GR 87) of proper design, acceptable quality, and approved materials of construction as well as metallurgy. It is also required that two hardened steel washers be used under the head of each nut and that a non–metallic-based lubricant (i.e., oil and graphite) be used on the nuts, bolts, and washers.

The flanges are assumed to be in good condition and in compliance with ASME B16.5 specifications. Special attention should be given to seating surface finish and flatness.

Only torque wrenches that have been calibrated should be used. The proper bolt tightening pattern must be followed (see Figure 4.6 for proper bolting pattern) with the desired ultimate torque value arrived at in a minimum of three equal increments. All bolts in the flange should then be checked in consecutive order in a counterclockwise direction.

The contact dimensions listed are taken from the inside diameter (ID) and outside diameter (OD) of the windings, which are different from the ASME ring gasket dimensions. No provisions have been made in these tables to account for vibration effects on the bolts. These tables are based on ambient conditions, without compensation for elevated temperatures. If conditions different from these exist, we suggest that further analysis be performed to determine the appropriate torque values.

Gasket Installation

In a flanged connection, all components must be correct to achieve a seal. The most common cause of leaky gasketed joints is improper installation procedures.

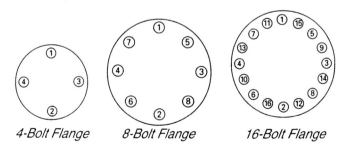

4-Bolt Flange   8-Bolt Flange   16-Bolt Flange

Figure 4-6. Installation sequence for 4-, 8-, and 16-bolt flanges.

Bolting Procedures

- Place the gasket on the flange surface to be sealed
- Bring the opposing flange into contact with the gasket
- Clean the bolts and lubricate them with a quality lubricant, such as an oil and graphite mixture
- Place the bolts into the bolt holes
- Finger-tighten the nuts
- Follow the bolting sequence in the diagrams above
- During the initial tightening sequence, do not tighten any bolts more than 30 percent of the recommended bolt stress. Doing so will cause cocking of the flange and the gasket will be crushed
- Upon reaching the recommended torque requirements, do a clockwise bolt-to-bolt torque check to make certain that the bolts have been stressed evenly
- Due to creep and stress relaxation, it is essential to pre-stress the bolts to ensure adequate stress load during operation

Hydrostatic Testing Precautions

If hydrostatic tests are to be performed at pressures higher than those for which the flange was rated, higher bolt pressures must be applied in order to get a satisfactory seal under the test conditions.

Use high-strength alloy bolts (ASTM B193 grade B7 is suggested) during the tests. They may be removed upon completion. Higher stress values required to seat the gasket during hydrostatic tests at higher than flange-rated pressures may cause the standard bolts to be stressed beyond their yield points.

Upon completion of hydrostatic testing, relieve all bolt stress by 50 percent of the allowable stress.

Begin replacing the high-strength alloy bolts (suggested for test conditions) one by one with the standard bolts while maintaining stress on the gasket.

After replacing all the bolts, follow the tightening procedure recommended in the bolting sequence diagrams (Figure 4-6).

Pre-Stressing Bolts for Thermal Expansion

Bolts should be pre-stressed to compensate for thermal expansion as well as for relaxation, creep, hydrostatic end pressure, and residual gasket loads.

A difference in the coefficient of thermal expansion between the materials of the flange and the bolts may change loads. In cases of serious thermal expansion, it may be necessary to apply a minimum of stress to the bolts and allow the pipe expansion to complete the compression of the gasket.

A gasket with a centering guide ring should be compressed to the guide ring. A gasket without a centering guide ring must be installed with precautions taken to prevent thermal expansion from crushing the gasket beyond its elastic limit.

Calculating Load Requirements

The load requirements can be calculated from two formulas that define the minimum load required to effect a seal on a particular gasket. The two formulas are

Wml and Wm2. When these formulas have been calculated, the larger load of the two is the load necessary to effect a seal.

Let:

π = 3.14
p = Maximum internal pressure
M   = Gasket factor "M" defined in Figure 4-7
        (M = 3 for spiral woud gaskets)
Y   = Seating stress "Y" defined in Figure 4-7
        (Y = 10,000 psi for spiral wound gaskets)
N   = Basic width of a gasket per chart in Figure 4-8
        (For raised face flanges see diagram 1a)
B_0 = Basic seating width of a gasket per chart, Figure 4-8
        (For raised face flanges, B_0 = N/2)
B_1 = Effective seating width of a gasket; must be determined.
ID = Inside diameter of gasket
OD = Outside diameter of gasket

For gaskets where the raised face is smaller than the OD of the gasket face, the OD is equal to the outer diameter of the raised face.
    Find:

ID = _____
OD = _____

"M" and "Y" data are to be used for flange designs only as specified in the ASME Boiler and Pressure Vessel Code Division 1, Section VIII, Appendix 2. They are not meant to be used as gasket seating stress values in actual service. Our bolt torque tables give that information and should be used as such.

"M" - Maintenance Factor

A factor that provides the additional preload needed in the flange fasteners to maintain the compressive load on a gasket after internal pressure is applied to a joint.

$$M = (W - A_2P)/A_1P$$

Where: W = Total Fastener force (lb. or N)

A_2 = Inside area of gasket (in.^{2} or mm^2)

P = Test pressure (psig or N/mm^2)

A_1 = Gasket area (in.^{2} or mm^2)

"Y" - Minimum Design Seating Stress

The minimum compressive stress in pounds per square inch (or bar) on the contact area of the gasket that is required to provide a seal at an internal pressure of 2 psig (0.14 bar).

$$Y = W/A_1$$

Gasket Design	Gasket Material	Gasket Factor "M"	Min. Design Seating Stress "Y" (psi)
Spiral wound metal, non-asbestos filled	 Stainless steel or MONEL®	3.00	10,000
Garlock CONTROLLED DENSITY® flexible graphite-filled spiral wound	 Stainless steel or MONEL®	3.00	7,500
Garlock EDGE®	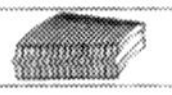 Stainless steel or MONEL®	2.00	5,000
Corrugated metal, non-asbestos or Corrugated metal-jacketed, non-asbestos filled	 Soft aluminum	2.50	2,900
	Soft copper or brass	2.75	3,700
	Iron or soft steel	3.00	4,500
	MONEL® or 4%-6% chrome	3.25	5,500
	Stainless steel	3.50	6,500
Corrugated metal	 Soft aluminum	2.75	3,700
	Soft copper or brass	3.00	4,500
	Iron or soft steel	3.25	5,500
	MONEL® or 4%-6% chrome	3.50	6,500
	Stainless steel	3.75	7,600
Flat metal-jacketed, non-asbestos filled	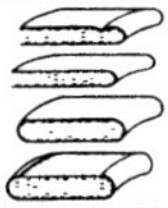 Soft aluminum	3.25	5,500
	Soft copper or brass	3.50	6,500
	Iron or soft steel	3.75	7,600
	MONEL®	3.50	8,000
	4%-6% chrome	3.75	9,000
	Stainless steel	3.75	9,000
Grooved metal	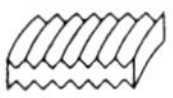 Soft aluminum	3.25	5,500
	Soft copper or brass	3.50	6,500
	Iron or soft steel	3.75	7,600
	MONEL® or 4%-6% chrome	3.75	9,000
	Stainless steel	4.25	10,100
Solid flat metal	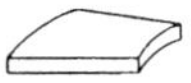 Soft aluminum	4.00	8,800
	Soft copper or brass	4.75	13,000
	Iron or soft steel	5.50	18,000
	MONEL® or 4%-6% chrome	6.00	21,800
	Stainless steel	6.50	26,000
Ring joint	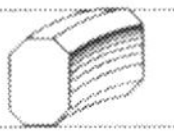 Iron or soft steel	5.50	18,000
	MONEL® or 4%-6% chrome	6.00	21,800
	Stainless steel	6.50	26,000

MONEL® is a registered trademark of International Nickel.

Figure 4-7. Gasket factors "M" and "Y."

Flange and Gasket Diagram		Basic Gasket Seating Width, B_0	
		Column 1 (Solid flat metal and ring joint gaskets)	**Column 2** (Spiral wound, metal jacketed, corrugated metal, grooved metal gaskets)
1a		$\dfrac{N}{2}$	$\dfrac{N}{2}$
1b*			
1c	$W \le N$	$\dfrac{W+T}{2}, \left[\dfrac{W+N}{4}\text{max.}\right]$	$\dfrac{W+T}{2}, \left[\dfrac{W+N}{4}\text{max.}\right]$
1d*	$W \le N$		
2	1/64" Nubbin $\quad W \le \dfrac{N}{2}$	$\dfrac{W+N}{4}$	$\dfrac{W+3N}{8}$
3	1/64" Nubbin $\quad W \le \dfrac{N}{2}$	$\dfrac{N}{4}$	$\dfrac{3N}{8}$
4*		$\dfrac{3N}{8}$	$\dfrac{7N}{16}$
5*		$\dfrac{N}{4}$	$\dfrac{3N}{8}$
6		$\dfrac{W}{8}$	

N = Width of gasket

W = Width of contact area (raised face or serrations)

T = Thickness of gasket

B_0 = Basic seating width of gasket

B_1 = Effective seating width of gasket

$B_1 = B_0$ if $B_0 \le 1/4$";
$B_1 = (\sqrt{B_0})/2$ if $B_0 > 1/4$"

* Where serrations do not exceed 1/64" depth and 1/32" spacing, choose 1b or 1d.

H_G = Gasket load reaction force

G = Diameter of gasket load reaction force

h_G = Distance from G to bolt circle diameter

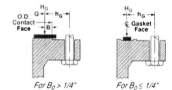

For $B_0 > 1/4$"　　　For $B_0 \le 1/4$"

Figure 4-8. Effective gasket sealing width.

Given the ID and OD, find the value of N. Then define B_0 in terms of N (see Figure 4-8):

$$N = \underline{\hspace{3cm}}$$
$$B_0 = \underline{\hspace{3cm}}$$

Determine if B_0 is greater or less than $1/4''$, then find B_1:

If $B_0 \leq 1/4''$, then $B_1 = B_0$;
If $B_0 > 1/4''$, then $B_1 = (\sqrt{B_0})/2$;
$$B_1 = \underline{\hspace{3cm}}$$

Using B_1, determine G:

$$G = OD - [(B_1)(2)]$$

Now, insert these values in the final equations to determine minimum required load:

$$Wm1 = [\pi(P)(G^2)/4] + [2(B_1)(\pi)(G)(M)(P)]$$
$$Wm2 = \pi(B_1)(G)(Y)$$

When Wm1 and Wm2 have been calculated, the larger of the two numbers is the minimum load required to seat a gasket. In most cases the available bolt load in a connection is greater than the minimum load on the gasket. If not, higher bolt stresses or changes in the gasket design are required for an effective seal.

NOTE: Flange design code suggestions for low-pressure applications calling for minimum seating stress (Y value) are sometimes inadequate to seat the gasket because the bolting and flange rigidity are insufficient to effect a proper seal. Care should be taken to ensure that flange conditions provide a suitable seating surface. For internal pressure to be contained, flange rotation and sufficient residual loads must also be considered in the flange design.

General Installation and Inspection Procedure

This segment covers recommended procedures relating to the preparation and inspection of a joint prior to the actual bolt-up. Obviously, high temperature piping joints in hydrogen-containing streams are less forgiving than those in more moderate service. Critical flanges are defined as joints in services in excess of 500°F and in sizes above six in. in diameter: *(Text continued on page 175)*

DATE: _____

RECORDER: _____

CREW: _____

LINE NO.: _____

Joint No. (in Flow Direction Sequence)	Flange One		Flange Two		Joint Alignment (3)					Gasket Condition			Bolts		Remarks
	Face (1) Condition	Warpage Check (2)	Face (1) Condition	Warpage Check (2)	a	b	a-b	c	d	New Gasket?	Proper Type?	Any Defects	Lubri-cated?	Length Checked?	
1															
2															
3															
4															
5															
6															

Notes:

1) Note any scratches, dents, weld spatter or scale. Note if facing was wire brushed. Note if facing was refaced or any other repairs made.

2) Use a straight edge and check for any deviation of facing from straight edge.

3) See sketches.

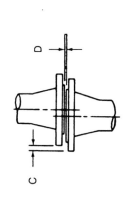

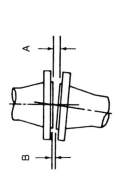

Figure 4-9. Typical flanged joint record form.

Table 4-5
Torque to Stress Bolts

The torque required to produce a certain stress in bolting is dependent on several conditions, including:

- Diameter and number of threads on bolt
- Condition of nut bearing surfaces
- Lubrication of bolt threads and nut bearing surfaces.

The tables below reflect the results of many tests to determine the relation between torque and bolt stress. Values are based on steel bolts that have been well-lubricated with a heavy graphite and oil mixture. A nonlubricated bolt has an efficiency of about 50 percent of a well-lubricated bolt. Also, different lubricants produce results that vary from 50 to 100 percent of the tabulated stress figures.

For Alloy Steel Stud Bolts (Load in pounds on stud bolts when torque load is applied)

Nominal Diameter of Bolt (inches)	Number of Threads (per inch)	Diameter at Root of Thread (inches)	Area at Root of Thread (sq. inch)	Stress					
				30,000 psi		45,000 psi		60,000 psi	
				Torque (ft lbs)	Compression (lbs)	Torque (ft lbs)	Compression (lbs)	Torque (ft lbs)	Compression (lbs)
1/4	20	0.185	0.027	4	810	6	1,215	8	1,620
5/16	18	0.240	0.045	8	1,350	12	2,025	16	2,700
3/8	16	0.294	0.068	12	2,040	18	3,060	24	4,080
7/16	14	0.345	0.093	20	2,790	30	4,185	40	5,580

1/2	13	0.400	0.126	30	3,780	45	5,670	60	7,560
9/16	12	0.454	0.162	45	4,860	68	7,290	90	9,720
5/8	11	0.507	0.202	60	6,060	90	9,090	120	12,120
4/3	10	0.620	0.302	100	9,060	150	13,590	200	18,120
7/8	9	0.731	0.419	160	12,570	240	18,855	320	25,140
1	8	0.838	0.551	245	16,530	368	24,795	490	33,060
1 1/8	8	0.963	0.728	355	21,840	533	32,760	710	43,680
1 1/4	8	1.088	0.929	500	27,870	750	41,805	1,000	55,740
1 3/8	8	1.213	1.155	680	34,650	1,020	51,975	1,360	69,300
1 1/2	8	1.338	1.405	800	42,150	1,200	63,225	1,600	84,300
1 5/8	8	1.463	1.680	1,100	50,400	1,650	75,600	2,200	100,800
1 3/4	8	1.588	1.980	1,500	59,400	2,250	89,100	3,000	118,800
1 7/8	8	1.713	2.304	2,000	69,120	3,000	103,680	4,000	138,240
2	8	1.838	2.652	2,200	79,560	3,300	119,340	4,400	159,120
2 1/4	8	2.088	3.423	3,180	102,690	4,770	154,035	6,360	205,380
2 1/2	8	2.338	4.292	4,400	128,760	6,600	193,140	8,800	257,520
2 3/4	8	2.588	5.259	5,920	157,770	8,880	236,655	11,840	315,540
3	8	2.838	6.324	7,720	189,720	11,580	264,580	15,440	379,440

Table continued on next page.

Table 4-5—cont'd
Torque to Stress Bolts

For Machine Bolts and Cold Rolled Steel Stud Bolts (Load in pounds on stud bolts when torque load is applied)

Nominal Diameter of Bolt (inches)	Number of Threads (per inch)	Diameter at Root of Thread (inches)	Area at Root of Thread (sq. inch)	Stress					
				7,500 psi		15,000 psi		30,000 psi	
				Torque (ft lbs)	Compression (lbs)	Torque (ft lbs)	Compression (lbs)	Torque (ft lbs)	Compression (lbs)
1/4	20	0.185	0.027	1	203	2	405	4	810
5/16	18	0.240	0.045	2	338	4	675	8	1,350
3/8	16	0.294	0.068	3	510	6	1,020	12	2,040
7/16	14	0.345	0.093	5	698	10	1,395	20	2,790
1/2	13	0.400	0.126	8	945	15	1,890	30	3,780
9/16	12	0.454	0.162	12	1,215	23	2,340	45	4,860
5/8	11	0.507	0.202	15	1,515	30	3,030	60	6,060
3/4	10	0.620	0.302	25	2,265	50	4,530	100	9,060
7/8	9	0.731	0.419	40	3,143	80	6,285	160	12,570
1	8	0.838	0.551	62	4,133	123	8,265	245	16,530
1 1/8	7	0.939	0.693	98	5,190	195	10,380	390	20,760
1 1/4	7	1.064	0.890	137	6,675	273	13,350	545	26,700
1 3/8	6	1.158	1.054	183	7,905	365	15,810	730	31,620
1 1/2	6	1.283	1.294	219	9,705	437	19,410	875	38,820
1 5/8	5 1/2	1.389	1.515	300	11,363	600	22,725	1,200	45,450
1 3/4	5	1.490	1.744	390	13,080	775	26,160	1,550	52,320
1 7/8	5	1.615	2.049	525	15,368	1,050	30,735	2,100	61,470
2	4 1/2	1.711	2.300	563	17,250	1,125	34,500	2,250	69,000

(Text continued from page 170)

- Indentify critical flanges and maintain records. A suitable record form is attached in Figure 4-9. A suggested identification procedure is to use the line identification number and proceed in the flow direction with joints #1, #2, etc.

Prior to Gasket Insertion

- Check condition of flange faces for scratches, dirt, scale, and protrusions. Wire brush clean as necessary. Deep scratches or dents will require refacing with a flange facing machine.
- Check that flange facing gasket dimension, gasket material and type, and bolting are per specification. Reject nonspecification situations. Improper gasket size is a common error.
- Check gasket condition. Only new gaskets should be used. Damaged gaskets (including loose spiral windings) should be rejected. The ID windings on spiral-wound gaskets should have at least three evenly spaced spot welds or approximately one spot weld every six in. of circumference (see API 601).
- Use a straightedge and check facing flatness. Reject warped flanges.
- Check alignment of mating flanges. Avoid use of force to achieve alignment. Verify that:
  1. The two flange faces are parallel to each other within $1/32$ in. at the extremity of the raised face
  2. Flange centerlines coincide within $1/8$ in.

Joints not meeting these criteria should be rejected.

Controlled Torque Bolt-Up of Flanged Connections

Experience shows that controlled torque bolt-up is warranted for certain flanged connections. These would typically include:

- All flanges (all ratings and sizes) with a design temperaure >900°F
- All flanges (all ratings) 12 in. diameter and larger with a design temperature >650°F
- All 6 in. diameter and larger 1,500 pound class flanges with a design temperature >650°F
- All 8 in. diameter and larger 900 pound class flanges with a design temperature >650°F
- All flanges not accessible from a maintenance platform and >50 ft above grade

Table 4-6
Flange and Bolt Dimensions for Standard Flanges

NPS (inches)	150 psi Dia. of Flange (inches)	No. of Bolts	Dia. of Bolts (inches)	Bolt Circle (inches)	300 psi Dia. of Flange (inches)	No. of Bolts	Dia. of Bolts (inches)	Bolt Circle (inches)
1/4	3 3/8	4	1/2	2 1/4	3 3/8	4	1/2	2 1/4
1/2	3 1/2	4	1/2	2 3/8	3 3/4	4	1/2	2 5/8
3/4	3 7/8	4	1/2	2 3/4	4 5/8	4	5/8	3 1/4
1	4 1/4	4	1/2	3 1/8	4 7/8	4	5/8	3 1/2
1 1/4	4 5/8	4	1/2	3 1/2	5 1/4	4	5/8	3 7/8
1 1/2	5	4	1/2	3 7/8	6 1/8	4	3/4	4 1/2
2	6	4	5/8	4 3/4	6 1/2	8	5/8	5
2 1/2	7	4	5/8	5 1/2	7 1/2	8	3/4	5 7/8
3	7 1/2	4	5/8	6	8 1/4	8	3/4	6 5/8
3 1/2	8 1/2	8	5/8	7	9	8	3/4	7 1/4
4	9	8	5/8	7 1/2	10	8	3/4	7 7/8
5	10	8	3/4	8 1/2	11	8	3/4	9 1/4
6	11	8	3/4	9 1/2	12 1/2	12	3/4	10 5/8
8	13 1/2	8	3/4	11 3/4	15	12	7/8	13
10	16	12	7/8	14 1/4	17 1/2	16	1	15 1/4
12	19	12	7/8	17	20 1/2	16	1 1/8	17 3/4
14	21	12	1	18 3/4	23	20	1 1/8	20 1/4
16	23 1/2	16	1	21 1/4	25 1/2	20	1 1/4	22 1/2
18	25	16	1 1/8	22 3/4	28	24	1 1/4	24 3/4
20	27 1/2	20	1 1/8	25	30 1/2	24	1 1/4	27
24	32	20	1 1/4	29 1/2	36	24	1 1/2	32

400 psi Dia. of Flange (inches)	No. of Bolts	Dia. of Bolts (inches)	Bolt Circle (inches)	600 psi Dia. of Flange (inches)	No. of Bolts	Dia. of Bolts (inches)	Bolt Circle (inches)
3 3/8	4	1/2	2 1/4	3 3/8	4	1/2	2 1/4
3 3/4	4	1/2	2 5/8	3 3/4	4	1/2	2 5/8
4 5/8	4	5/8	3 1/4	4 5/8	4	5/8	3 1/4
4 7/8	4	5/8	3 1/2	4 7/8	4	5/8	3 1/2
5 1/4	4	5/8	3 7/8	5 1/4	4	5/8	3 7/8
6 1/8	4	3/4	4 1/2	6 1/8	4	3/4	4 1/2
6 1/2	8	5/8	5	6 1/2	8	5/8	5
7 1/2	8	3/4	5 7/8	7 1/2	8	3/4	5 7/8
8 1/4	8	3/4	6 5/8	8 1/4	8	3/4	6 5/8
9	8	7/8	7 1/4	9	8	7/8	7 1/4
10	8	7/8	7 7/8	10 3/4	8	7/8	8 1/2
11	8	7/8	9 1/4	13	8	1	10 1/2
12 1/2	12	7/8	10 5/8	14	12	1	11 1/2
15	12	1	13	16 1/2	12	1 1/8	13 3/4
17 1/2	16	1 1/8	15 1/4	20	16	1 1/4	17
20 1/2	16	1 1/4	17 3/4	22	20	1 1/4	19 1/4
23	20	1 1/4	20 1/4	23 3/4	20	1 3/8	20 3/4
25 1/2	20	1 3/8	22 1/2	27	20	1 1/2	23 3/4
28	24	1 3/8	24 3/4	29 1/4	20	1 5/8	25 3/4
30 1/2	24	1 1/2	27	32	24	1 5/8	28 1/2
36	24	1 3/4	32	37	24	1 7/8	33

NPS (inches)	900 psi				1500 psi			
	Dia. of Flange (inches)	No. of Bolts	Dia. of Bolts (inches)	Bolt Circle (inches)	Dia. of Flange (inches)	No. of Bolts	Dia. of Bolts (inches)	Bolt Circle (inches)
$1/2$	$4^3/_4$	4	$3/_4$	$3^1/_4$	$4^3/_4$	4	$3/_4$	$3^1/_4$
$3/_4$	$5^1/_8$	4	$3/_4$	$3^1/_2$	$5^1/_8$	4	$3/_4$	$3^1/_2$
1	$5^7/_8$	4	$7/_8$	4	$5^7/_8$	4	$7/_8$	4
$1^1/_4$	$6^1/_4$	4	$7/_8$	$4^3/_8$	$6^1/_4$	4	$7/_8$	$4^3/_8$
$1^1/_2$	7	4	1	$4^7/_8$	7	4	1	$4^7/_8$
2	$8^1/_2$	8	$7/_8$	$6^1/_2$	$8^1/_2$	8	$7/_8$	$6^1/_2$
$2^1/_2$	$9^5/_8$	8	1	$7^1/_2$	$9^5/_8$	8	1	$7^1/_2$
3	$9^1/_2$	8	$7/_8$	$7^1/_2$	$10^1/_2$	8	$1^1/_8$	8
4	$11^1/_2$	8	$1^1/_8$	$9^1/_4$	$12^1/_4$	8	$1^1/_4$	$9^1/_2$
5	$13^3/_4$	8	$1^1/_4$	11	$14^3/_4$	8	$1^1/_2$	$11^1/_2$
6	15	12	$1^1/_8$	$12^1/_2$	$15^1/_2$	12	$1^3/_8$	$12^1/_2$
8	$18^1/_2$	12	$1^3/_8$	$15^1/_2$	19	12	$1^5/_8$	$15^1/_2$
10	$21^1/_2$	16	$1^3/_8$	$18^1/_2$	23	12	$1^7/_8$	19
12	24	20	$1^3/_8$	21	$26^1/_2$	16	2	$22^1/_2$
14	$25^1/_4$	20	$1^1/_2$	22	$29^1/_2$	16	$2^1/_4$	25
16	$27^3/_4$	20	$1^5/_8$	$24^1/_2$	$32^1/_2$	16	$2^1/_2$	$27^3/_4$
18	31	20	$1^7/_8$	27	36	16	$2^3/_4$	$30^1/_2$
20	$33^3/_4$	20	2	$29^1/_2$	$38^3/_4$	16	3	$32^3/_4$
24	41	20	$2^1/_2$	$35^1/_2$	46	16	$3^1/_2$	39

2500 psi			
Dia. of Flange (inches)	No. of Bolts	Dia. of Bolts (inches)	Bolt Circle (inches)
$5^1/_4$	4	$3/_4$	$3^1/_2$
$5^1/_2$	4	$3/_4$	$3^3/_4$
$6^1/_4$	4	$7/_8$	$4^1/_4$
$7^1/_4$	4	1	$5^1/_8$
8	4	$1^1/_8$	$5^3/_4$
$9^1/_4$	8	1	$6^3/_4$
$10^1/_2$	8	$1^1/_8$	$7^3/_4$
12	8	$1^1/_4$	9
14	8	$1^1/_2$	$10^3/_4$
$16^1/_2$	8	$1^3/_4$	$12^3/_4$
19	8	2	$14^1/_2$
$21^3/_4$	12	2	$17^1/_4$
$26^1/_2$	12	$2^1/_2$	$21^1/_4$
30	12	$2^3/_4$	$24^3/_8$

WARNING: Properties/applications shown throughout this table are typical. Your specific application should not be undertaken without independent study and evaluation for suitability. For specific application recommendations consult the manufacturer. Failure to select the proper sealing products could result in property damage and/or serious personal injury. Performance data published in this table have been developed from field testing, customer field reports and/or in-house testing.

While the utmost care has been used in compiling this material, we assume no responsibility for errors.

In addition, it is generally appropriate to apply the above criteria to flanged connections on equipment and other components such as:

- Valve bonnets, where the valve is positioned to include the above referenced design temperature/size/flange rating category
- Flanged equipment closures where they qualify for inclusion in the above categories
- All flanged connections which will eventually be covered with low temperature insulation within the above reference criteria

Adherence to the following procedure is recommended for controlled torquing of line flanges, bonnet joints, ect., when specified.

Preparation

- Thoroughly clean the flange faces and check for scars. Defects exceeding the permissible limits given in Table 4-7 should be repaired.

Table 4-7
Flange Face Damage/Acceptance Criteria

Type	Gasket Type Used	Damage	Critical Defect	Permissible Limits
1	Ring Joint	Scratch-like	Across seating surface	1–2 mils deep-one seating surface only
		Smooth depression		3 mils deep-one seating surface only
2	Spiral wound in tongue and groove joint	Scratch-like	$>\frac{1}{2}$ of tongue/ groove width	1 mil maximum
3	Spiral wound in raised face joint	Scratches, Smooth depressions & gen'l metal loss due to rusting.	$>\frac{1}{2}$ of seated width (min of 1/4″ intact surface left).	Up to $\frac{1}{2}$ of serrated finish depth
4	Asbestos	″	$>\frac{1}{2}$ of seated width	Up to $\frac{1}{2}$ of serrated finish depth

For gasket types 1 and 2 refacing required if more than 3–5 (permissible) defects found. Seating surface taken as center 50 percent of groove face.

- Check studs and nuts for proper size, conformance with piping material specifications, cleanliness, and absence of burrs
- Gaskets should be checked for size and conformance to specifications. Metal gaskets should have grease, rust, and burrs completely removed.
- Check flange alignment. Out-of-alignment of parallelism should be limited to the tolerance given in Figure 4-2.
- Number the studs and nuts to aid in identification and to facilitate applying crisscross bolt-up procedure
- Coat stud and nut thread, and nut and flange bearing surfaces with a liberal amount of bolt thread compound

Equipment

For studs larger than $1^{1}/_{2}$ in. in diameter, use "Select-A-Torq" hydraulic wrench (Model 5000 A) supplied by N-S-W Corp. of Houston, Texas, the "Hydra-Tork" wrench system (Model HT-6) supplied by Torque System, Inc., the "Hytorc" (Figure 4-10), tensioners by Hydratight-Sweeney (Figure 4-11), or one of many available Furmanite "Plarad" devices (Figure 4-12). Torque wrenches can be used on small flanges, with stud diameters less than $1^{1}/_{2}$ in. The torque wrenches should be calibrated at least once per week.

Figure 4-10. "Hytorc" stud tensioner.

Figure 4-11. Tensioners by Hydrotight-Sweeney.

Hot Bolting and Leakage Control

Hot bolting during startup and during process runs has been found to be an important factor in minimizing flange leakage. During heat-up and because of temperature changes, the bolts and gaskets deform permanently. This causes a loss of bolt stress after the temperature changes have smoothed out. Hot bolting helps correct this.

1. *Plarad 'K' tool loosening cap nuts on a
steam chest cover.*

2 3 4 5 —
   *Typical Torque wrench applications on
   bolted connections at a chemical plant.*

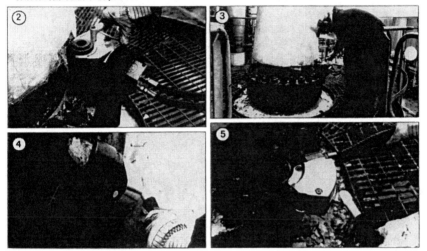

Figure 4-12. Furmanite "Plarad" hydraulic tensioning devices in action.

Hot Bolting Procedure

The objective of hot bolting is to restore the original bolt stress which has dropped due to yielding and/or creep of the flange joint components. If possible, this should be done with a bolt tensioning device. Hot bolting should start at the point of leakage and proceed in a crisscross pattern as described previously. Seized bolts sometimes present a problem when hot bolting. In such cases, it is necessary to use a wrench on both nuts.

Using Bolt Tensioners

There exists considerable experience with the use of various bolt tensioners for hot bolting. These procedures typically involve first running a die over the stud projections to facilitate subsequent installation of the tensioner heads. Mechanics are instructed to leave the heads in place for the minimum time necessary so as to prevent leakage of hydraulic fluid at the seals. Past procedures called for immersion of heads in water between applications; however, this is no longer necessary.

Using Hammer and Wrench or Torque Wrench

If leaks occur, it may be necessary to employ a 7 lb or heavier hammer to stop the leak. Tightening should first be done where the leakage has originated and the crisscross pattern should be used from there. Joints with spiral-wound gaskets can be tightened only to the limit of the steel centering ring thickness. Further tightening is fruitless if a spiral-wound gasket has already been tightened to this point.

If Hot Bolting Does Not Stop Leak

If leakage cannot be stopped by tightening, the line must be isolated and the joint broken to determine the cause:

- Examine flange facings for damage, distortion (warping), or foreign matter
- Check flange alignment, cut and realign piping if necessary
- Check gasket for proper material, dimensions, and type. Use a new gasket for reassembly of the joint.
- Check gasket deformation to determine if it was centered. This is best done by noting the position of the gasket *before* it is withdrawn and examining it immediately after withdrawal.
- Reassemble the joint
- If leakage persists, piping support and flexibility must be examined. It may be necessary to revise the support system or install spring hangers to lower bending moments.
- If leakage occurs during rainstorms, it will be necessary to install sheet metal rain shields, which may cover the top 180° of the flange, to prevent such leakage. These should be located about four inches away from the flange surface and should have sufficient width to cover the bolts plus two inches on each side.
- If leakage occurs during sudden changes in process temperatures, examine the process sequence to determine if steps can be taken to minimize rapid heat-up or cooling of lines. It may only be necessary to open a valve more slowly.

Recommendations for the Installation, Fabrication, Testing, and Cleaning of Air, Gas or Steam Piping*

The importance of starting any compressor with clean piping, particularly on the intake to any cylinder, cannot be over-emphasized. This is particularly important with multi-stage high-pressure compressors where

---

*Refer to appendices at the end of this chapter for typical checklists.

special metallic packing is required and parts are much more expensive than in a low-pressure compressor. Any dirt, rust, welding beads or scale carried into the compressor will cause scored packing rings, piston rods, cylinder bores, and pitted, Leaky valves.

It is important that the piping be fabricated with sufficient flange joints so that it can be dismantled easily for cleaning and testing. It is far better to clean and test piping in sections before actual erection than after it is in place.

If it is necessary to conduct the final test when the piping is in position, care should be taken to provide vents at the high spots so that air or gas will not be trapped in the piping. Make provision for complete drainage after the test is completed. These connections should be planned in advance.

When piping is cleaned in sections before erection, it is possible to do a thorough job of eliminating all acid. This is difficult to do with piping erected and in position, because carry-over of acid into the cylinders is almost certain to occur when the machine is started. This can cause extensive damage.

The use of chill-rings for butt welds in piping is recommended. This prevents welding beads from getting into the pipe to carry through, not only on the original starting, but later on during operation.

After hydrostatic tests have been made and the pipe sections have been cleaned as thoroughly as possible on the inside, the piping should be pickled by this procedure:

1. Pickle for 14 hours with hydrochloric acid. Circulate the acid continuously by means of a small pump. Use a five to 12 percent solution of hydrochloric acid, depending upon the condition of the pipe.
2. Neutralize the caustic.
3. Blow hot air through for several hours.
4. Fill with mineral seal oil and drain.
5. Blow out with hot air.
6. Pipe is now ready to use. If the pipe section is not to be assembled immediately, seal the ends tightly until ready for use. Then, before installation, pull through a swab saturated with carbon tetrachloride.

Even though this procedure has been carefully followed; on reciprocating compressor piping, a temporary filter (such as Type PT American Filter, Type PS Air-Maze, or equal) should be installed in the suction line to the suction bottle to remove particles 230 microns* (0.009 in. diameter) or larger. Provision must be made in the piping to check the pressure drop across the filter and to remove the filter cell for cleaning. Filter cell should be removed and left out only when the inlet line is free of welding beads, pipe scale, and other extraneous matter.

---

* 140 microns (0.0055 in. diameter) for nonlubricated cylinders.

On large piping (where a man can work inside), the pickling procedure can be omitted if the piping is cleaned mechanically with a wire brush, vacuumed and then thoroughly inspected for cleanliness. Time and trouble taken in the beginning to ensure that the piping is clean will shorten the break-in period, and may save a number of expensive shutdowns.

Pickling Procedure for Reciprocating Compressor Suction Piping: Method I

General Recommendations

1. The job should be executed by experienced people.
2. Operators must wear adequate safety equipment (gloves and glasses).
3. Accomplish entire pickling operation in as short a time as possible.

Preliminary Work

1. Install an acid-resistant pump connected to a circulating tank.
2. Provide $1\frac{1}{2}$ in. (or greater) acid resistant hoses for the connections (prepare suitable assembly sketch).
3. For ensuring the filling of the system, flow must go upward and vents must be installed.
4. Provide method for heating the solutions (e.g., a steam coil).

Pretreatment

Pretreatment is required only when traces of grease are present.

1. Fill the system with water at 90°C (194°F).
2. Add 2 percent sodium hydroxide and 0.5 percent sodium metasilicate (or sodium orthosilicate if cheaper). If these compounds are not available and only a small amount of grease is present 2 percent of NaOH and 3 percent of Na_2CO_3 may be used.
3. Circulate for 20–30 minutes at 90°C (194°F).
4. Dump the solution and wash with water until pH = 7.

Acid Treatment

1. Fill the system with water at 50°C (122°F).
2. Add 4 percent of Polinon 6A® and circulate to ensure its complete distribution.

3. Add hydrochloric acid to reach the concentration of 7 percent.
4. Circulate intermittently for about 45 minutes or more until the pickling has been accomplished.

Notes:

1. In order to avoid corrosion:
   (a) Keep the flow rate lower than 1 m/sec.
   (b) Take samples of the solution and check for the Fe^{+++} content: if $[Fe^{+++}] > 0.4$ percent dump solution.
2. In order to determine when the system has been adequately pickled, put a piece of oxidized steel in the circulation tank and inspect it frequently.

Neutralization

1. Add sodium hydroxide for neutralizing the acid, and water to avoid a temperature rise.
2. Circulate for 15–30 minutes.
3. Dump the solution and wash with water until pH = 7.

Note:

The concentration must be calculated on the overall volume of the solution.

Passivation

1. Fill the system with water at 40°C (104°F).
2. Add 0.5 percent of citric acid and circulate to ensure complete mixing.
3. Check the pH of the solution: if pH < 3.5, slowly add ammonia to raise pH to 3.5.
4. Circulate for 15–20 minutes.
5. Slowly add ammonia to raise pH to 6 in 10 minutes.
6. Add 0.5 percent sodium nitrite (or ammonium persulfate).
7. Circulate for 10 minutes.
8. Add ammonia to raise pH to 9.
9. Circulate for 45 minutes.

10. Stop the pump and hold the solution in the system for at least three hours.
11. Dump the solution.

Cleaning of Large Compressor Piping: Method II

Cleaning of the piping may be done by commercial companies with mobile cleaning equipment or by the following recommended cleaning procedure. After hydrostatic tests have been made and the pipe sections have been cleaned as thoroughly as possible on the inside, the piping should be pickled by the following (or equivalent) procedure:

1. Remove all grease, dirt, oil, or paint by immersing in a hot, caustic bath. The bath may be a solution of eight ounces of sodium hydroxide to one gallon of water with the solution temperature 180°–200°F. The time of immersion is at least thirty minutes, depending on the condition of the material.
2. Remove pipe from caustic and *immediately rinse with cold water*.
3. Place pipe in an acid pickling bath. Use a 5 to 12 percent solution of hydrochloric (muriatic) acid, depending upon the condition of the pipe. Rodine inhibitor should be added to the solution to prevent the piping from rusting quickly after removal from the acid bath. The temperature of the bath should be 140°–165°F. The time required in the acid bath to remove scale and rust will vary, depending on the solution strength and condition of piping; however, six hours should be a minimum. The normal time required is about 12 to 14 hours.
4. Remove pipe from acid bath and *immediately wash with cold water* to remove all traces of acid.
5. Without allowing piping to dry, immerse in a hot neutral solution. A one to two ounce soda ash per gallon of water solution may be used to maintain a pH of 9 or above. The temperature of the solution should be 160°–170°F. Litmus paper may be used to check the wet piping surface to determine that an acidic condition does not exist. If acidic, then repeat neutral solution treatment.
6. Rinse pipe with cold water, *drain thoroughly and blow out with hot air until dry*.
7. *Immediate steps must be taken to prevent rusting*, even if piping will be placed in service shortly. Generally, a dip or spray coating of light water displacement mineral oil will suffice; however, if piping is to be placed in outdoor storage for more than several weeks, a hard-coating water displacement type rust preventative should be applied.

8. Unless piping is going to be placed in service immediately, suitable gasketed closures must be placed on the ends of the piping and all openings to prevent entrance of moisture or dirt. Use of steel plate discs and thick gaskets is recommended for all flanges. Before applying closures, the flange surfaces should be coated with grease.

9. Before installation, *check that no dirt or foreign matter has entered piping and that rusting has not occurred.* If in good condition, then pull through a swab saturated with carbon tetrachloride.

10. For nonlubricated (NL) units where oil coating inside piping is not permissible (due to process contamination), even for the starting period, consideration should be given to one of the following alternatives:

   (a) Use of nonferrous piping materials, such as aluminum.
   (b) Application of a plastic composition or other suitable coating after pickling to prevent rusting.
   (c) After rinsing with water in step six, immerse piping in a hot phosphoric bath. The suggested concentration is three to six ounces of iron phosphate per gallon of water, heated to 160°–170°F, with pH range of 4.2 to 4.8. The immersion time is three to five minutes or longer, depending on density of coating required. *Remove and dry thoroughly, blowing out with hot air.*

   CAUTION: Hydrochloric acid in contact with the skin can cause burns. If contacted, *acid should be washed off immediately with water.* Also, if indoors, adequate ventilation, including a vent hood, should be used. When mixing the solution, always *add the acid to the water, never the water to the acid.*

On large piping (where a man can work inside), the pickling procedure can be omitted if the piping is cleaned mechanically with a wire brush, vacuumed and then *thoroughly inspected for cleanliness.* Time and trouble taken in the beginning to insure that the piping is clean will shorten the break-in period, and may save a number of expensive shut downs.

Temporary Line Filters

When first starting, it is advisable to use a temporary line filter in the intake line near the compressor to catch any dirt, chips, or other foreign material that may have been left in the pipe. But clean the pipe first. Do not depend on a temporary line filter. If the gas or air being compressed may, at times, contain dust, sand, or other abrasive particles, a gas scrubber or air cleaner must be *installed permanently and serviced regularly.*

Even though the previous cleaning procedure has been carefully followed on the compressor piping, a temporary filter (such as Type PT American Filter or equal) should be installed in the suction line to the suction bottle to remove particles 230 microns (0.009 in.) in diameter or larger. If the compressor is an "NL" (nonlubricated) design, the filter should be designed to remove particles 140 microns (0.0055 in.) in diameter or larger. Provision must be made in the piping to check the pressure drop across the filter and to remove the filter cell for cleaning. If the pressure drop across the filter exceeds 5 percent of the upstream line pressure, remove the filter, *clean thoroughly* and reinstall. The filter cell should be removed and left out only when the inlet line is free of welding beads, pipe scale, and other extraneous matter.

Appendix 4-A

Detailed Checklist for Rotating Equipment: Machinery Piping*

* Source: P.C. Monroe, as originally presented at First International Pump Symposium, and Short Course, Texas A & M University, Houston, Texas, May 21–24, 1984.

DATE/BY

1. CHECK PUMP DISCHARGE FLANGE FOR LEVEL IN TWO DIRECTIONS 90° APART. MAXIMUM OUT OF LEVEL 0.002"/FOOT. _____

2. CHECK PARALLELISM OF SUCTION AND DISCHARGE PIPE FLANGES TO PUMP FLANGES. MAXIMUM OUT OF PARALLELISM 0.030" AT GASKET SURFACE. _____

3. CHECK CONCENTRICITY OF SUCTION AND DISCHARGE PIPING TO PUMP FLANGES. FLANGE BOLTS TO SLIP INTO HOLES BY HAND. _____

4. CHECK TO SEE IF PROPER PIPE SUPPORTS ARE INSTALLED. _____

5. CHECK FOR 5 PIPE DIAMETERS OF STRAIGHT PIPE BEFORE SUCTION AND AFTER DISCHARGE FLANGES OF PUMP. _____

6. TIGHTEN PIPING FLANGE BOLTS IN CRISS-CROSS PATTERN TO PROPER TORQUE. COAT BOLT THREADS WITH ANTI-SEIZE COMPOUND BEFORE TIGHTENING. _____

7. CHECK PUMP SHAFT DEFLECTION USING FACE AND RIM METHOD DURING FLANGE BOLT TIGHTENING. MAXIMUM ALLOWED DEFLECTION IS 0.002" TIR. _____

NOTE: PREFERRED METHOD OF PIPING PUMP IS TO BRING THE PIPING TO THE PUMP. AFTER PUMP IS GROUTED AND ALIGNED, SUCTION AND DISCHARGE FLANGES WITH GASKETS ARE 4-BOLTED TO PUMP. PIPING IS THEN FIELD FITTED AND WELDED TO THE FLANGES.

8. PROVIDE A DROP OUT SPOOL PIECE AT THE SUCTION
    FLANGE FOR A CONICAL TYPE TEMPORARY MECHANICAL
    STRAINER. SEE FIGURE 1.   _____

9. TEMPORARY MECHANICAL STRAINER INSTALLED.   _____

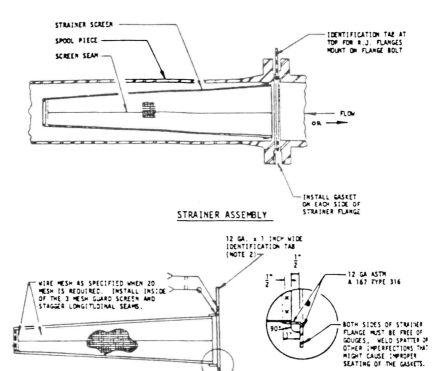

STRAINER SCREEN
SPOOL PIECE
SCREEN SEAM

IDENTIFICATION TAB AT
TOP FOR R.J. FLANGES
MOUNT ON FLANGE BOLT

FLOW
OR

INSTALL GASKET
ON EACH SIDE OF
STRAINER FLANGE

STRAINER ASSEMBLY

12 GA. x 1 INCH WIDE
IDENTIFICATION TAB
(NOTE 2)

WIRE MESH AS SPECIFIED WHEN 20
MESH IS REQUIRED. INSTALL INSIDE
OF THE 3 MESH GUARD SCREEN AND
STAGGER LONGITUDINAL SEAMS.

12 GA ASTM
A 167 TYPE 316

BOTH SIDES OF STRAINER
FLANGE MUST BE FREE OF
GOUGES, WELD SPATTER OR
OTHER IMPERFECTIONS THAT
MIGHT CAUSE IMPROPER
SEATING OF THE GASKETS.

SCREEN SECTION   **DETAIL A**

SEE DETAIL "A"

FIGURE 1

Appendix 4-B

Specifications for Cleaning Mechanical Seal Pots and Piping for Centrifugal Pumps

I. SCOPE

    1.   THIS SPECIFICATION COVERS THE INSPECTION, CLEANING, AND PRESERVATION OF MECHANICAL SEAL POTS AND THEIR ASSOCIATED PIPING ON NEW PUMPS PRIOR TO STARTUP AND COMMISSIONING, AND ON EXISTING PUMPS WHICH HAVE THEIR MECHANICAL SEAL PIPING MODIFIED.

    2.   THIS WORK SHALL BE DONE CONCURRENTLY WITH CLEANING AND SETTING OF MECHANICAL SEALS ON NEW PUMPS.

II. MATERIALS

    1.   THE FOLLOWING MATERIALS WILL BE REQUIRED FOR THIS WORK:

        FRESH WATER
        SODIUM HYDROXIDE
        SODIUM METASILICATE
        TRISODIUM PHOSPHATE
        *NON-IONIC DETERGENT
        *HYDROCHLORIC ACID
        *AMMONIUM BIFLOURIDE
        *ARMOHIB 28 OR EQUIVALENT
        *CITRIC ACID
        *SODA ASH
        *SODIUM NITRITE
        10 MICRON FILTER SCREENS

*THESE CHEMICALS ARE NOT USED TO CLEAN STAINLESS STEEL PIPING OR SEAL POTS.

    2.   QUANTITIES WILL VARY BASED ON THE SIZE AND AMOUNT OF PIPING TO BE CLEANED.

III. DISASSEMBLY

    1.   ALL INSTRUMENTATION IN THE SEAL PIPING SHALL BE REMOVED PRIOR TO DISASSEMBLY OF THE PIPING AND SEAL POTS.

2. THE INSTRUMENTATION SHALL BE TAGGED AND STORED IN
    A SAFE MANNER.

3. THE SEAL PIPING SHALL BE BROKEN AT THE TWO CLOSEST
    FLANGED OR UNION CONNECTIONS TO THE PUMP OR GLAND
    AND TAKEN TO THE SHOP IN ONE PIECE.

4. IF THE PIPING CAN BE BROKEN DOWN INTO SHORTER SEGMENTS
    FOR EASIER CLEANING IT MAY BE DONE IN THE SHOP.

5. OPENINGS ON THE PUMP AND/OR GLAND SHALL BE PLUGGED
    OR CAPPED TO PREVENT ENTRY OF DIRT INTO THE PUMP
    WHILE THE PIPING IS IN THE SHOP.

IV. CLEANING

1. ALL PIPING AND SEAL POTS SHALL BE FLUSHED WITH 170°F
    TO 200°F ALKALINE SOLUTION CONTAINING THREE (3) PER-
    CENT NaOH, ONE (1) PERCENT SODIUM METASILICATE, ONE
    (1) PERCENT TRISODIUM PHOSPHATE AND ONE TENTH (0.1)
    PERCENT NON-IONIC DETERGENT IN FRESH WATER. THIS
    SOLUTION IS TO BE CIRCULATED FOR AT LEAST SIX (6)
    HOURS. THE PIPING AND POTS SHALL THEN BE DRAINED
    AND FLUSHED WITH FRESH WATER UNTIL THE pH LEVEL IS
    LESS THAN 8.0. WATER FOR STAINLESS STEEL PIPING AND
    SEAL POTS SHALL CONTAIN LESS THAN 50 PPM CHLORIDE.

2. STAINLESS STEEL PIPING AND SEAL POTS SHALL BE FURTHER
    FLUSHED WITH 160°F FRESH WATER CONTAINING LESS THAN
    50 PPM CHLORIDE AT A HIGH ENOUGH VELOCITY TO REMOVE
    ALL DIRT. THE PIPING RUNS SHALL BE POSITIONED AND
    SUPPORTED SO AS TO HAVE NO DEAD SPOTS WITH THE WATER
    ENTERING AT THE LOWEST POINT IN THE PIPING AND EXITING
    AT THE HIGHEST POINT. THE CIRCULATED WATER SHALL BE
    FILTERED THROUGH TEN (10) MICRON FILTER SCREENS AND
    RETURNED TO AN OPEN STORAGE TANK BEFORE BEING RECIR-
    CULATED. WHEN THE FILTER SCREENS REMAIN CLEAN FOR
    FIVE (5) MINUTES, CIRCULATION MAY BE TERMINATED AND
    THE PIPING AND SEAL POT AIR DRIED.

3. CARBON STEEL PIPING AND POTS SHALL BE FLUSHED IN THE
    SAME MANNER AS THE STAINLESS STEEL PIPING AND SEAL
    POTS EXCEPT THAT SIX (6) TO TEN (10) PERCENT INHIBITED
    HYDROCHLORIC ACID BY WEIGHT WITH TWENTY-FIVE HUNDRETHS
    (0.25) PERCENT AMMONIUM BIFLORIDE BY WEIGHT INHIBITED
    WITH TWO (2) GALLONS PER 1,000 GALLONS ARMOHIB 28,
    RODINE 213, OR EQUIVALENT IN FRESH WATER SHALL BE USED.

4. THE ACID SOLUTION SHALL BE HEATED TO 160°F AND
    CIRCULATED AT A SLOW RATE FOR SIX (6) HOURS OR UNTIL
    THE REACTION IS COMPLETE.

5. SAMPLES OF THE ACID SOLUTION SHALL BE TAKEN AT HOURLY
    INTERVALS AND TESTED FOR ACID STRENGTH AND TOTAL IRON
    CONCENTRATION. IF THE ACID CONCENTRATION DROPS BELOW
    THREE (3) PERCENT, ACID SHALL BE ADDED TO RAISE THE
    CONCENTRATION TO FIVE (5) PERCENT. CLEANING SHALL
    CONTINUE UNTIL IRON AND ACID CONCENTRATION REACH
    EQUILIBRIUM.

6. AFTER ACID FLUSHING HAS BEEN COMPLETED, THE PIPING
    AND SEAL POTS SHALL BE DRAINED AND FLUSHED WITH FRESH
    WATER CONTAINING ONE TENTH (0.1) PERCENT BY WEIGHT
    CITRIC ACID.

7. THE PIPING AND SEAL POTS SHALL THEN BE NEUTRALIZED WITH ONE (1) PERCENT BY WEIGHT SODA ASH Na_2CO_3 PLUS FIVE TENTHS (0.5) PERCENT BY WEIGHT SODIUM NITRITE $NaNO_2$ SOLUTION IN FRESH WATER CIRCULATED FOR TWO HOURS HEATED TO 175°F TO 190°F.

8. AFTER NEUTRALIZATION THE PIPING AND SEAL POT SHALL BE RINSED WITH FRESH WATER AND AIR DRIED.

V. PRESERVING

1. IMMEDIATELY AFTER DRYING, THE PIPING AND SEAL POTS SHALL BE FLUSHED OR SPRAYED WITH THE SAME OIL AS WILL BE USED IN THE SEAL SYSTEM.

2. AFTER THE OIL TREATMENT THE PIPING SHALL BE REASSEMBLED IF IT WAS DISASSEMBLED IN THE SHOP. ALL OPENINGS SHALL BE PLUGGED OR CAPPED TO PREVENT ENTRY OF ANY CONTAMINATING MATTER.

3. AS SOON AS ALL WORK ON THE MECHANICAL SEALS HAS BEEN COMPLETED, THE PIPING AND SEAL POTS SHALL BE REINSTALLED AND THE LINES FILLED WITH PROPER SEAL OIL.

4. THE INSTRUMENTATION SHALL BE REINSTALLED AT THIS TIME.

5. AFTER FILLING THE LINES WITH OIL AND REINSTALLING THE INSTRUMENTATION THE PUMP SHALL BE TAGGED TO INDICATE THAT THE SEAL SYSTEM IS READY FOR SERVICE.

Appendix 4-C

Detailed Checklist for Rotating Equipment: Pump Piping

1. Valves accessible. _____

2. Instrumentation viewable and accessible. _____

3. Valves, glands, and instrumentation tagged. _____

4. Seal pots and piping adequately supported.

   Maximum unsupported spans:

   1/2 inch line - 2' 0" _____
   3/4 inch line - 2' 6" _____
   1 inch line - 3' 0" _____

5. Seal flush piping per API plan _____. Sketch attached. _____

6. Auxiliary flush piping per API plan _____. Sketch attached. _____

7. Cooling water piping per API plan _____. Sketch attached. _____

8. Seal pot(s) and piping free of scale and dirt. _____

9. Seal pot(s) filled to middle of sight glass. _____

10. Pipe thread seal weld covers all threads, 1/4 inch minimum weld.

11. SUCTION LINE STRAINER INSTALLED _____. MESH.   _____

12. SUCTION LINE STRAINER TABBED.   _____

13. HORIZONTAL SUCTION LINE REDUCER, IF USED, IS
     ECCENTRIC, FLAT SIDE UP.   _____

14. 3/4 INCH VALVED CONNECTION ON EACH SIDE OF
     STRAINER.   _____

15. PIPE FLANGE RATINGS: SUCTION   INCH,   LB.
                          DISCHARGE   INCH,   LB.   _____

16. DISCHARGE LINE INCREASER, IF USED, IS
     CONCENTRIC.   _____

17. DISCHARGE CHECK VALVE INSTALLED, FLOW
     DIRECTION CORRECT.   _____

18. VALVED BYPASS AROUND CHECK VALVE. (FOR HOT
     SERVICES).   _____

19. CHECK VALVE DRILLED IF NO BYPASS IS INSTALLED
     IN HOT SERVICES.   _____

20. ONE INCH VALVED WARM-UP LINE BETWEEN DISCHARGE
     AND SUCTION LINES.   _____

21. DOUBLE BLOCK VALVED LINES HAVE BODY DRAIN VALVES
     IN BLOCK VALVES OR VALVED DRAIN IN SPOOL BETWEEN
     BLOCK VALVES.   _____

22. PIPE NIPPLES THREADED ON ENDS ONLY.   _____

23. NO PIPE BUSHINGS.   _____

24. CARBON STEEL PIPE SCH 160, 1/2 INCH NPS
     MINIMUM.   _____

25. STAINLESS STEEL PIPE SCH 80S, 1/2 INCH NPS
     MINIMUM.   _____

26. PIPE NIPPLES AT BRANCH CONNECTIONS GREATER
     THAN 2 INCH LONG BUT LESS THAN 4 INCH LONG.   _____

27. PIPING HIGH POINTS HAVE VALVED VENTS.   _____

28. LOW POINTS, GAUGE GLASSES AND PRESSURE GAUGES
     HAVE VALVED DRAINS.   _____

29. BLOCK VALVES INSTALLED BETWEEN INSTRUMENT AND
     LINE.   _____

Part II

Alignment and Balancing

Chapter 5

Machinery Alignment*

For most rotating machines used in the process industries, the trend is toward higher speeds, higher horsepowers per machine, and less sparing. The first of these factors increases the need for precise balancing and alignment. This is necessary to minimize vibration and premature wear of bearings, couplings, and shaft seals. The latter two factors increase the economic importance of high machine reliability, which is directly dependent on minimizing premature wear and breakdown of key components. Balancing, deservedly, has long received attention from machinery manufacturers and users as a way to minimize vibration and wear. Many shop and field balancing machines, instruments, and methods have become available over the years. Alignment, which is equally important, has received proportionately less notice than its importance justifies.

Any kind of alignment, even straightedge alignment, is better than no alignment at all. Precise, two-indicator alignment is better than rough alignment, particularly for machines 3,600 rpm and higher. It can give greatly improved bearing and seal life, lower vibration, and better overall reliability. It does take longer, however, especially the first time it is done to a particular machine, or when done by inexperienced personnel. The process operators and mechanical supervisors must be made aware of this time requirement. If they insist on having the job done in a hurry, they should do so with full knowledge of the likelihood of poor alignment and reduced machine reliability. Figure 5-1 shows a serious machinery failure

*Main source: Malcolm G. Murray, Jr., *Alignment Manual for Horizontal, Flexibly-Coupled Rotating Machines* available from publisher, Murray and Garig Tool Works, 220 East, Texas Avenue, Baytown, Texas 77520; Tel. (281) 427-5923. Adapted by permission. Certain portions of this chapter, e.g., laser-optic alignment and some of the alignment tolerance criteria, are from other sources.

Figure 5-1. Machinery damage caused by bearing seizure. Bearing seizure was the result of gear coupling damage, and gear coupling damage was caused by excessive misalignment, caused by piping forces.

which started with piping-induced misalignment, progressed to coupling distress, bearing failure, and finally, total wreck.

Prealignment Requirements

The most important requirement is to have someone who knows what he is doing, and cares enough to do it right. Continuity is another important factor. Even with good people, frequent movement from location to location can cause neglect of things such as tooling completeness and prealignment requirements.

The saying that "you can't make a silk purse out of a sow's ear" also applies to machinery alignment. Before undertaking an alignment job, it is prudent to check for other deficiencies which would largely nullify the benefits or prevent the attainment and retention of good alignment. Here is a list of such items and questions to ask oneself:

Foundation   Adequate size and good condition? A rule of thumb calls for concrete weight equal to three times machine weight for rotating machines, and five times for reciprocating machines.

Grout   Suitable material, good condition, with no voids remaining beneath baseplate? Tapping with a small hammer can detect hollow spots, which can then be filled by epoxy injection or other means. This is a lot of trouble, though, and often is not necessary if the lack of grout is not causing vibration or alignment drift.

Baseplate   Designed for adequate rigidity? Machine mounting pads level, flat, parallel, coplanar, clean? Check with straightedge and feeler gauge. Do this upon receipt of new pumps, to make shop correction possible—and maybe collect the cost from the pump manufacturer. Shims clean, of adequate thickness, and of corrosion- and crush-resistant material? If commercial pre-cut shims are used, check for actual versus marked thicknesses to avoid a soft foot condition. Machine hold-down bolts of adequate size, with clearance to permit alignment corrective movement? Pad height leaving at least 2 in. jacking clearance beneath center at each end of machine element to be adjusted for alignment? If jackscrews are required, are they mounted with legs sufficiently rigid to avoid deflection? Are they made of type 316 stainless steel, or other suitable material, to resist field corrosion? Water or oil cooled or heated pedestals are usually unnecessary, but can in some cases be used for onstream alignment thermal compensation.

Piping   Is connecting piping well fitted and supported, and sufficiently flexible, so that no more than 0.003 in. vertical and horizontal (measured separately—not total) movement occurs at the flexible coupling when the last pipe flanges are tightened? Selective flange bolt tightening may be required, while watching indicators at the coupling. If pipe flange angular misalignment exists, a "dutchman" or tapered filler piece may be necessary. To determine filler piece dimensions, measure flange gap around circumference, then calculate as follows:

$$\frac{1}{8} \text{ in.} + (\text{Max. Gap} - \text{Min. Gap}) \left[\frac{\text{Gasket O.D.}}{\text{Flange O.D.}} \right] =$$

Maximum Thickness of
Tapered Filler Piece

$^1/_8$ in. = Dutchman Minimum Thickness (180° from Maximum Thickness). Dutchman OD and ID same as gasket OD and ID.

Spiral wound gaskets may be helpful, in addition to or instead of a tapered filler piece. Excessive parallel offset at the machine flange connection cannot be cured with a filler piece. It may be possible to absorb it by offsetting several successive joints slightly, taking advantage of clearance between flange bolts and their holes. If excessive offset remains, the piping should be bent to achieve better fit. For the "stationary" machine element, the piping may be connected either before or after the alignment is done—provided the foregoing precautions are taken, and final alignment remains within acceptable tolerances. In some cases, pipe expansion or movement may cause machine movement leading to misalignment and increased vibration. Better pipe supports or stabilizers may be needed in such situations. At times it may be necessary to adjust these components with the machine running, thus aligning the machine to get minimum vibration. Sometimes, changing to a more tolerant type of coupling, such as elastomeric, may help.

Coupling Installation Some authorities recommend installation on typical pumps and drivers with an interference fit, up to 0.0005 in. per in. of shaft diameter. In our experience, this can give problems in subsequent removal or axial adjustment. If an interference fit is to be used, we prefer a light one—say 0.0003 in. to 0.0005 in. overall, regardless of diameter. For the majority of machines operating at 3,600 rpm and below, you can install couplings with 0.0005 in. overall diametral clearance, using a setscrew over the keyway. For hydraulic dilation couplings and other nonpump or special categories, see manufacturers' recommendations or appropriate section of this text. Many times, high-performance couplings require interference fits as high as 0.0025 in. per in. of shaft diameter.

Coupling cleanliness, and for some types, lubrication, are important and should be considered. Sending a repaired machine to the field with its lubricated coupling-half unprotected, invites lubricant contamination, rusting, dirt accumulation, and premature failure. Lubricant should be chosen from among those recommended by the coupling manufacturer or a reputable oil company. Continuous running beyond two years is inadvisable without inspecting a grease lubricated coupling, since the centrifuging effects are likely to cause caking and loss of lubricity. Certain lubricants, e.g., Amoco and Koppers coupling greases, are reported to eliminate this problem, but visual external inspection is still advisable to detect leakage. Continuous lube couplings are subject to similar problems, although such remedies as anti-sludge holes can be used to allow longer runs at higher speeds. By far the best remedy is *clean* oil, because even small amounts of water will promote sludge formation. Spacer length can be important, since parallel misalignment accommodation is directly proportional to such length.

Alignment Tolerances

Before doing an alignment job, we must have tolerances to work toward. Otherwise, we will not know when to stop. One type of "tolerance" makes *time* the determining factor, especially on a machine that is critical to plant operation, perhaps the only one of its kind. The operations superintendent may only be interested in getting the machine back on the line, *fast*. If his influence is sufficient, the job may be hurried and done to rather loose alignment tolerances. This can be unfortunate, since it may cause excessive vibration, premature wear, and early failure. This gets us back to the need for having the tools and knowledge for doing a good alignment job efficiently. So much for the propaganda—now for the tolerances.

Tolerances must be established before alignment, in order to know when to stop. Various tolerance bases exist. One authority recommends $^1/_2$-mil maximum centerline offset per in. of coupling length, for hot running misalignment. A number of manufacturers have graphs which recommend tolerances based on coupling span and speed. A common tolerance in terms of face-and-rim measurements is 0.003-in, allowable face gap difference and centerline offset. This ignores the resulting accuracy variation due to face diameter and spacer length differences, but works adequately for many machines.

Be cautious in using alignment tolerances given by coupling manufacturers. These are sometimes rather liberal and, while perhaps true for the coupling itself, may be excessive for the coupled machinery.

A better guideline is illustrated in Figure 5-2, which shows an upper, absolute misalignment limit, and a lower, "don't exceed for good long-term operation limit." The real criterion is the running vibration. If

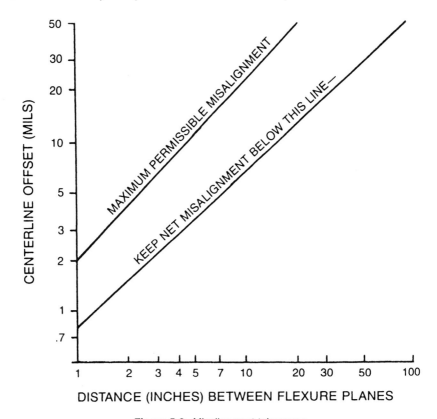

Figure 5-2. Misalignment tolerances.

excessive, particularly at twice running frequency and axially, further alignment improvement is probably required. Analysis of failed components such as bearings, couplings, and seals can also indicate the need for improved alignment.

Figure 5-2 can be applied to determine allowable misalignment for machinery equipped with nonlubricated metal disc and diaphragm couplings, up to perhaps 10,000 rpm. If the machinery is furnished with gear-type couplings, Figure 5-2 should be used up to 3,600 rpm only. At speeds higher than 3,600 rpm, gear couplings will tolerate with impunity only those shaft misalignments which limit the sliding velocity of engaging gear teeth to less than perhaps 120 in. per minute. For gear couplings, this velocity can be approximated by $V = (\pi DN) \tan \alpha$, where

       D = gear pitch diameter, in.
       N = revolutions per minute

2 tan α = total indicator reading obtained at hub outside diameter, divided by distance between indicator planes on driver and driven equipment couplings.

Say, for example, we were dealing with a 3,560 rpm pump coupled to a motor driven via a 6-in. pitch diameter gear coupling. We observe a total indicator reading of 26 mils in the vertical plane and a total indicator reading of 12 mils in the horizontal plane. The distance between the flexing member of the coupling, i.e., flexing member on driver and flexing member on driven machine, is 10 in. The total net indicator reading is $[(26)^2 + (12)^2]^{1/2}$ = 28.6 mils. Tan α ($\frac{1}{2}$)(28.6)/10) = 1.43 mils/in., or 0.00143 in./in. The sliding velocity is therefore $[(\pi)(6)(3560)(0.00143)]$ = 96 in. per minute. Since this is below the maximum allowable sliding velocity of 120 in. per minute, the installation would be within allowable misalignment.

Choosing an Alignment Measurement Setup

Having taken care of the preliminaries, we are now ready to choose an alignment setup, or arrangement of measuring instruments. Many such setups are possible, generally falling into three broad categories: face-and-rim, reverse-indicator, and face-face-distance. The following sketches show several of the more common setups, numbered arbitrarily for ease of future reference. Note that if measurements are taken with calipers or ID micrometers, it may be necessary to reverse the sign from that which would apply if dial indicators are used.

Figures 5-3 through 5-8 show several common arrangements of indicators, jigs, etc. Other arrangements are also possible. For example, Figures 5-3 and 5-4 can be done with jigs, either with or without breaking the coupling. They can also sometimes be done when no spacer is present, by using right-angle indicator extension tips. Figures 5-6 and 5-7 can be set up with both extension arms and indicators on the same side, rather than 180° opposite as shown. In such cases, however, a sign reversal will occur in the calculations. Also, we can indicate on back of face, as for connected metal disc couplings. Again, a sign reversal will occur.

In choosing the setup to use, personal preference and custom will naturally influence the decision, but here are some basic guidelines to follow.

Reverse-Indicator Method

This is the setup we prefer for most alignment work. As illustrated in Figure 5-9, it has several advantages:

Setup 1—
Two-Indicator Face-and-Rim

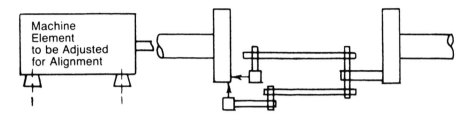

Setup 1A—
Two Indicator
Face-and-Rim with Stationary
Shaft Not Rotatable by Hand

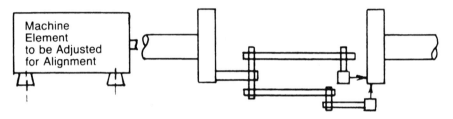

Figure 5-3. Two-indicator face-and-rim alignment method.

Setup 2—
Three Indicator Face-and-Rim

Figure 5-4. Three-indicator face-and-rim alignment method.

Setup 3—
Close-Coupled Face-and-Rim

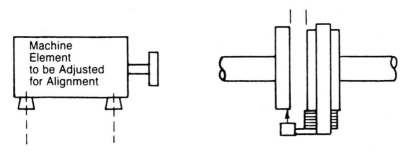

Figure 5-5. Close-coupled face-and-rim alignment method.

Setup 4—
Reverse-Indicator
Using Clamp-on Jigs with
Extended Posts

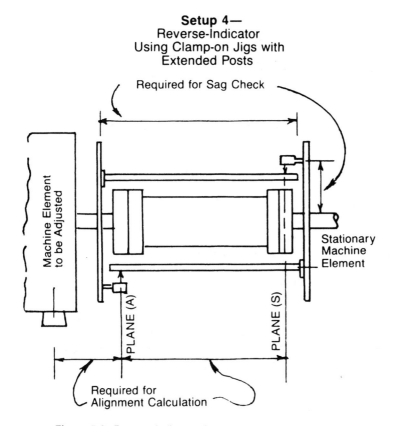

Figure 5-6. Reverse-indicator alignment using clamp-on jigs.

Setup 5

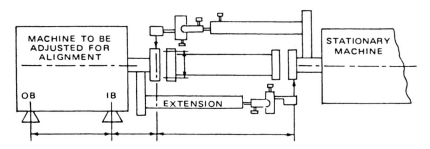

Figure 5-7. Reverse-indicator alignment using face-mounted brackets or any other brackets which hold the indicators as shown.

Setup 6

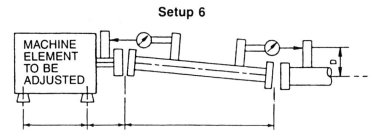

Figure 5-8. Two-indicator face-face-distance alignment method.

1. Accuracy is not affected by axial movement of shafts in sleeve bearings.
2. Both shafts turn together, either coupled or with match marks, so coupling eccentricity and surface irregularities do not reduce accuracy of alignment readings.
3. Face alignment, if desired, can be derived quite easily without direct measurement.
4. Rim measurements are easy to calibrate for bracket sag. Face sag, by contrast, is considerably more complex to measure.
5. Geometric accuracy is usually better with reverse-indicator method in process plants, where most couplings have spacers.
6. With suitable clamp-on jigs, the reverse-indicator method can be used quite easily for measuring without disconnecting the coupling or removing its spacer. This saves time, and for gear couplings, reduces the chance for lubricant contamination.
7. For the more complex alignment situations, where thermal growth and/or multi-element trains are involved, reverse-indicator can be

Figure 5-9. Reverse-indicator setup.

used quite readily to draw graphical plots showing alignment conditions and moves. It is also useful for calculating optimum moves of two or more machine elements, when physical limits do not allow full correction to be made by moving a single element.

8. When used with jigs and posts, single-axis leveling is sufficient for ball-bearing machines, and two-axis leveling will suffice for sleeve-bearing machines.

9. For long spans, adjustable clamp-on jigs are available for reverse-indicator application, without requiring coupling spacer removal. Face-and-rim jigs for long spans, by contrast, are usually nonadjustable custom brackets requiring spacer removal to permit face mounting.

10. With the reverse-indicator setup, we mount only one indicator per bracket, thus reducing sag as compared to face-and-rim, which mounts two indicators per bracket. (Face-and-rim can do it with one per bracket if we use two brackets, or if we remount indicators and rotate a second time, but this is more trouble.)

There are some limitations of the reverse-indicator method. It should not be used on close-coupled installations, unless jigs can be attached behind the couplings to extend the span to 3 in. or more. Failure to observe this limitation will usually result in calculated moves which overcorrect for the misalignment.

Both coupled shafts *must* be rotatable, preferably by hand, and preferably while coupled together. If only one shaft can be rotated, or if neither can be rotated, the reverse-indicator method cannot be used.

If the coupling diameter exceeds available axial measurement span, geometric accuracy will be poorer with reverse-indicator than with face-and-rim.

If required span exceeds jig span capability, either get a bigger jig or change to a different measurement setup such as face-face-distance. Cooling tower drives would be an example of this.

Face-and-Rim Method

This is the "traditional" setup which is probably the most popular, although it is losing favor as more people learn about reverse-indicator.

Advantages of face-and-rim:

1. It can be used on large, heavy machines whose shafts cannot be turned.
2. It has better geometric accuracy than reverse-indicator, for large diameter couplings with short spans.
3. It is easier to apply on short-span and small machines than is reverse-indicator, and will often give better accuracy.

Limitations of face-and-rim:

1. If used on a machine in which one or both shafts cannot be turned, some runout error may occur, due to shaft or coupling eccentricity.
2. If used on a sleeve bearing machine, axial float error may occur. One method of avoiding this is to bump the turned shaft against the axial stop each time before reading. Another way is to use a second face indicator 180° around from the first, and take half the algebraic difference of the two face readings after 180° rotation from zero start. Figure 5-10 illustrates this alignment method. Two 2-in. tubular graphite jigs are used for light weight and high rigidity.
3. If used with jigs and posts, two or three axis leveling is required, for ball and sleeve bearing machines respectively. Reverse-indicator requires leveling in one less axis for each.
4. Face-and-rim has lower geometric accuracy than reverse-indicator, for spans exceeding coupling or jig diameter.

Figure 5-10. Face-and-rim indicator setup using lightweight, high-rigidity tubular graphite fiber-reinforced epoxy jigs.

5. Face sag is often insignificant, but it can occur on some setups, and result in errors if not accounted for. Calibration for face sag is considerably more complex than for rim sag.
6. For long spans, face-and-rim jigs are usually custom-built brackets requiring spacer removal to permit face mounting. Long-span reverse-indicator jigs, by contrast, are available in adjustable clamp-on models not requiring spacer removal.
7. Graphing the results of face-and-rim measurements is more complex than with reverse-indicator measurements.

Face-Face-Distance Method

Advantages of face-face-distance:

1. It is usable on long spans, such as cooling tower drives, without elaborate long-span brackets or consideration of bracket sag.
2. It is the basis for thermal growth measurement in the Indikon proximity probe system, and again is unaffected by long axial spans.
3. It is sometimes a convenient method for use with diaphragm couplings such as Lucas Aerospace (Utica, New York), allowing mounting of indicator holders on spacer tube, with indicator contact points on diaphragm covers.

Limitations of face-face-distance:

1. It has no advantage over the other methods for anything except long spans.
2. It cannot be used for installations where no coupling spacer is present.
3. Its geometric accuracy will normally be lower than either of the other two methods.
4. It may or may not be affected by axial shaft movement in sleeve bearings, but this can be avoided by the same techniques as for face-and-rim.

Laser-Optic Alignment

In the early 1980s, by means of earth-bound laser beams and a reflector mounted on the moon, man has determined the distance between earth and the moon to within about 6 inches.

Such accuracy is a feature of optical measurement systems, as light travels through space in straight lines, and a bundled laser ray with particular precision.

Thus, critical machinery alignment, where accuracy of measurement is of paramount importance, is an ideal application for a laser-optic alignment system.

The inherent problems of mechanical procedure and sequence of measuring have been solved by Prüftechnik Dieter Busch, of Ismaning, Germany, whose OPTALIGN® system (Figure 5-11) comprises a semiconductor laser emitting a beam in the infrared range (wavelength 820 mm), along with a beamfinder incorporating an infrared detector. The laser beam is refracted through a prism and is caught by a receiver/detector.

These light-weight, nonbulky devices are mounted on the equipment shafts, and only a cord-connected microcomputer module is external to the beam emission and receiver/detector devices.

The prism redirects the beam and allows measurement of parallel offset in one plane and angularity in another, thus simultaneously controlling both. In one 90° rotation of the shafts all four directional alignment corrections are determined.

With the data automatically obtained from the receiver/detector, the microcomputer instantaneously yields the horizontal and vertical adjustment results for the alignment of the machine to be moved.

Physical contact between measuring points on both shafts is no longer required, as this is now bridged by the laser beam, eliminating the possibilities for error arising from gravitational hardware sag as well as from

Figure 5-11. Optalign® laser-optic alignment system.

sticky dial indicators, etc. The system's basic attachment is still carried out with a standard quick-fit bracketing system, or with any other suitable attachment hardware.

If the reader owns an OPTALIGN® or the newer "smartALIGN®" (Figure 5-12) system, he does not have to be concerned with sag. Other reader must continue the checkout process.

Checking for Bracket Sag

Long spans between coupling halves may cause the dial indicator fixture to sag measurably because of the weight of the fixture and the dial

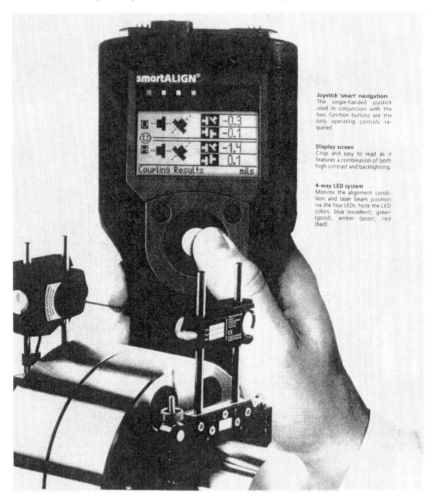

Figure 5-12. SmartALIGN® system. (Source: Prüftechnik, A. G., Ismaning, Germany.)

indicators. Although sag may be minimized by proper bracing, sag effects should still be considered in vertical alignment. To determine sag, install the dial indicators on the alignment fixture in the same orientation and relative position as in the actual alignment procedure with the fixture resting on a level surface as shown in Figure 5-13. With a small sling and scale, lift the indicator end of the fixture so that the fixture is in the horizontal position. Note the reading on the scale. Assume for example that the scale reading was 7.5 lbs. Next, mount the alignment fixture on the coupling

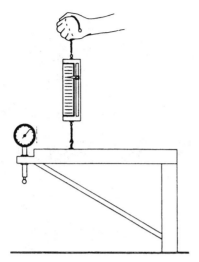

Figure 5-13. Testing for bracket sag.

hub with the dial indicator plunger touching the top vertical rim of the opposite coupling hub. Set the dial indicator to zero. Next, locate the sling in the same relative position as before and, while observing the scale, apply an upward force so as to repeat the previous scale reading (assumed 7.5 lbs in our example). Note the dial indicator reading while holding the upward force. Let us assume for example that we observe a dial indicator reading of −0.004 in. Using this specific methodology, sag error applies equally to the top and bottom readings. Therefore, the sag correction to the total indicator reading is double the indicated sag and must be *algebraically* subtracted from the bottom vertical parallel reading, i.e., −(2) (−0.004) = +0.008 correction to bottom reading.

This method is a clever one for face-mounted brackets. For clamp-on brackets, however, it would be easier and more common to attach them to a horizontal pipe on sawhorses, and roll top to bottom. Figure 5-14 shows this conventional method which, except for the sag compensator device, is almost universally employed. The sag compensator feature incorporates a weight-beam scale which applies an upward force when the indicator bracket is located at the top of the machine shaft, and an equal, but opposite, force when the indicator bracket and shaft combination is rotated to the down position, 180° removed.

In any event, let us assume that we obtain readings of 0 and +0.160 in. at the top and bottom vertical parallels respectively. We correct for sag in the following manner:

Figure 5-14. Sag compensator.

Using the first method of sag determination,

we observe bottom parallel reading	+ 0.160 in.
Sag correction − 2(−0.004) = + 0.008	+ 0.008 in.
Corrected bottom parallel reading	+ 0.168 in.

Bracket Sag Effect on Face Measurements

Bracket sag is generally thought to primarily affect rim readings, with little effect on face readings. Often this is true, but some risk may be incurred by assuming this without a test. Unlike rim sag, face sag effect depends not only on jig or bracket stiffness, but on its geometry.

Determining face sag effect is fairly easy. First get rim sag for span to be used (we are referring here to the full indicator deflection due to sag when the setup is rotated from top to bottom). This may be obtained by trial, with rim indicator only, or from a graph of sags compiled for the bracket to be used. Then install a setup with rim indicator only, on calibration pipe or on actual field machine, and "lay on" the face indicator and accessories, noting additional rim indicator deflection when this is done. Double this additional deflection, and add it to the rim sag found previously, if both the face and rim indicators are to be used simultaneously. If the face and the rim indicators are to be used separately, to reduce sag, use the original rim sag in the normal manner, and use this same original rim sag as shortly to be described in determining face sag effect—in this latter case utilizing a rim indicator installed temporarily with the face indicator for this purpose. If the face indicator is a different type (i.e., different weight) from the rim indicator, obtain rim sag using this face indicator on the rim, and use this figure to determine face sag effect.

Now install face and rim setup on the actual machine, and zero the indicators. With indicators at the top, deflect bracket upward an amount equal to the appropriate rim sag, reading on the rim indicator, and note the face indicator reading. The face sag correction with indicators at bottom would be this amount with opposite sign. If zeroing the setup at the bottom, the face sag correction at the top would be this amount with same sign (if originally determined at top, as described).

Face Sag Effect—Examples

Example 1

Face and rim indicators are to be used together as shown in Figure 5-3. Assume you will obtain the following from your sag test:

Rim sag with rim indicator only = 0.004 in.

Rim sag with two indicators = 0.007 in.

Mount the setup on the machine in the field, and with indicators at top, deflect the bracket upward 0.007 in. as measured on the rim indicator. When this is done, the face indicator reads plus 0.002 in. Face sag correction at the bottom position would therefore be minus 0.002 in. If you wish to zero at the bottom for alignment, but otherwise have data as noted, the face sag correction at the top would be plus 0.002 in.

Example 2

Face and rim indicators are to be used separately to reduce sag. Both indicators are the same type and weight. Other basic data are also the same.

Install face indicator and temporary rim indicator on the machine in the field, and place in top position. Zero indicators and deflect upward 0.004 in. as measured on rim indicator. Face indicator reads plus 0.0013 in. Face sag correction at the bottom would therefore be minus 0.0013 in. If zeroing at the bottom for alignment, but otherwise the same as above, face sag correction at top would be plus 0.0013 in.

Example 3

This will determine sag for "3-Indicator Face-and-Rim Setup" shown in Figure 5-4.

Set up the jig to the same geometry as for field installation but with rim indicator only and roll 180° top to bottom on pipe to get total single indicator rim sag ____ (Step 1).

Zero rim indicator on top and add or "lay on" face indicator, noting rim indicator deflection that occurs ____ (Step 2). Double this ____ (Step 3).

Add it to original total single indicator rim sag (Step 1). ____ (Step 4).

This figure, preceded by a plus sign, will be the sag correction for the rim indicator readings taken at bottom.

With field measurement setup as shown, zero all indicators, and deflect the indicator end of the upper bracket upward an amount equal to the total rim sag (Step 4). Note the face sag effect by reading the face indicator. This amount, with opposite sign, is the face sag correction to apply to the readings taken at the lower position ____ (Step 5).

Now deflect the upper bracket back down from its "total rim sag" deflection an amount equal to Step 3.

The amount of sag remaining on the face indicator, preceded by the *same* sign, is the sag correction for the single face indicator being read at the top position ____ (Step 6).

All of the foregoing refers of course to *bracket* sag. In long machines, we will also have *shaft* sag. This is mentioned only in passing, since there is no need to do anything about it at this time. Our "point-by-point" alignment will automatically take care of shaft sag. For initial leveling of large turbogenerators, etc., especially if using precision optical equipment, shaft sag must be considered. Manufacturers of such machines know this, and provide their erectors with suitable data for sag compensation. Further discussion of shaft sag is beyond the scope of this text.

Leveling Curved Surfaces

It is common practice to set up the "rim" dial indicators so their contact tips rest directly on the surface of coupling rims or shafts. If gross misalignment is not present, and if coupling and/or shaft diameters are large, which is usually the case, accuracy will often be adequate. If, however, major misalignment exists, and/or the rim or shaft diameters are small, a significant error is likely to be present. It occurs due to the measurement surface curvature, as illustrated in Figures 5-15 and 5-16.

This error can usually be recognized by repeated failure of top-plus-bottom (T + B) readings to equal side-plus-side (S + S) readings within

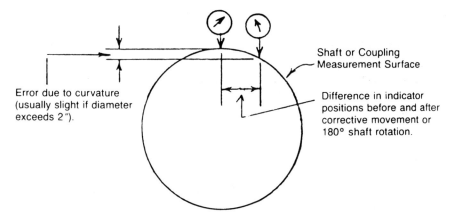

Figure 5-15. Error can be induced due to curvature effect on misaligned components.

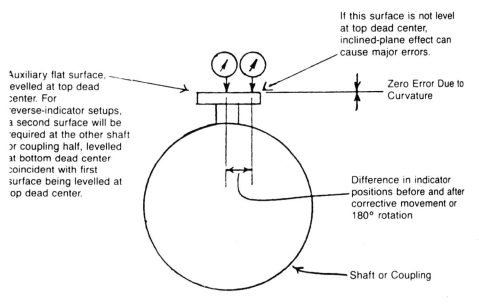

Figure 5-16. Auxiliary flat surface added to avoid curvature-induced measurement error.

one or two thousandths of an inch, and by calculated corrections resulting in an improvement which undershoots or overshoots and requires repeated corrections to achieve desired tolerance. A way to minimize this error is to use jigs, posts, and accessories which "square the circle." Here we attach flat surfaces or posts to the curved surfaces, and level them at top and bottom dead center. This corrects the error as shown in Figure 5-14.

For this method to be fully effective, rotation should be performed at accurate 90° quadrants, using inclinometer or bubble-vial device.

In most cases, however, this error is not enough to bother eliminating—it is easier just to make a few more corrective moves, reducing the error each time.

Jig Posts

The preceding explanation showed a rudimentary auxiliary surface, or "jig post," used for "squaring the circle." A more common reason for using jig posts is to permit measurement without removing the spacer on a concealed hub gear coupling. If jig posts are used, it is important that they be used properly. In effect, we must ensure that the surfaces contacted by the indicators meet these criteria:

- As already shown, they must be leveled in coordination at top and bottom dead centers, to avoid inclined plane error
- If any axial shaft movement can occur, as with sleeve bearings, the surfaces should also be made parallel to their shafts. This can be done by leveling axially at the top, rotating to the bottom, and rechecking. If bubble is not still level, tilt the surface back toward level for a half correction.
- If face readings are to be taken on posts, the post face surfaces should be machined perpendicular to their rim surfaces. In addition to this, and to Steps 1 and 2 just described, rotate shafts so posts are horizontal. Using a level, adjust face surfaces so they are vertical. Rotate 180° and recheck with level. If not still vertical, tilt back toward vertical to make a half correction on the bubble. This will accomplish our desired objective of getting the face surface perpendicular to the shaft in all measurement planes.

The foregoing assumes use of tri-axially adjustable jig posts. If such posts are not available, it may be possible to get good results using accurately machined nonadjustable posts. If readings and corrections do not turn out as desired, however, it could pay to make the level checks as described—they might pinpoint the problem and suggest a solution such as using a nonpost measurement setup.

Interpretation and Data Recording

Due to sag as well as geometry of the machine installation, it is difficult and deceptive to try second-guessing the adequacy of alignment solely from the "raw" indicator readings. It is necessary to correct for sag, then note the "interpreted" readings, then plot or calculate these to see the overall picture—including equivalent face misalignment if primary readings were reverse-indicator on rims only. Sometimes thermal offsets must be included, which further complicates the overall picture.

As a way to systematically consider these factors and arrive at a solution, it is helpful to use prepared data forms and stepwise calculation.

Suppose we are using the two-indicator face-rim method shown in Figure 5-3; let's call it "Setup #1." To start, prepare a data sheet as shown in Figure 5-17. Next, measure and fill in the "basic dimensions" at the top. Then, fill in the orientation direction, which is *north* in our example. Next, take a series of readings, zeroing at the top, and returning for final readings which should also be zero or nearly so. Now do a further check: Add the top and bottom readings algebraically (T + B), and add the side readings (S + S). The two sums should be equal, or nearly so. If the checks are poor, take a new set of readings. Do the checks *before* accounting for bracket sag. Now, fill in the known or assumed bracket sag. If the bracket does not sag (optimist!), fill in zero. Combine the sag algebraically with the vertical rim reading as shown, and get the net reading using (+) or (−) as appropriate to accomplish the sag correction. A well-prepared form will have this sign printed on it. If it does not, mentally figure out what must be done to "un-sag" the bracket in the final position, and what sign would apply when doing so.

Now we are ready to interpret our data in the space provided on the form. To do this, first take half of our net rim reading:

$$\frac{-0.011}{2} = -0.0055$$

This is because we are looking for centerline rather than rim offset. Since its sign is minus, we can see from the indicator arrangement sketch that the machine element to be adjusted is *higher* than the stationary element, at the plane of measurement. This assumes the use of a conventional American dial indicator, in which a positive reading indicates contact point movement into the indicator.

By the same reasoning, we can see that the bottom face distance is 0.007 in. wider than the top face distance.

Going now to the horizontal readings, we make the north rim reading zero by adding −0.007 in. to it. To preserve the equality of our algebra, we

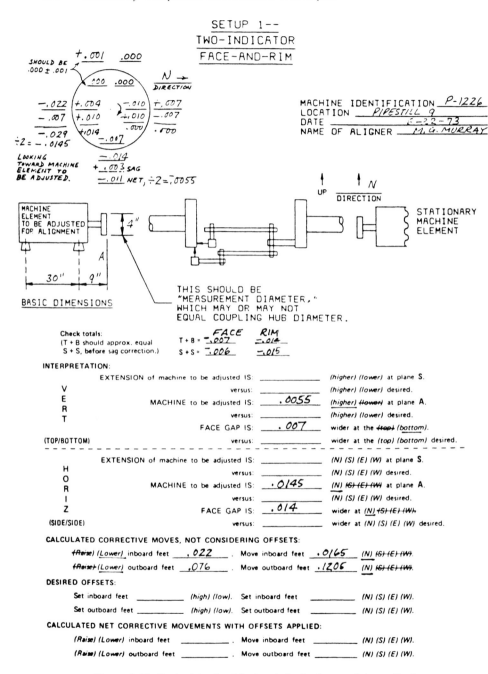

Figure 5-17. Basic data sheet for two-indicator face-and-rim method.

also add −0.007 in. to the south rim reading, giving us −0.029 in. Taking half of this, we find that the machine element to be adjusted is 0.0145 in. north of the stationary element at the plane of measurement.

Finally, we do a similar operation on our horizontal face readings, and determine that the north face distance is wider by 0.014 in.

The remaining part of the form provides space to put the calculated corrective movements. Although these have been filled in for our example, let's leave them for the time being, since we are not yet ready to explain the calculation procedure. We will show you how to get these numbers later. If you think you already know how, go ahead and try—the results may be interesting.

You have now seen the general idea about data recording and interpretation. By doing it systematically, on a prepared form corresponding to the actual field setup, you can minimize errors. If you are interrupted, you will not have to wonder what those numbers meant that you wrote down on the back of an envelope an hour ago. We will defer consideration of the remaining setups, until we have explained how to calculate alignment corrective movements. We will then take numerical examples for all the setups illustrated, and go through them all the way.

Calculating the Corrective Movements

Many machinists make alignment corrective movements by trial and error. A conscientious person can easily spend two days aligning a machine this way, but by knowing how to calculate the corrections, the time can be cut to two hours or less.

Several methods, both manual and electronic, exist for doing such calculations. All, of course, are based on geometry, and some are rather complicated and difficult to follow. For those interested in such things, see References 1–15. Years ago, the alignment specialist made use of programmable calculator solutions. Perhaps he used popular calculators such as the TI 59 and HP 67. By recording the alignment measurements on a prepared form, and entering these figures in the prescribed manner into the calculator, the required moves came out as answers. A variation of this was the TRS 80 pocket computer which had been programmed to do alignment calculations via successive instructions to the user telling him what information to enter.

By far the simplest calculator is the one described earlier in conjunction with the laser-based OPTALIGN® and smartALIGN® systems.

The foregoing electronic systems are popular, and have advantages in speed, accuracy, and ease of use. They have disadvantages in cost, usability under adverse field and hazardous area conditions, pilferage, sensitivity to damage from temperature extremes and rough handling, and availability to the field machinist at 2:00 A.M. on a holiday weekend. They also, for the most part, work mainly with numbers, and the answers may require acceptance on blind faith. By contrast, graphical methods inherently aid visualization by showing the relationship of adjacent shaft centerlines to scale.

Manual calculation methods have the advantage of low investment (pencil and paper will suffice, but even the simplest calculator will be faster). They have the disadvantage, some say, of requiring more thinking than the programmed electronic solutions, particularly to choose the plus and minus signs correctly.

The graphical methods, which "old-timers" prefer, have the advantage of aiding visualization and avoiding confusion. Their accuracy will sometimes be less than that of the "pure" mathematical methods, but usually not enough to matter. Investment is low—graph paper and plotting boards are inexpensive. Speed is high once proficiency is attained, which usually does not take long.

In this text, we will emphasize the graphical approach. Before doing so, let's highlight some common manual mathematical calculations.

Nelson[11] published an explanation of one rather simple method a number of years ago. A shortened explanation is given in Figure 5-18. For our given example, this would work out as follows:

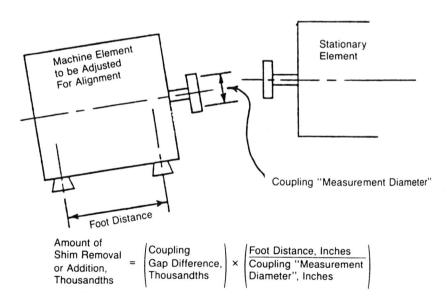

Figure 5-18. Basic mathematical formula used in determining alignment corrections.

Gap difference: 0.007 in.
Foot distance: 30 in.
Coupling measurement diameter: 4 in.

$$(0.007) \times \frac{(30)}{(4)} = 0.0525\,\text{in.} - \text{say } 0.053\,\text{in. shim addition beneath}$$

inboard feet, or removal beneath
outboard feet, or a combination
of the two, for a total of
0.053 in. correction.

Then, using rim measurements, determine parallel correction, and add or remove shims equally at all feet. Now do horizontal alignment similarly, and repeat as necessary.

Nelson's method is easy to understand, and it works. It is basically a four-step procedure in this order:

1. Vertical angular correction.
2. Vertical parallel correction.
3. Horizontal angular correction.
4. Horizontal parallel correction.

It has three disadvantages, however. First, it requires four steps, whereas the more complex mathematical methods can combine angular and parallel data, resulting in a two-step correction. Secondly, it is quite likely that initial angular correction will subsequently have to be partially "un-done," when making the corresponding parallel correction. Nobody likes to cut and install shims, then end up removing half of them. Finally, it is designed only for face-and-rim setups, and does not apply to the increasingly popular reverse-indicator technique.

We will now show two additional examples, wherein the angular and parallel correction are calculated at the same time, for an overall two-step correction. Frankly, we ourselves no longer use these methods, nor do we still use Nelson's method, but are including them here for the sake of completeness. Graphical methods, as shown later, are easier and faster. In particular, the alignment plotting board should be judged extremely useful. Readers who are not interested in the mathematical method may wish to skip to our later page, where the much easier graphical methods are explained. But, in any event, here is the full mathematical treatment.

In our first example, we will reuse the data already given in our setup No. 1 data sheet.

First, we will solve for vertical corrections:

Using Nelson's method, we found it necessary to make a 0.053 in. shim correction. Let us arbitrarily say this will be a shim addition beneath the inboard feet. At the coupling face, we then get a rise of:

$$\left(\frac{39}{30}\right)(0.053) = 0.069 \text{ in.}$$

Since we were already 0.0055 in. too high here, this puts us

$$0.069 \text{ in.} + 0.0055 \text{ in.} = 0.0745 \text{ in.}$$

too high. Therefore, subtract 0.0745 in. (call it 0.075 in.) at all feet. Thus our net shim change will be:

Inboard feet: 0.053 in. − 0.075 in. = 0.022 in. shim removal

Outboard feet: 0.075 in. shim removal

For the horizontal corrections, we proceed similarly:

$$(0.014) \times \frac{30}{4} = 0.105 \text{ in.}$$ Outboard (west) feet must move north, inboard (east) feet must move south, or a combination of the two, for angular alignment.

Let us say the outboard feet move north 0.105 in. This makes the coupling face move south, pivoting about the inboard feet:

$$\frac{9}{30}(0.105) = 0.0315 \text{ in.}$$

Since it was already 0.0145 in. too far north, it is now:

$$0.0315 - 0.0145 = 0.0170 \text{ in.}$$

too far south, as are the feet. Therefore, net correction will be:

Move outboard feet 0.105 in. + 0.017 in. = 0.122 in. north
Move inboard feet   0.017 in. north

It can be seen that our answers agree closely with those on the data sheet, which were obtained graphically. The differences are not large enough to cause us trouble in the actual field alignment correction.

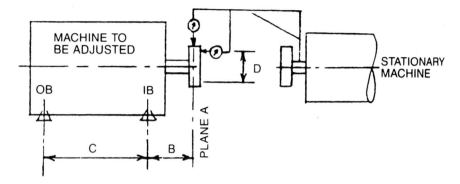

Figure 5-19. Machine sketch for face-and-rim alignment method.

Alternatively, this first example could be solved with another similar "formula method." To begin with, we draw the machine sketch, Figure 5-19. Then, we proceed by jotting down the relevant formulas:

$$\text{Correction at IB} = \pm\left(\frac{\text{Face Gap}}{\text{Difference}}\right)\left(\frac{B}{D}\right) \pm \begin{array}{c}\text{Net Parallel}\\\text{Offset of Shaft}\\\text{Centerlines at}\\\text{Plane A}\end{array}$$

$$\text{Correction at OB} = \pm\left(\frac{\text{Face Gap}}{\text{Difference}}\right)\left(\frac{B+C}{D}\right) \pm \begin{array}{c}\text{Net Parallel}\\\text{Offset of Shaft}\\\text{Centerlines at}\\\text{Plane A}\end{array}$$

For our example, the solutions would be as follows:

Vertical

$$\text{At IB,} -(0.007)\left(\frac{9}{4}\right)-\left(\frac{0.011}{2}\right) = -0.0157 - 0.0055 = -0.0212$$
$$\text{say lower IB } 0.021\,\text{in.}$$

$$\text{At OB,} -(0.007)\left(\frac{9+30}{4}\right)-\left(\frac{0.011}{2}\right) = -0.0683 - 0.0055 = -0.0738$$
$$\text{say lower OB } 0.074\,\text{in.}$$

Horizontal

At IB, $-(0.014)\left(\dfrac{9}{4}\right)N \pm \left(\dfrac{0.029}{2}\right)S$

$\qquad = -0.0316\,N - 0.0145\,S$
$\qquad = 0.0171\,N;$ say move IB 0.017 in. north

At OB, $-(0.014)\left(\dfrac{9+30}{4}\right)N \pm \left(\dfrac{0.029}{2}\right)S$

$\qquad = -0.137\,N - 0.0145\,S$
$\qquad = 0.1225\,N$
$\qquad =$ say move OB 0.122 in. or 123 in. north

As you can see, the values found this way are close to those found earlier. The main problem people have with applying these formulas is choosing between plus and minus for the terms. The easiest way, in our opinion, is to visualize the "as found" conditions, and this will point the way that movement must proceed to go to zero misalignment. For example, our bottom face distance is wide—therefore we need to *lower* the feet (pivoting at plane A) which we denote with a minus sign. The machine element to be adjusted is *higher* at plane A—so we need to lower it some more, which takes another minus sign. For the horizontal, our north face distance is wider, so we need to move the feet north (again pivoting at plane A). The machine element to be adjusted is *north* at plane A, so we need to move it south. Call north plus or minus, so long as you call south the opposite sign. Not really hard, but a lot of people have trouble with the concept, which is why we prefer to concentrate on graphical methods, where direction of movement becomes more obvious. We will get into this shortly, but first let's do a reverse-indicator problem mathematically.

For our reverse-indicator example, we will use the setup shown earlier as Figure 5-6. Also, we must now refer to the appropriate data sheet, Figure 5-20. Finally, we resort to some triangles, Figures 5-21 and 5-22, to assist us in visualizing the situation.

Figure 5-19 represents the elevation view. Solving, we obtain:

$0.007\,\text{in.} - (0.012\,\text{in.} - 0.007\,\text{in.})\left(\dfrac{12}{14}\right)$

$\qquad = 0.0027\,\text{in. too high at inboard feet.}$

$0.007\,\text{in.} - (0.012\,\text{in.} - 0.007\,\text{in.})\left(\dfrac{12+26}{14}\right)$

$\qquad = 0.0066\,\text{in. too low at outboard feet.}$

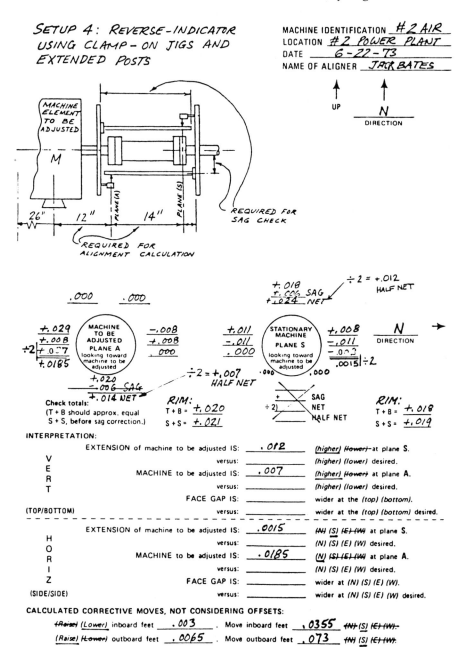

Figure 5-20. Data sheet reverse-indicator alignment method.

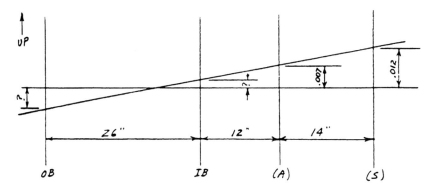

Figure 5-21. Elevation triangles for reverse-indicator alignment example.

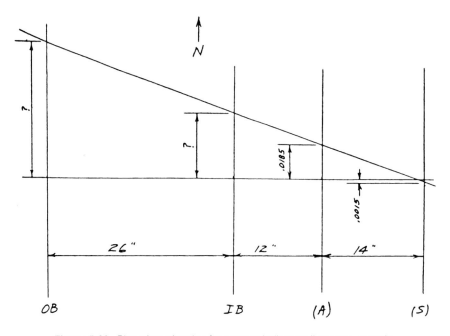

Figure 5-22. Plan view triangles for reverse-indicator alignment example.

Figure 5-20 represents the plan view. Here,

$$0.0185 + (0.0185 + 0.0015)\left(\frac{12}{14}\right)$$

$$= 0.0357 \text{ in. too far north at inboard feet.}$$

$$0.0185 + (0.0185 + 0.0015)\left(\frac{12 + 26}{14}\right)$$

$$= 0.0729 \text{ in. too far north at outboard feet.}$$

Summarizing, we should:

Lower inboard feet 0.003 in.
Lower outboard feet 0.0065 in., say 0.007 in.
Move inboard feet south 0.036 in.
Move outboard feet south 0.073 in.

These results obviously agree closely with our graphical results. Again, the same results could have been obtained mathematically. To begin with, we have to provide a machine sketch, Figure 5-23. Then:

$$\frac{\text{Correction}}{\text{at Inboard}} = \pm \left[\frac{\text{Centerline}}{\text{Offset at S}} \pm \frac{\text{Centerline}}{\text{Offset at A}} \right] \left(\frac{B+C}{B} \right)$$

$$\pm \frac{\text{Centerline}}{\text{Offset at S.}}$$

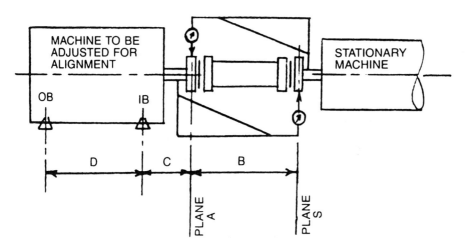

Figure 5-23. Machine sketch for reverse-indicator alignment example.

$$\begin{array}{l} \text{Correction} \\ \text{at Outboard} \end{array} = \pm \left[\begin{array}{l} \text{Centerline} \\ \text{Offset at S} \end{array} \pm \begin{array}{l} \text{Centerline} \\ \text{Offset at A} \end{array} \right] \left(\frac{B+C+D}{B} \right) \\ \quad\quad\quad\quad\quad \pm \begin{array}{l} \text{Centerline} \\ \text{Offset at S.} \end{array}$$

Using numbers from our example:

$$+\left[0.012 - 0.007\right]\left(\frac{14+12}{14}\right) - 0.012 = 0.0093 - 0.012 = -0.0027$$

say lower IB 0.003 in.

$$+\left[0.012 - 0.007\right]\left(\frac{14+12+26}{14}\right) - 0.12 = +0.0066\,\text{in.}$$

say raise OB 0.007 in.

$$-\left[0.0015 + 0.0185\right]\left(\frac{14+12}{14}\right)$$
$$+0.0015 = -0.0371 + 0.0015 = -0.0356$$

say move IB 0.036 in. south

$$-\left[0.0015 + 0.0185\right]\left(\frac{14+12+26}{14}\right)$$
$$+0.0015 = -0.074 + 0.0015 = -0.0725$$

say move OB 0.072 in. or 0.073 in. south

Again, the answers come out all right if you get the signs right, but the visualization is difficult unless you make scale drawings or graphical plots representing the "as found" conditions.

The Graphical Procedure for Reverse Alignment*

As mentioned earlier, the reverse dial indicator method of alignment is probably the most popular method of measurement, because the dial indicators are installed to measure the relative position of two shaft centerlines. This section emphasizes this method because of the ease of graphically illustrating the shaft position.

What Is Reverse Alignment?

Reverse alignment is the measurement of the axis or the centerline of one shaft to the relative position of the axis of an opposing shaft center-

---

*Courtesy of A-Line Mfg., Inc., Liberty Hill, Texas (Tel. 877-778-5454).

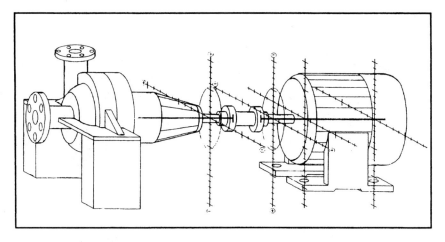

Figure 5-24. Centerline measurement—both vertical and horizontal.

line. This measurement can be projected the full length of both shafts for proper positioning if you need to allow for thermal movement. The measurement also shows the position of the shaft centerlines at the coupling flex planes, for the purpose of selecting an allowable tolerance. The centerline measurements are taken in both horizontal and vertical planes (Figure 5-24).

Learning How to Graph Plot

Graphical alignment is a technique that shows the relative positon of the two shaft centerlines on a piece of square grid graph paper.

First we must view the equipment to be aligned in the same manner that appears on the graph plot. In this example we view the equipment with the "**FIXED**" on the left and the "**MOVABLE**" on the right (Figure 5-25). This remains the same view both vertically and horizontally. Mark these sign conventions on graph paper, as shown in Figure 5-26.

Example   **Scale: Each Square ↔ = 1.0″**
                    Scale: Each Square ↕ = 0.001″

Next, measure:

A. Distance between indicators
B. Distance between indicator and front foot
C. Distance between feet

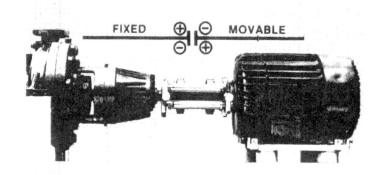

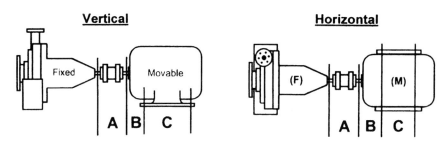

Figure 5-25. Views of equipment to be aligned.

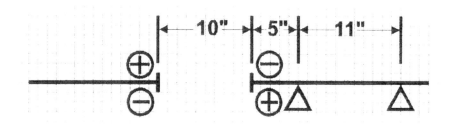

Figure 5-26. Choose convenient sign convention on graph paper.

The direction of indicator movements is shown in Figure 5-27. Choose dial indicators that read 0.001-inch (or "one mil"), and become familiar with the logic of dial indicator sweeps (Figure 5-28). Note that this illustration shows the true arc of measurement. The centerline of the opposing shaft to be 0.004″ lower and 0.002″ to the right of the centerline of the shaft being measured.

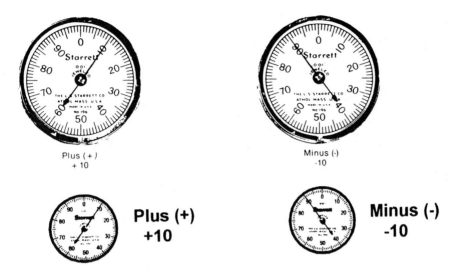

Figure 5-27. Direction of indicator movements.

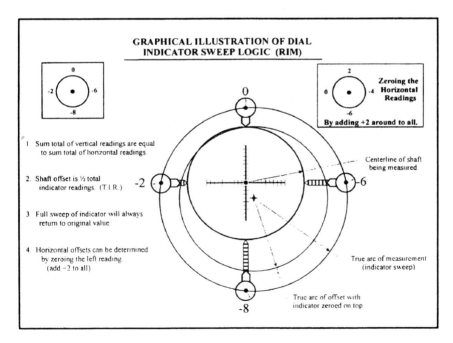

Figure 5-28. Graphical illustration of dial indicator sweep logic. Measurements are made on coupling rim.

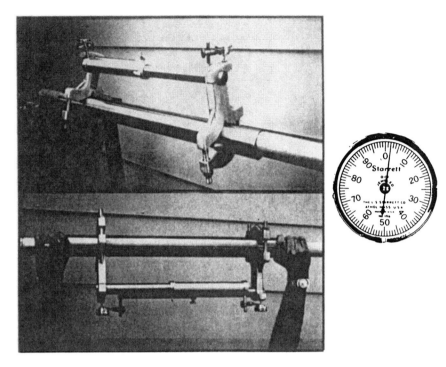

Figure 5-29. Sag check. Example: 0.002″ sag. Position indicator to read +2.

The **most important factors to remember** about the logic of the dial indicator sweep are:

1. The plus and minus sign show direction.
2. The number value shows how far (distance).
3. The offset is $^1/_2$ the total indicator reading (TIR).

Sag Check

To perform this check (Figure 5-29), clamp the brackets on a sturdy piece of pipe the same distance they will be when placed on the equipment. Zero both indicators on top, then rotate to bottom. The difference between the top and bottom reading is the sag.

Sag will always have a negative value, so when allowing for sag on the vertical move always start with a plus (+) reading.

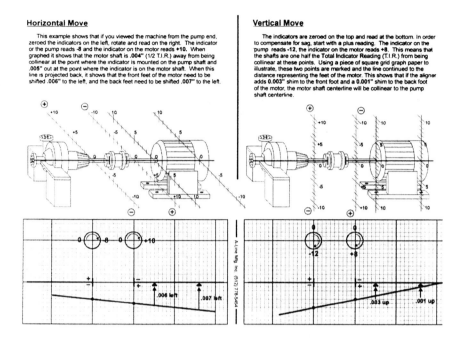

Horizontal Move

This example shows that if you viewed the machine from the pump end, zeroed the indicators on the left, rotate and read on the right. The indicator or the pump reads -8 and the indicator on the motor reads +10. When graphed it shows that the motor shaft is .004" (1/2 T.I.R.) away from being collinear at the point where the indicator is mounted on the pump shaft and .005" out at the point where the indicator is on the motor shaft. When this line is projected back, it shows that the front feet of the motor need to be shifted .006" to the left, and the back feet need to be shifted .007" to the left.

Vertical Move

The indicators are zeroed on the top and read at the bottom. In order to compensate for sag, start with a plus reading. The indicator on the pump reads -12, the indicator on the motor reads +8. This means that the shafts are one half the Total Indicator Reading (T.I.R.) from being collinear at these points. Using a piece of square grid graph paper to illustrate, these two points are marked and the line continued to the distance representing the feet of the motor. This shows that if the aligner adds 0.003" shim to the front foot and a 0.001" shim to the back foot of the motor, the motor shaft centerline will be collinear to the pump shaft centerline.

Figure 5-30. Horizontal and vertical moves explained.

Making the Moves

The next step is "making your moves," as illustrated in Figure 5-30. The correct account of movement will have been predefined as discussed later in this segment.

Using the reverse method of centerline measurement, the tolerance window (Figure 5-31) can be visually illustrated on a piece of square grid graph paper. Each horizontal square will represent 1 inch, each vertical square will represent 1 one-thousandth of an inch (0.001").

Figure 5-31 shows a typical pump and motor arrangement with the coupling flex planes 8" apart. An allowable tolerance of $^1/_2$ thousandths (0.0005") per inch of coupling separation is selected. This is typical for equipment operating at speeds up to 10,000 rpm. The aligner will now apply the tolerance window to the graph paper 0.004" above and 0.004" below the fixed centerline at the same location where the flexing elements are shown in the figure.

After the adjustment has been made and a new set of indicator readings have been taken, if the movable centerline stays within the tolerance window at both flex planes, the alignment is now within tolerance.

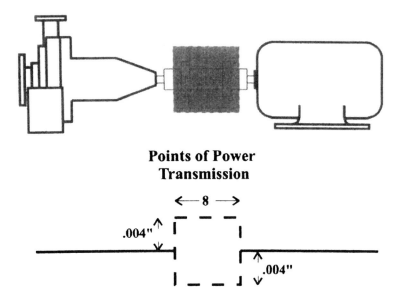

**Points of Power
Transmission**

**Allowable alignment when moveable C/L
is within tolerance box**

Figure 5-31. Tolerance window ("tolerance box").

Thermal movement calculations need to be applied to ensure that the machine *can move into* tolerance and *not move out* of tolerance.

It should be noted that the generally accepted value is $^1/_2$ thousandths per inch (0.0005″) deviation from colinear for each inch of distance between the coupling flex planes. This is probably too close a tolerance for general purpose pumps, but is not difficult to obtain. Since unwanted loads (thermal and other) are difficult to predict, the tighter tolerance gives a margin of safety.

Summary of Graphical Procedure

Figures 5-32 through 5-38 give a convenient summary of the graphical procedure.

The "Optimum Move" Alignment Method

At times, as in mixing alcohol with water and measuring volumes, the whole can be less than the sum of its parts. A parallel situation exists in

(Text continued on page 245)

(A) Preparation

1. Remove all existing shims under one foot, clean the base thoroughly, scrape away all rust, file off any nicks and burrs. Thread the base bolts into the base, retap holes if necessary. If old shims are to be used, clean them thoroughly. Replace damaged shims. Always use a minimum amount of shims.

(B) Installation of brackets.

1. Clean mounting surface, file off all nicks and burrs.
2. Check indicators for sticking and loose needle.
3. Aim indicator stem directly toward center line of shaft.

(C) Measurement.

(D) Layout on graph paper.

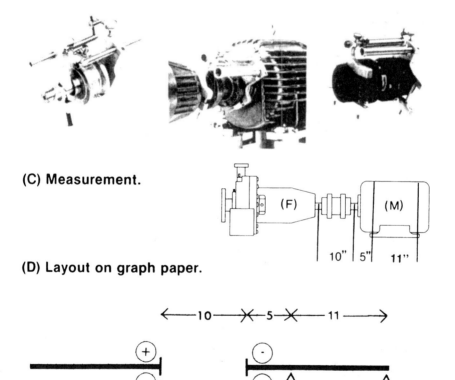

Figure 5-32. Getting set up for the graphical procedure.

SCALE ↔ = 1"
      ↕ = .001"

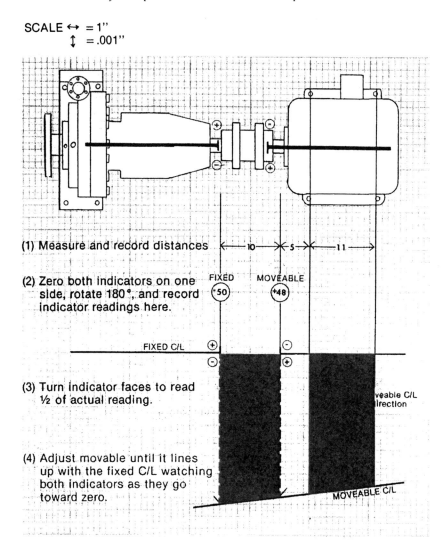

(1) Measure and record distances

(2) Zero both indicators on one
    side, rotate 180°, and record
    indicator readings here.

(3) Turn indicator faces to read
    ½ of actual reading.

(4) Adjust movable until it lines
    up with the fixed C/L watching
    both indicators as they go
    toward zero.

If alignment cannot be achieved because equipment is boltbound, it is now
time to solve the problem.
If proper alignment is achieved, tighten all base bolts

Figure 5-33. Preliminary horizontal move.

(F) Check for soft foot.

1. Move indicators to 12 o'clock position, get some depression on the indicators, and zero.

2. Loosen one base bolt. If indicator moves away from zero, place whatever amount of shims that will slide under that foot. Retighten base bolt.

3. Repeat this procedure to the remaining feet.

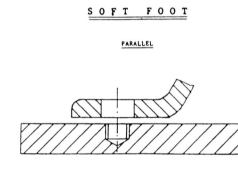

(G) Vertical move

Place both indicators on top and position dials to zero. If compensating for sag, start with a plus (+) reading. (remember example: .002" sag, start with indicator reading + 2)
Rotate to bottom and record readings here.

Figure 5-34. Preparing for the vertical move includes soft foot check.

SCALE ↔ = 1"

(1) Zero both indicators on top
or start at a plus reading to
compensate for sag (if allowing
for .002" sag, start indicator reading + 2

(2) Measure And record distances   10" 5" 11"

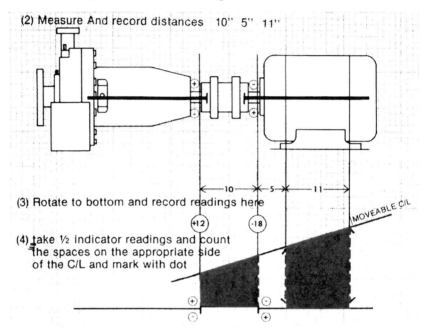

(3) Rotate to bottom and record readings here

(4) take ½ indicator readings and count
the spaces on the appropriate side
of the C/L and mark with dot

(5) With straight edge draw a line connecting
these two dots and extend the line over
the part of the graph representing
the feet

(6) Count the spaces at the feet of Moveable C/L
To the line on the same plane as the fixed C/L
is the shim Adjustment.

After shim adjustment is satisfactory, go back and complete horizontal
move.

IMPORTANT: Be sure to check for correct coupling spacing.

Figure 5-35. Calculate the vertical move.

EXAMPLE 1.

Pump stays the same, motor grows .003" at both feet.

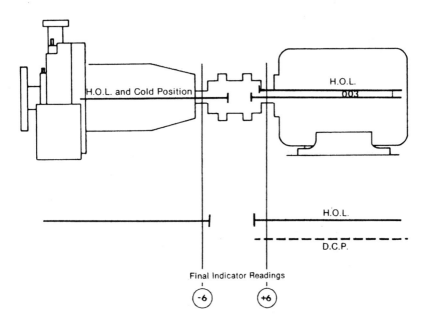

Figure 5-36. Thermal growth considerations, parallel. Thermal movements in machinery can be graphically illustrated when the aligner knows the precalculated heat movements.

Example 2.

Pump will rise .001" at front reference point, .004" at back reference point. The motor will rise .003" at both reference points.

Step 1. Calculate thermal movement from ambient to running position.

    2. Set up graph using H.O.L. as target and layout D.C.P.

    3. Use ½ indicator reading at the fixed D.C.P. to layout the A.C.P. of the motor. (D1)

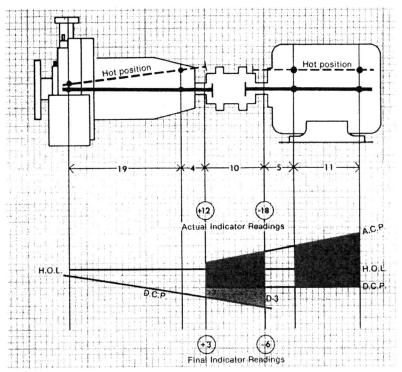

H.O.L. - Hot Operating Line.
D.C.P. - Desired Cold Position.
A.C.P. - Actual Cold Position.

D-1 - ½ indicator readings taken from the projected C/L of the fixed.
D-2 - Total shim adjustment.
D-3 - Final indicator reading.

Figure 5-37. Thermal growth considerations, angular.

Misalignment is the deviation of relative shaft position from a colinear axis of rotation, measured at the points of power transmission, when the equipment is running at normal operating conditions.

A good rule of thumb for general purpose equipment (3600 RPMs and less) is .0005" (½ thousandth) per inch of coupling span.

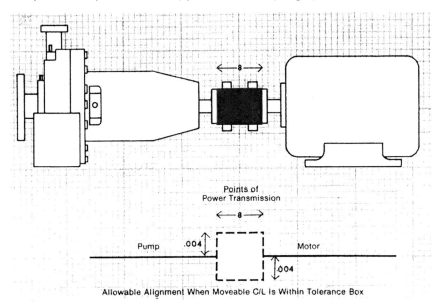

Figure 5-38. Defining the "tolerance box."

(Text continued from page 238)

the method we are about to illustrate[16]. In effect, we will see that by making optimum movements of both elements to be aligned, the maximum movement required at any point is a great deal less than if either element were to be moved by itself. Figure 5-39 shows an electric motor-driven centrifugal pump with severe vertical misalignment. The numbers are actual, from a typical job, and were not made up for purposes of this text.

As can be seen, regardless of whether we chose to align the motor to the pump or vice versa, we needed to lower the feet considerably—from 0.111 to 0.484 in. As it happened, the motor feet had only 0.025 in. total shimming, and the pump, as usual, had no shimming at all.

Some would shim the pump "straight up" to get it higher than the motor, and then raise the motor as required. This, in fact, was first attempted by our machinists. They had raised the pump about ³/₈ in., at which point the piping interfered, and the pump was still not high enough. By inspection of

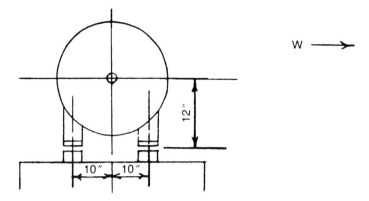

Figure 5-39. Horizontal movement by vertical adjustment: electric motor example.

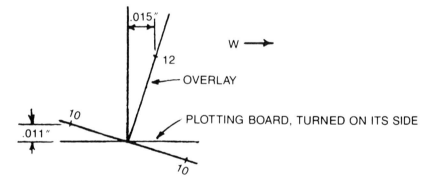

Figure 5-40. Plotting board solution for electric motor movement exercise of Figure 5-39.

Figures 5-41 and 5-42 it can be seen that they would have needed to raise it 0.484 in. (or 0.459 in. if all outboard motor shims had been removed).

Figure 5-42 shows the solution used to achieve alignment without radical shimming or milling. As can be seen, our maximum shim addition was 0.050 in., which is much lower than the values found earlier for single-element moves. We could have reduced this shimming slightly by removing our 0.025 in. existing shims from beneath the outboard feet of the motor, but chose not to do so, leaving some margin for single-element trim adjustments. As it turned out, the trimming went the other way, with 0.012 in. and 0.014 in. additions required beneath the motor inboard and outboard, respectively. This reflects such factors as heel-and-toe effect causing variation in foot pivot centers. This is normal for

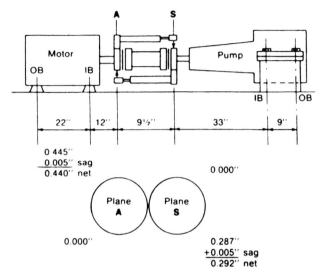

Interpreting, we see that the extension of the motor shaft centerline is 0.146 in. higher than the pump shaft centerline at plane "S" and, the motor shaft centerline is 0.220 in. higher than the extension of the pump shaft centerline at plane "A."

Solving by plotting board we get the following; if the motor is moved:

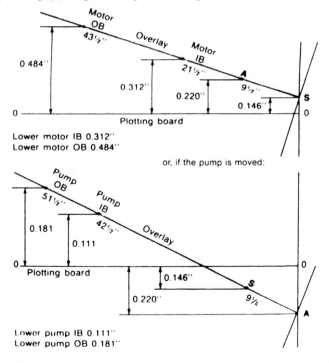

Figure 5-41. Motor-pump vertical misalignment with single element move solutions.

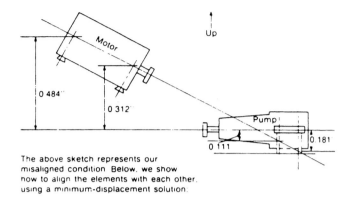

The above sketch represents our
misaligned condition. Below. we show
how to align the elements with each other.
using a minimum-displacement solution:

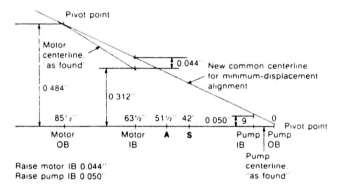

The alignment dimensions are scaled much larger than the linear dimensions
for ease in working the problem. This makes the new common centerline
have an exaggerated tilt as shown here. although it would not be apparent
in the actual field installation.

Figure 5-42. Plotting board or graph paper plot showing optimum two-element move.

situations such as this with short foot centers and long projections to
measurement planes.

Several variations on the foregoing example are worth noting, and are
shown in Figure 5-43. The basic approach is the same for all though, and
is easy to apply once the principle is understood.

We have, to this point, made no mention of thermal growth. If this is
to be considered, the growth data may be superimposed on the basic misalignment plots, or included prior to plotting, before proceeding with the
optimum-move solution. Also, of course, there are valid nongraphical
methods of handling the alignment solutions shown here—but we find the
graphical approach easier for visualization, and accurate enough if done
carefully.

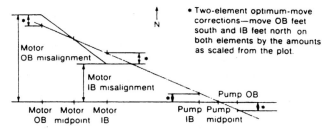

• Two-element optimum-move
corrections—move OB feet
south and IB feet north on
both elements by the amounts
as scaled from the plot.

For horizontal corrections, a midpoint plot, asshown above, is often useful for minimizing movements.

If, however, we cannot move south (or down for vertical misalignment), our optimum-move plot would look like one of the following:

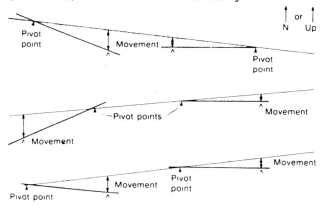

With three or more elements, the same principles apply. For example, the following would give us minimum movements assuming two-directional capability:

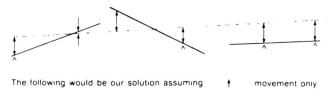

The following would be our solution assuming ↑ movement only

Figure 5-43. Various possibilities in plotting minimum displacement alignment.

Thermal Growth—Twelve Ways to Correct for It

Thermal growth of machines may or may not be significant for alignment purposes. In addition, movement due to pipe effects, hydraulic forces and torque reactions may enter the picture. Relative growth of the two or more elements is what concerns us, not absolute growth referenced to a fixed benchmark (although the latter could have an indirect effect if piping forces are thereby caused). Vibration, as measured by seismic or proximity probe instrumentation, can give an indication of whether thermal growth is causing misalignment problems due to differences between ambient and operating temperatures. If no problem exists, then a "zero-zero" ambient alignment should be sufficient. Our experience has been that such zero-zero alignment is indeed adequate for the majority of electric motor driven pumps. Zero-zero has the further advantage of simplicity, and of being the best starting point when direction of growth is unknown. Piping is often the "tail that wags the dog," causing growth in directions that defy prediction. For these reasons, we favor zero-zero unless we have other data that appear more trustworthy, or unless we are truly dealing with a predictable hot pump thermal expansion situation.

If due to vibration or other reasons it is decided that thermal growth correction should be applied, several approaches are available, as follows:

1. Pure guesswork, or guesswork based on experience.
2. Trial-and-error.
3. Manufacturers' recommendations.
4. Calculations based on measured or assumed metal temperatures, machine dimensions, and handbook coefficient of thermal expansion.
5. Calculations based on "rules-of-thumb," which incorporate the basic data of 4.
6. Shut down, disconnect coupling, and measure before machines cool down.
7. Same as 6, except use clamp-on jigs to get faster measurements without having to break the coupling.
8. Make mechanical measurements of machine housing growth during operation, referenced to baseplate or foundation, or between machine elements. (Essinger.)
9. Same as 8, except use eddy current shaft proximity probes as the measuring elements, with electronic indication and/or recording. (Jackson; Dodd/Dynalign; Indikon.)
10. Measure the growth using precise optical instrumentation.
11. Make machine and/or piping adjustments while running, using vibration as the primary reference.

12. Laser measurement represents another possibility. The OPTA-LIGN® method mentioned earlier also covers hot alignment checks.

Let us now examine the listed techniques individually.

Guesswork. Guesswork is rarely reliable. Guesswork based on experience, however, may be quite all right—although perhaps in such cases it isn't really guesswork. If a certain thermal growth correction has been found satisfactory for a given machine, often the same correction will work for a similar machine in similar service.

Trial-and-Error. Highly satisfactory, if you have plenty of time to experiment and don't damage anything while doing so. Otherwise, to be avoided.

Manufacturers' Recommendations. Variable. Some will work well, others will not. Climatic, piping, and process service differences can, at times, change the growth considerably from manufacturers' predictions based on their earlier average experience.

Calculations Based on Measured or Assumed Metal Temperatures, Machine Dimensions, and Handbook Coefficients of Thermal Expansion. Again, results are variable. An infrared thermometer is a useful tool here, for scanning a machine for temperature. This method ignores effects due to hydraulic forces, torque reactions, and piping forces.

Calculations Based on Rules of Thumb. Same comment as previous paragraph.

Shut Down, Disconnect Coupling, and Measure before Machines Cool Down. About all this can be expected to do is give an indication of the credulity of the person who orders it done. In the time required to get a set of measurements by this method, most of the thermal growth and all of the torque and hydraulic effect will have vanished.

Same as Previous Paragraph Except Use Clamp-On Jigs to Get Faster Measurements Without Having to Break the Coupling. This method, used in combination with backward graphing, should give better results than 6, but how much better is questionable. Even with "quick" jigs, a major part of the growth will be lost. Furthermore, shrinkage will be occurring during the measurement, leading to inconsistencies. Measurement of torque and hydraulic effects will also be absent by this method. Some training courses advocate this technique, but we do not. If used, however, three sets of data should be taken, at close time intervals—not two sets as some texts recommend. The cooling, hence shrinkage, occurs at a variable rate, and three points are required to establish a curve for backward graphing.

Make Mechanical Measurements of Machine Housing Growth During Operation, Referenced to Baseplate or Foundation, or Between Machine Elements. This method can be used for machines with any type of coupling, including continuous-lube. Essinger[5] describes one variation, using baseplate or foundation reference points, and measurement between these and bearing housing via a long stroke indicator having Invar 36 extensions subject to minimum expansion-contraction error. Hot and cold data are taken, and a simple graphic triangulation method gives vertical and horizontal growth at each plane of measurement. This method is easy to use, where physical obstructions do not prevent its use. Bear in mind that base plate thermal distortion may affect results. It is reasonably accurate, except for some machines on long, elevated foundations, where errors can occur due to unequal growth along the foundation length. In such cases, it may be possible to apply Essinger's method between machine cases, without using foundation reference points. A further variation is to fabricate brackets between machine housings and use a reverse-indicator setup, except that dial calipers may be better than regular dial indicators which would be bothered by vibration and bumping.

Same as Previous Paragraph, But Use Eddy Current Shaft Proximity Probes as the Measuring Elements, with Electronic Indicating and/or Recording. Excepting the PERMALIGN® method, this one lends itself the best to keeping a continuous record of machine growth from startup to stabilized operation. Due to the complexity and cost of the instrumentation and its application, this technique is usually reserved for the larger, more complex machinery trains. Judging by published data, the method gives good results, but it is not the sort of thing that the average mechanic could be fully responsible for, nor would it normally be justified for an average, two-element machinery train. In some cases, high machine temperatures can prevent the use of this method. The Dodd bars offer the advantage over the Jackson method that cooled posts are not needed and thermal distortion of base plate does not affect results. The Indikon system also has these advantages, and in addition can be used on unlimited axial spans. It is, however, more difficult to retrofit to an existing machine.

Measure the Growth Using Precise Optical Instrumentation. This method makes use of the precise tilting level and jig transit, with optical micrometer and various accessories. By referencing measurements to fixed elevations or lines of sight, movement of machine housing points can be determined quite accurately, while the machine is running. As with the previous method, this system is sophisticated and expensive, with delicate equipment, and requires personnel more knowledgeable than the average mechanic. It is therefore reserved primarily for the more complex machinery trains. It has given good results at times, but has also given erroneous or questionable data in other instances. The precise tilting level has additional use in soleplate and shaft leveling, which are not difficult to learn.

Several consultants offer optical alignment services. For the plant having only infrequent need for such work, it is usually more practical to engage such a consultant than to attempt it oneself.

Make Machine and/or Piping Adjustments While Running, Using Vibration as the Primary Reference. Baumann and Tipping[2] describe a number of horizontal onstream alignments, apparently made with success. Others are reluctant to try such adjustments for fear of movement control loss that could lead to damage. We have, however, frequently adjusted pipe supports and stabilizers to improve pump alignment and reduce vibration while the pump was running.

Laser Measurements

With the introduction of the modern, up-to-date PERMALIGN® system, laser-based alignment verification has been extended to cover hot alignment checks. Figure 5-44 illustrates how the PERMALIGN® is mounted onto both coupled machines to monitor alignment. The measurements are then taken when the monitor (shown mounted on the left-hand machine) emits a laser beam, which is reflected by the prism mounted on the other machine (shown on the right). The reflected beam reenters the monitor and strikes a position detector inside. When either machine moves, the reflected beam moves as well, changing its position in the detector. This detector information is then processed so that the amount of machine movement is shown immediately in terms of $1/100$ mm or mils in the display, located directly below the monitor lens. Besides displaying detector X and Y co-ordinates, the LCD also indicates system temperature and other operating information.

Thermal Growth Estimation by Rules of Thumb

We will now describe several "rules of thumb" for determining growth. Frankly, we have little faith in any of them, but are including them here for the sake of completeness.

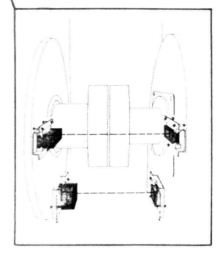

Figure 5-44. Hot alignment of operating machines being verified by laser-optic means (courtesy Prüftechnik A.G., Ismaning, Germany).

The following is for "foot-mounted horizontal, end suction centrifugal pumps driven by electric motors":

For liquids 200°F and below, set motor shaft at same height as pump shaft.

For liquids above 200°F, set pump shaft 0.001 in. lower, per 100°F of temperature above 200°F per in. distance between pump base and shaft centerline.

Example: 450°F liquid; pump dimension from base to centerline is 10 in.

$$\frac{(450-200)}{100}(0.001)(10) = 0.025 \text{ in.} \quad \text{Therefore, set pump } 0.025 \text{ in. low}$$

$$(\text{or set motor } 0.025 \text{ in. high}).$$

The following applies to "foot mounted pumps or turbines":

$$\text{Thermal growth (mils)} = 6 \times \frac{(T_o - T_a)}{100} \times L$$

Where L = Distance from base to shaft centerline, feet
        T_o = Operating temperature, °F
        T_a = Ambient temperature, °F

For centerline mounted pumps, we are told to change the coefficient from 6 to 3. Another rule tells us to use the coefficient 3 for foot mounted pumps!

Yet another source tells us to use the following formula:

$$\text{Thermal growth, inches,} = 0.008 \times \frac{(T_o - T_a)}{100} \times \frac{L}{3},$$

$$\text{for centerline mounted pumps.}$$

For foot mounted pumps, use L in place of $\dfrac{L}{3}$.

Another rule of thumb says to neglect thermal growth in centerline mounted pumps when fluid temperature is below 400°F, and to cool the pedestal when fluid temperature exceeds 400°F. This rule is somewhat unrealistic, since the benefits of omitting the cooling clearly outweigh the advantages of including it!

Yet another rule tells us to allow for 0.0015 in. growth per in. of height from base to shaft centerline, for any steam turbine—regardless of steam or ambient temperatures. Another chart goes into elaborate detail, recommending various differences in centerline height between turbine and pump based on machine types and service conditions, but without considering their dimensions.

For electric motor growth, we have the following:

(Foot to shaft centerline, in.) (6×10^{-6}) (nameplate temp rise, °F) = motor vertical growth, in. This is inconvenient, since motor temperature rise is normally given in degrees centigrade. In case you have forgotten how to convert, °F = (°C × 9/5) + 32.

Another rule says to use half of the above figure.

Then there is the rule that advises using 7 L, where L represents distance from base to shaft centerline in feet, and the answer comes out in thousandths of an inch. Yet another source says to use 4 L. These rules all assume uniform vertical expansion from one end to the other. However, on motors having single end fans, the expansion will be greater at the air outlet end. Angular misalignment caused by this difference can exceed parallel misalignment caused by overall growth! The same can be true of certain other machines with a steep temperature gradient from one end to the other, such as blowers, compressors, and turbines.

The rules just cited were found in various published or filmed instructions from major pump manufacturers, oil refining companies and, in one case, a technical magazine published for the electric power industry. Their inconsistency, and their failure to recognize certain growth phenomena, make their accuracy rather questionable. This is especially true where piping growth can affect machine alignment.

Finally, the reader may wish to review either ref. 17 or 18, which give quick updates on shaft alignment technology.

References

1. *Alignment Procedure, Revised Edition.* Buffalo, New York, Joy Manufacturing Company, 1970. (This describes and illustrates a mathematical formula progressive calculation approach to determining corrective movements based on reverse-indicator measurements.)
2. Baumann, Nelson P. and Tipping, William E., Jr., "Vibration Reduction Techniques for High-Speed Rotating Equipment—ASME Paper 65-WA/PWR-3." New York: The American Society of Mechanical Engineers, 1965.
3. Dodd, V. R., *Total Alignment.* The Petroleum Publishing Company, Tulsa, 1975.
4. Dreymala, James, *Factors Affecting and Procedures of Shaft Alignment.* Dreyco Mechanical Services, Houston, 1974.
5. Essinger, Jack N., "Alignment of Turbomachinery—A Review of Techniques Employing Dial Indicators." Paper presented at Second Symposium on Compressor Train Reliability Improvement, Manufacturing Chemists Association, Houston, Texas, April 4, 1972.

Similar information was published in *Hydrocarbon Processing*, September 1973.

6. Gibbs, C. R. and Wren, J. R., "Aligning Horizontal Machine Sets." *Allis-Chalmers Engineering Review.* About 1968—exact date not known.

7. Jackson, Charles, "How to Align Barrel-Type Centrifugal Compressors." *Hydrocarbon Processing* (September 1971) (Corrected Reprint).

8. Jackson, Charles, "Start Cold for Good Alignment of Rotating Equipment." *The Oil and Gas Journal*, March 11, 1974, Pages 124–130.

9. Jackson, Charles, "Techniques for Alignment of Rotating Equipment." *Hydrocarbon Processing*, LV (January 1976), Pages 81–86.

10. King, W. F. and Petermann, J. E., "Align Shafts, Not Couplings!" *Allis-Chalmers Electrical Review.* Second Quarter 1951, Pages 26–29.

11. Nelson, Carl A., "Orderly Steps Simplify Coupling Alignment." *Plant Engineering*, June 1967, Pages 176–178.

12. "Service Memo SD-5-69; Reverse Reading Coupling Alignment." Houston: Dresser Industries, Inc., Machinery Group, 1969.

13. Durkin, Tom, "Aligning Shafts." *Plant Engineering*, January 11, 1979, Pages 86–90, and February 8,1979, Pages 102–105.

14. Zatezalo, John, "A Machinery Alignment System for Industry." Pittsburgh: IMS–Industrial Maintenance Systems, Inc., 1981.

15. Hamar, Martin R., "Laser Alignment in Industry–ASTME Paper MR68–408." Dearborn, Michigan: The American Society of Tool and Manufacturing Engineers, 1968.

16. Murray, Malcolm G., "Out of Room? Use Minimum Movement Machinery Alignment." *Hydrocarbon Processing*, Houston, January 1979, Pages 112–114.

17. Bloch, Heinz P., "Updating Shaft Alignment Knowledge." *Maintenance Technology*, April 2004.

18. Bloch, Heinz P., "Update Your Shaft Alignment Knowledge." *Chemical Engineering*, September 2004.

Chapter 6
Balancing of Machinery Components*

This chapter contains some of the theoretical aspects of balancing and balancing machines, to give a better understanding of the process of balancing a rotor and of the working principles of balancing machines[1,2].

Definition of Terms

Definitions of many terms used in balancing literature and in this text are contained in Appendix A. Commonly used synonyms for some of these standard terms are also included. For further information on terminology, refer to ISO Standard No. 1925 (see Appendix 6C).

Purpose of Balancing

An unbalanced rotor will cause vibration and stress in the rotor itself and in its supporting structure. Balancing of the rotor is therefore necessary to accomplish one or more of the following:

1. Increase bearing life.
2. Minimize vibration.
3. Minimize audible and signal noises.
4. Minimize operating stresses.
5. Minimize operator annoyance and fatigue.

---

6. Minimize power losses.
7. Increase quality of product.
8. Satisfy operating personnel.

Unbalance in just one rotating component of an assembly may cause the entire assembly to vibrate. This induced vibration in turn may cause excessive wear in bearings, bushings, shafts, spindles, gears, etc., substantially reducing their service life. Vibration sets up highly undesirable alternating stresses in structural supports and frames that may eventually lead to their complete failure. Performance is decreased because of the absorption of energy by the supporting structure. Vibrations may be transmitted through the floor to adjacent machinery and seriously impair its accuracy or proper functioning.

The Balancing Machine as a Measuring Tool

A balancer or balancing machine is necessary to detect, locate, and measure unbalance. The data furnished by the balancer permit changing the mass distribution of a rotor, which, when done accurately, will balance the rotor. Balance is a zero quantity, and therefore is detected by observing an absence of unbalance. The balancer measures only unbalance, never balance.

Centrifugal force acts upon the entire mass of a rotating element, impelling each particle outward and away from the axis of rotation in a radial direction. If the mass of a rotating element is evenly distributed about its shaft axis, the part is "balanced" and rotates without vibration. However, if an excess of mass exists on one side of a rotor, the centrifugal force acting upon this heavy side exceeds the centrifugal force exerted by the light side and pulls the entire rotor in the direction of the heavy side. Figure 6-1 shows the side view of a rotor having an excess mass m on one side. Due to centrifugal force exerted by m during rotation, the entire rotor is being pulled in the direction of the arrow F.

Causes of Unbalance

The excess of mass on one side of a rotor shown in Figure 6-1 is called unbalance. It may be caused by a variety of reasons, including:

1. Tolerances in fabrication, including casting, machining, and assembly.
2. Variation within materials, such as voids, porosity, inclusions, grain, density, and finishes.

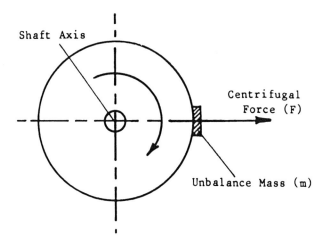

Figure 6-1. Unbalance causes centrifugal force.

3. Nonsymmetry of design, including motor windings, part shapes, location, and density of finishes.
4. Nonsymmetry in use, including distortion, dimensional changes, and shifting of parts due to rotational stresses, aerodynamic forces, and temperature changes.

Often, balancing problems can be minimized by symmetrical design and careful setting of tolerances and fits. Large amounts of unbalance require large corrections. If such corrections are made by removal of material, additional cost is involved and part strength may be affected. If corrections are made by addition of material, cost is again a factor and space requirements for the added material may be a problem.

Manufacturing processes are the major source of unbalance. Unmachined portions of castings or forgings which cannot be made concentric and symmetrical with respect to the shaft axis introduce substantial unbalance. Manufacturing tolerances and processes which permit any eccentricity or lack of squareness with respect to the shaft axis are sources of unbalance. The tolerances, necessary for economical assembly of several elements of a rotor, permit radial displacement of assembly parts and thereby introduce unbalance.

Limitations imposed by design often introduce unbalance effects which cannot be corrected adequately by refinement in design. For example, electrical design considerations impose a requirement that one coil be at a greater radius than the others in a certain type of universal motor armature. It is impractical to design a compensating unbalance into the armature.

Fabricated parts, such as fans, often distort nonsymmetrically under service conditions. Design and economic considerations prevent the adaptation of methods which might eliminate this distortion and thereby reduce the resulting unbalance.

Ideally, rotating parts always should be designed for inherent balance, whether a balancing operation is to be performed or not. Where low service speeds are involved and the effects of a reasonable amount of unbalance can be tolerated, this practice may eliminate the need for balancing. In parts which require unbalanced masses for functional reasons, these masses often can be counterbalanced by designing for symmetry about the shaft axis.

A rotating element having an uneven mass distribution, or unbalance, will vibrate due to the excess centrifugal force exerted during rotation by the heavier side of the rotor. Unbalance causes centrifugal force, which in turn causes vibration. When at rest, the excess mass exerts no centrifugal force and, therefore, causes no vibration. Yet, the actual unbalance is still present.

Unbalance, therefore, is independent of rotational speed and remains the same, whether the part is at rest or is rotating (provided the part does not deform during rotation). Centrifugal force, however, varies with speed. When rotation begins, the unbalance will exert centrifugal force tending to vibrate the rotor. The higher the speed, the greater the centrifugal force exerted by the unbalance and the more violent the vibration. Centrifugal force increases proportionately to the square of the increase in speed. If the speed is doubled, the centrifugal force quadruples; if the speed is tripled, the centrifugal force is multiplied by nine.

Units of Unbalance

Unbalance is measured in ounce-inches, gram-inches, or gram-millimeters, all having a similar meaning, namely a mass multiplied by its distance from the shaft axis. An unbalance of $100 \, g \cdot in.$, for example, indicates that one side of the rotor has an excess mass equivalent to 10 grams at a 10 in. radius, or 20 grams at a 5 in. radius (see Figure 6-2).

In each case, the mass, when multiplied by its distance from the shaft axis, amounts to the same unbalance value, namely 100 gram-inches. A given mass will create different unbalances, depending on its distance from the shaft axis. To determine the unbalance, simply multiply the mass by the radius.

Since a given excess mass at a given radius represents the same unbalance regardless of rotational speed, it would appear that it could be corrected at any speed, and that balancing at service speeds is unnecessary.

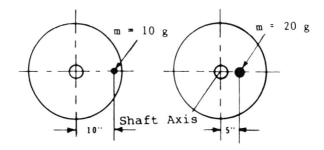

Figure 6-2. Side view of rotors with 100 g · in. unbalance.

This is true for rigid rotors as listed in Table 6-5. However, not all rotors can be considered rigid, since certain components may shift or distort unevenly at higher speeds. Thus they may have to be balanced at their service speed.

Once the unbalance has been corrected, there will no longer be any significant disturbing centrifugal force and, therefore, no more unbalance vibration. A small residual unbalance will usually remain in the part, just as there is a tolerance in any machining operation. Generally, the higher the service speed, the smaller should be the residual unbalance.

In many branches of industry, the unit of gram · inch (abbreviated g · in.) is given preference because it has proven to be the most practical. An ounce is too large for many balancing applications, necessitating fractions or a subdivision into hundredths, neither of which has become very popular.

Types of Unbalance

The following paragraphs explain the four different types of unbalance as defined by the internationally accepted ISO Standard No. 1925 on balancing terminology. For each of the four mutually exclusive cases an example is shown, illustrating displacement of the principal axis of inertia from the shaft axis caused by the addition of certain unbalance masses in certain distributions to a perfectly balanced rotor.

Static Unbalance

Static unbalance, formerly also called force unbalance, is illustrated in Figure 6-3 below. It exists when the principal axis of inertia is displaced parallel to the shaft axis. This type of unbalance is found primarily in

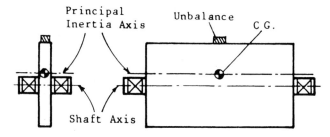

Figure 6-3. Static unbalance.

narrow, disc-shaped parts such as flywheels and turbine wheels. It can be corrected by a single mass correction placed opposite the center-of-gravity in a plane perpendicular to the shaft axis, and intersecting the CG.

Static unbalance, if large enough, can be detected with conventional gravity-type balancing methods. Figure 6-3A shows a concentric rotor with unbalance mass on knife edges. If the knife-edges are level, the rotor will turn until the heavy or unbalanced spot reaches the lowest position. Figure 6-3B shows an equivalent condition with an eccentric rotor. The rotor with two equal unbalance masses equidistant from the CG as shown in Figure 6-3C is also out of balance statically, since both unbalance masses could be combined into one mass located in the plane of the CG.

Static unbalance can be measured more accurately by centrifugal means on a balancing machine than by gravitational means on knife-edges or rollers. Static balance is satisfactory only for relatively slow-revolving, disc-shaped parts or for parts that are subsequently assembled onto a larger rotor which is then balanced dynamically as an assembly.

Couple Unbalance

Couple unbalance, formerly also called moment unbalance, is illustrated in Figure 6-4 and 6-4A. It is that condition for which the principal axis of inertia intersects the shaft axis at the center of gravity. This arises when two equal unbalance masses are positioned at opposite ends of a rotor and spaced 180° from each other. Since this rotor will not rotate when placed on knife-edges, a dynamic method must be employed to detect couple unbalance. When the workpiece is rotated, each end will vibrate in opposite directions and give an indication of the rotor's uneven mass distribution.

Couple unbalance is sometimes expressed in gram · inch · inches or gram · in.^{2} (or ounce-in.^{2}), wherein the second in. dimension refers to the distance between the two planes of unbalance.

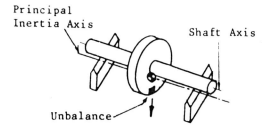

Figure 6-3A. Concentric disc with static unbalance.

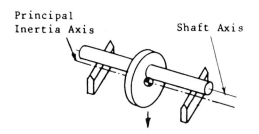

Figure 6-3B. Eccentric disc, therefore static unbalance.

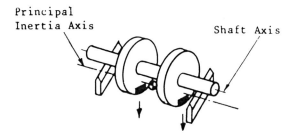

Figure 6-3C. Two discs of equal mass and identical static unbalance, aligned to give statically unbalanced assembly.

It is important to note that couple unbalance cannot be corrected by a single mass in a single correction plane. At least two masses are required, each in a different transverse plane (perpendicular to the shaft axis) and 180° opposite to each other. In other words, a couple unbalance needs another couple to correct it. In the example in Figure 6-4B, for instance, correction could be made by placing two masses at opposite angular positions on the main body of the rotor. The axial location of the correction

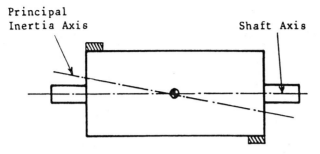

Figure 6-4. Couple unbalance.

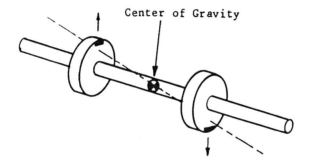

Figure 6-4A. Discs of Figure 6-3C, realigned to cancel static unbalance, now have couple unbalance.

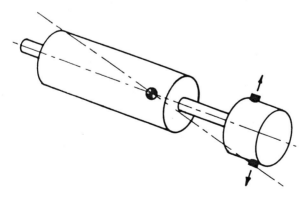

Figure 6-4B. Couple unbalance in outboard rotor component.

couple does not matter as long as its value is equal in magnitude but opposite in direction to the unbalance couple.

Quasi-Static Unbalance

Quasi-static unbalance, Figure 6-5, is that condition of unbalance for which the central principal axis of inertia intersects the shaft axis at a point other than the center of gravity. It represents the specific combination of static and couple unbalance where the angular position of one couple component coincides with the angular position of the static unbalance. This is a special case of dynamic unbalance.

Dynamic Unbalance

Dynamic unbalance, Figure 6-6, is that condition in which the central principal axis of inertia is neither parallel to, nor intersects the shaft axis.

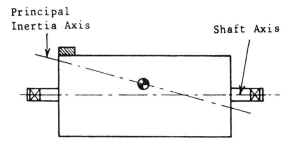

Figure 6-5. Quasi-static unbalance.

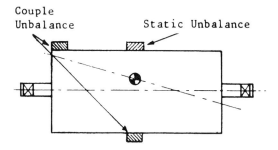

Figure 6-5A. Couple plus static unbalance results in quasi-static unbalance provided one couple mass has the same angular position as the static mass.

Quasi-Static Unbalance

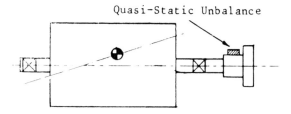

Figure 6-5B. Unbalance in coupling causes quasi-static unbalance in rotor assembly.

Unbalance masses not
diametrically opposed.

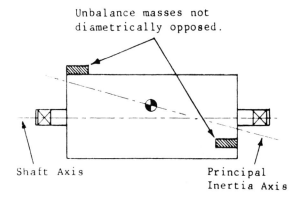

Shaft Axis   Principal
                                    Inertia Axis

Figure 6-6. Dynamic unbalance.

It is the most frequently occurring type of unbalance and can only be corrected (as is the case with couple unbalance) by mass correction in at least two planes perpendicular to the shaft axis.

Another example of dynamic unbalance is shown in Figure 6-6A.

Motions of Unbalanced Rotors

In Figure 6-7, a rotor is shown spinning freely in space. This corresponds to spinning above resonance in soft bearings. In Figure 6-7A only static unbalance is present and the center line of the shaft sweeps out a cylindrical surface. Figure 6-7B illustrates the motion when only couple unbalance is present. In this case, the centerline of the rotor shaft sweeps out two cones which have their apexes at the center-of-gravity of the rotor. The effect of combining these two types of unbalance when they occur in the same axial plane (quasi-static unbalance) is to move the apex of the cones away from the center-of-gravity. In the case of dynamic unbalance

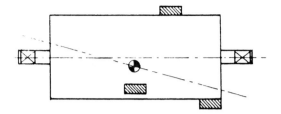

Figure 6-6A. Couple unbalance plus static unbalance results in dynamic unbalance.

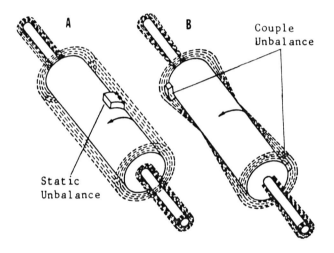

Figure 6-7. Effect of unbalance on free rotor motion.

there will be no apex and the shaft will move in a more complex combination of the motions shown in Figure 6-7.

Effects of Unbalance and Rotational Speed

As has been shown, an unbalanced rotor is a rotor in which the principal inertia axis does not coincide with the shaft axis.

When rotated in its bearings, an unbalanced rotor will cause periodic vibration of, and will exert a periodic force on, the rotor bearings and their supporting structure. If the structure is rigid, the force is larger than if the structure is flexible (except at resonance). In practice, supporting structures are neither entirely rigid nor entirely flexible but somewhere in between. The rotor-bearing support offers some restraint, forming a

spring-mass system with damping, and having a single resonance frequency. When the rotor speed is below this frequency, the principal inertia
axis of the rotor moves outward radially. This condition is illustrated in
Figure 6-8A.

If a soft pencil is held against the rotor, the so-called high spot is marked
at the same angular position as that of the unbalance. When the rotor speed
is increased, there is a small time lag between the instant at which the
unbalance passes the pencil and the instant at which the rotor moves out
enough to contact it. This is due to the damping in the system. The angle
between these two points is called the "angle of lag" (see Figure 6-8B).
As the rotor speed is increased further, resonance of the rotor and its supporting structure will occur; at this speed the angle of lag is 90° (see Figure
6-8C). As the rotor passes through resonance, there are large vibration
amplitudes and the angle of lag changes rapidly. As the speed is increased

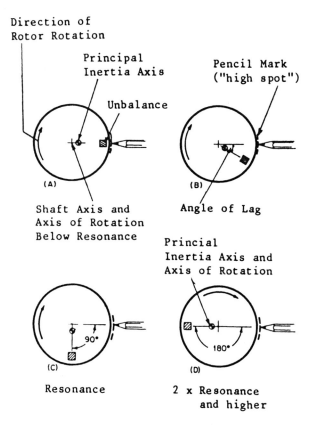

Figure 6-8. Angle of lag and migration of axis of rotation.

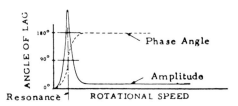

Figure 6-9. Angle of lag and amplitude of vibration versus rotational speed.

further, vibration subsides again; when increased to nearly twice resonance speed, the angle of lag approaches 180° (see Figure 6-8D). At speeds greater than approximately twice resonance speed, the rotor tends to rotate about its principal inertia axis at constant amplitude of vibration; the angle of lag (for all practical purposes) remains 180°.

In Figure 6-8 a soft pencil is held against an unbalanced rotor. In (A) a high spot is marked. Angle of lag between unbalance and high spot increases from 0° (A) to 180° in (D) as rotor speed increases. The axis of rotation has moved from the shaft axis to the principal axis of inertia.

Figure 6-9 shows the interaction of rotational speed, angle of lag, and vibration amplitude as a rotor is accelerated through the resonance frequency of its suspension system.

Correlating CG Displacement with Unbalance

One of the most important fundamental aspects of balancing is the direct relationship between the displacement of center-of-gravity of a rotor from its journal axis, and the resulting unbalance. This relationship is a prime consideration in tooling design, tolerance selection, and determination of balancing procedures.

For a disc-shaped rotor, conversion of CG displacement to unbalance, and vice versa, is relatively simple. For longer workpieces it can be almost as simple, if certain approximations are made. First, consider a disc-shaped rotor.

Assume a perfectly balanced disc, as shown in Figure 6-10, rotating about its shaft axis and weighing 999 ounces. An unbalance mass m of one ounce is added at a ten in. radius, bringing the total rotor weight W up to 1,000 ounces and introducing an unbalance equivalent to 10 ounce·in. This unbalance causes the CG of the disc to be displaced by a distance e in the direction of the unbalance mass.

Since the entire mass of the disc can be thought to be concentrated in its center-of-gravity, it (the CG) now revolves at a distance e about the

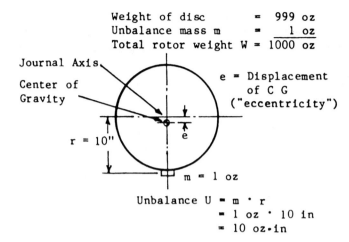

Figure 6-10. Disc-shaped rotor with displaced center of gravity due to unbalance.

shaft axis, constituting an unbalance of U = We. Substituting into this formula the known values for the rotor weight, we get:

$10 \text{ oz} \cdot \text{in.} = 1,000 \text{ oz} \cdot \text{e}$

Solving for e we find

$$e = \frac{10 \text{ oz} \cdot \text{in.}}{1,000 \text{ oz}} = 0.01 \text{ in.}$$

In other words, we can find the displacement e by the following formula:

$$e \text{ (in.)} = \frac{U \text{ (oz} \cdot \text{in.)}}{W \text{ (oz)}}$$

For example, if a fan is first balanced on a tightly fitting arbor, and subsequently installed on a shaft having a diameter 0.002 in. smaller than the arbor, the total play resulting from the loose fit may be taken up in one direction by a set screw. Thus the entire fan is displaced by one half of the play or 0.001 in. from the axis about which it was originally balanced. If we assume that the fan weighs 100 pounds, the resulting unbalance will be:

$U = 100 \text{ lb} \cdot 16 \text{ oz}/\text{lb} \cdot 0.001 \text{ in.} = 1.6 \text{ oz} \cdot \text{in.}$

The same balance error would result if arbor and shaft had the same diameter, but the arbor (or the shaft) had a total indicated runout (TIR) of 0.002 in. In other words, the displacement is always only one half of the total play or TIR.

The CG displacement e discussed above equals the shaft displacement only if there is no influence from other sources, a case seldom encountered. Nevertheless, for balancing purposes, the theoretical shaft respectively CG displacement is used as a guiding parameter.

On rotors having a greater length than a disc, the formula e = U/W for finding the correlation between unbalance and displacement still holds true if the unbalance happens to be static only. However, if the unbalance is anything other than static, a somewhat more complicated situation arises.

Assume a balanced roll weighing 2,000 oz, as shown in Figure 6-11, having an unbalance mass m of 1 oz near one end at a radius r of 10 in. Under these conditions the displacement of the center-of-gravity (e) no longer equals the displacement of the shaft axis (d) in the plane of the bearing. Since shaft displacement at the journals is usually of primary interest, the correct formula for finding it looks as follows (again assuming that there is no influence from bearings and suspension):

$$d = \frac{mr}{W+m} + \frac{mrjh}{Iz - Ix}$$

Where:

d = Displacement of principal inertia axis from shaft axis in plane of bearing
W = Rotor weight

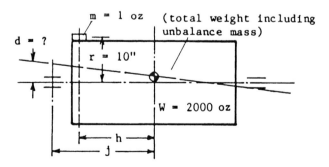

Figure 6-11. Roll with unbalance.

m = Unbalance mass
r = Radius of unbalance
h = Distance from center-of-gravity to plane of unbalance
j = Distance from center-of-gravity to bearing plane
Ix = Moment of inertia around transverse axis
Iz = Polar moment of inertia around journal axis

Since neither the polar nor the transverse moments of inertia are known, this formula is impractical. Instead, a widely accepted approximation may be used.

The approximation lies in the assumption that the unbalance is static (see Figure 6-12). Total unbalance is thus 20 oz · in. Displacement of the principal inertia axis from the bearing axis (and the eccentricity e of CG) in the rotor is therefore:

$$e = \frac{20 \, oz \cdot in.}{2,000 \, oz} = 0.01 \, in.$$

If the weight distribution is not equal between the two bearings but is, say, 60 percent on the left bearing and 40 percent on the right bearing, then the unbalance in the left plane must be divided by 60 percent of the rotor weight to arrive at the approximate displacement in the left bearing plane, whereas the unbalance in the right plane must be divided by 40 percent of the rotor weight.

An assumed unbalance of 10 oz · in. in the left plane (close to the bearing) will thus cause an approximate eccentricity in the left bearing of:

$$e = \frac{10 \, oz \cdot in.}{2,000 \, oz \cdot 0.6} = 0.00833 \, in.$$

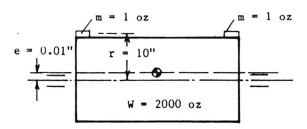

(total weight including unbalance masses)

Figure 6-12. Symmetric rotor with static unbalance.

and in the right bearing of:

$$e = \frac{10\,oz \cdot in.}{2,000\,oz \cdot 0.4} = 0.0125\,in.$$

Quite often the reverse calculation is of interest. In other words, the unbalance is to be computed that results from a known displacement. Again the assumption is made that the resulting unbalance is static.

For example, assume an armature and fan assembly weighing 2,000 lbs and having a bearing load distribution of 70 percent at the armature (left) end and 30 percent at the fan end (see Figure 6-13). Assume further that the assembly has been balanced on its journals and that the rolling element bearings added afterwards have a total indicated runout of 0.001 in., causing an eccentricity of the shaft axis of $\frac{1}{2}$ of the TIR or 0.0005 in.

Question: How much unbalance does the bearing runout cause in each side of the rotor?

Answer: In the armature end

$$U = 1,400\,lb \cdot 16\,oz/lb \cdot 0.0005\,in. = 11.2\,oz \cdot in.$$

In the fan end

$$U = 600\,lb \cdot 16\,oz/lb \cdot 0.0005\,in. = 4.8\,oz \cdot in.$$

When investigating the effect of bearing runout on the balance quality of a rotor, the unbalance resulting from the bearing runout should be added to the residual unbalance to which the armature was originally balanced on the journals; only then should the sum be compared with the recommended balance tolerance. If the sum exceeds the recommended toler-

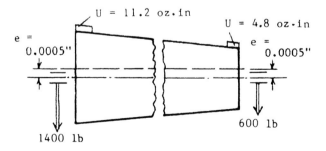

Figure 6-13. Unbalance resulting from bearing runout in an asymmetric rotor.

ance, the armature will either have to be balanced to a smaller residual unbalance on its journals, or the entire armature/bearing assembly will have to be rebalanced in its bearings. The latter method is often preferable since it circumvents the bearing runout problem altogether, although field replacement of bearings will be more problematic.

Balancing Machines

The purpose of a balancing machine is to determine by some technique both the magnitude of unbalance and its angular position in each of one, two, or more selected correction planes. For single-plane balancing this can be done statically, but for two- or multi-plane balancing, it can be done only while the rotor is spinning. Finally, all machines must be able to resolve the unbalance readings, usually taken at the bearings, into equivalent values in each of the correction planes.

On the basis of their method of operation, balancing machines and equipment can be grouped in three general categories:

1. Gravity balancing machines.
2. Centrifugal balancing machines.
3. Field balancing equipment.

In the first category, advantage is taken of the fact that a body free to rotate always seeks that position in which its center-of-gravity is lowest. Gravity balancing machines, also called nonrotating balancing machines, include horizontal ways or knife-edges, roller stands, and vertical pendulum types (Figure 6-14). All are capable of only detecting and/or indicating static unbalance.

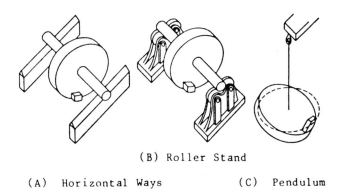

(B) Roller Stand

(A)   Horizontal Ways   (C)   Pendulum

Figure 6-14. Static balancing devices.

In the second category, the amplitude and phase of motions or reaction forces caused by once-per-revolution centrifugal forces resulting from unbalance are sensed, measured, and displayed. The rotor is supported by the machine and rotated around a horizontal or vertical axis, usually by the drive motor of the machine. A centrifugal balancing machine (also called a rotating balancing machine) is capable of measuring static unbalance (single plane machine) or static and couple unbalance (two-plane machine). Only a two-plane rotating balancing machine can detect couple and/or dynamic unbalance.

Field balancing equipment, the third category, provides sensing and measuring instrumentation only; the necessary measurements for balancing a rotor are taken while the rotor runs in its own bearings and under its own power. A programmable calculator or handheld computer may be used to convert the vibration readings (obtained in several runs with test masses) into magnitude and phase angle of the required correction masses.

Gravity Balancing Machines

First, consider the simplest type of balancing—usually called "static" balancing, since the rotor is not spinning.

In Figure 6-14A, a disc-type rotor on a shaft is shown resting on knife-edges. The mass added to the disc at its rim represents a known unbalance. In this illustration, and those which follow, the rotor is assumed to be balanced without this added unbalance mass. In order for this balancing procedure to work effectively, the knife-edges must be level, parallel, hard, and straight.

In operation, the heavier side of the disc will seek the lowest level—thus indicating the angular position of the unbalance. Then, the magnitude of the unbalance usually is determined by an empirical process, adding mass to the light side of the disc until it is in balance, i.e., until the disc does not stop at the same angular position.

In Figure 6-14B, a set of balanced rollers or wheels is used in place of the knife edges. Rollers have the advantage of not requiring as precise an alignment or level as knife edges; also, rollers permit run-out readings to be taken.

In Figure 6-14C, another type of static, or "nonrotating", balancer is shown. Here the disc to be balanced is supported by a flexible cable, fastened to a point on the disc which coincides with the center of the shaft axis slightly above the transverse plane containing the center-of-gravity. As shown in Figure 6-14C, the heavy side will tend to seek a lower level than the light side, thereby indicating the angular position of the unbalance. The disc can be balanced by adding mass to the diametrically opposed side of the disc until it hangs level. In this case, the center-of-gravity is moved until it is directly under the flexible support cable.

Static balancing is satisfactory for rotors having relatively low service speeds and axial lengths which are small in comparison with the rotor diameter. A preliminary static unbalance correction may be required on rotors having a combined unbalance so large that it is impossible in a dynamic, soft-bearing balancing machine to bring the rotor up to its proper balancing speed without damaging the machine. If the rotor is first balanced statically by one of the methods just outlined, it is usually possible to decrease the initial unbalance to a level where the rotor may be brought up to balancing speed and the residual unbalance measured. Such preliminary static correction is not required on hard-bearing balancing machines.

Static balancing is also acceptable for narrow, high speed rotors which are subsequently assembled to a shaft and balanced again dynamically. This procedure is common for single stages of jet engine turbines and compressors.

Centrifugal Balancing Machines

Two types of centrifugal balancing machines are in general use today, soft-bearing and hard-bearing machines.

Soft-Bearing Balancing Machines

The soft-bearing balancing machine derives its name from the fact that it supports the rotor to be balanced on bearings which are very flexibly suspended, permitting the rotor to vibrate freely in at least one direction, usually the horizontal, perpendicular to the rotor shaft axis (see Figure 16-15). Resonance of rotor and bearing system occurs at one half or less of the lowest balancing speed so that, by the time balancing speed is reached, the angle of lag and the vibration amplitude have stabilized and can be measured with reasonable certainty (see Figure 6-16A).

Bearings (and the directly attached support components) vibrate in unison with the rotor, thus adding to its mass. Restriction of vertical motion does not affect the amplitude of vibration in the horizontal plane, but the added mass of the bearings does. The greater the combined rotor-and-bearing mass, the smaller will be the displacement of the bearings, and the smaller will be the output of the devices which sense the unbalance.

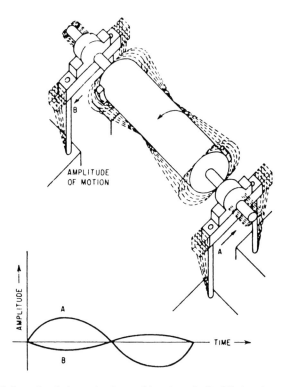

Figure 6-15. Motion of unbalanced rotor and bearings in flexible-bearing, centrifugal balancing machines.

As far as the relationship between unbalance and bearing motion is concerned, the soft-bearing machine is faced with the same complexity as shown in Figure 6-11.

Therefore, a direct indication of unbalance can be obtained only after calibrating the indicating elements for a given rotor by use of test masses which constitute a known amount of unbalance.

For this purpose the soft-bearing balancing machine instrumentation contains the necessary circuitry and controls so that, upon proper calibration for the particular rotor to be balanced, an exact indication of amount-of-unbalance and its angular position is obtained. Calibration varies between parts of different mass and configuration, since displacement of the principal axis of inertia in the balancing machine bearings is dependent upon rotor mass, bearing and suspension mass, rotor moments of inertia, and the distance between bearings.

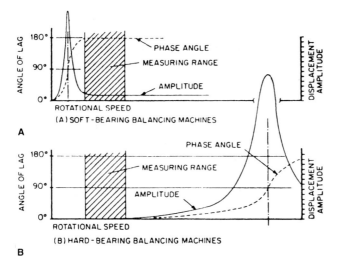

Figure 6-16. Phase angle and displacement amplitude versus rotational speed in soft-bearing and hard-bearing balancing machines.

Hard-Bearing Balancing Machines

Hard-bearing balancing machines are essentially of the same construction as soft-bearing balancing machines, except that their bearing supports are significantly stiffer in the transverse horizontal direction. This results in a horizontal resonance for the machine which occurs at a frequency several orders of magnitude higher than that for a comparable soft-bearing balancing machine. The hard-bearing balancing machine is designed to operate at speeds well below this resonance (see Figure 6-16B) in an area where the phase angle lag is constant and practically zero, and where the amplitude of vibration—though small—is directly proportional to centrifugal forces produced by unbalance.

Since the force that a given amount of unbalance exerts at a given speed is always the same, no matter whether the unbalance occurs in a small or large, light or heavy rotor, the output from the sensing elements attached to the balancing machine bearing supports remains proportional to the centrifugal force resulting from unbalance in the rotor. The output is not influenced by bearing mass, rotor mass, or inertia, so that a permanent relation between unbalance and sensing element output can be established.

Centrifugal force from a given unbalance rises with the square of the balancing speed. Output from the pick-ups rises proportionately with the third power of the speed due to a linear increase from the rotational frequency superimposed on a squared increase from centrifugal force. Suitable integrator circuitry then reduces the pickup signal inversely proportional to the cube of the balancing speed increase, resulting in a constant unbalance readout. Unlike soft bearing balancing machines, the use of calibration masses is not required to calibrate the machine for a given rotor.

Angle of lag is shown as a function of rotational speed in Figure 6-16A for soft-bearing balancing machines whose balancing speed ranges start at approximately twice the resonance speed of the supports; and in Figure 6-16B for hard-bearing balancing machines. Here the resonance frequency of the combined rotor-bearing support system is usually more than three times greater than the maximum balancing speed.

For more information on hard-bearing and other types of balancing machines, see articles on advantages of hard-bearing machines and on balancing specific types of rotors. (Reprints are available through Schenck Trebel Corporation.)

Both soft- and hard-bearing balancing machines use various types of sensing elements at the rotor-bearing supports to convert mechanical vibration into an electrical signal. These sensing elements are usually velocity-type pickups, although certain hard-bearing balancing machines use magnetostrictive or piezo-electric pickups.

Measurement of Amount and Angle of Unbalance

Three basic methods are used to obtain a reference signal by which the phase angle of the amount-of-unbalance indication signal may be correlated with the rotor. On end-drive machines (where the rotor is driven via a universal-joint driver or similarly flexible coupling shaft) a phase reference generator, directly coupled to the balancing machine drive spindle, is used. On belt-drive machines (where the rotor is driven by a belt over the rotor periphery) or on air-drive or self-drive machines, a stroboscopic lamp flashing once per rotor revolution, or a scanning head (photoelectric cell with light source) is employed to obtain the phase reference.

Whereas the scanning head only requires a single reference mark on the rotor to obtain the angular position of unbalance, the stroboscopic light necessitates attachment of an angle reference disc to the rotor, or placing an adhesive numbered band around it. Under the once-per-revolution flash of the strobe light the rotor appears to stand still so that an angle reading can be taken opposite a stationary mark.

With the scanning head, an additional angle indicating circuit and instrument must be employed. The output from the phase reference sensor

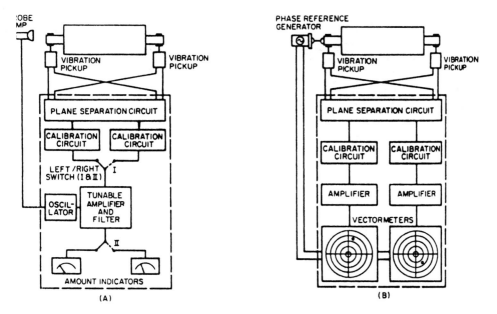

Figure 6-17. Block diagram of typical balancing machine instrumentations. (A) Amount of unbalance indicated on analog meters, angle by strobe light. (B) Combined amount and angle indication on Vector meters, simultaneously in two correction planes.

(scanning head) and the pickups at the rotor-bearing supports are processed and result in an indication representing the amount-of-unbalance and its angular position.

In Figure 6-17 block diagrams are shown for typical balancing instrumentations.

Figure 6-17A illustrates an indicating system which uses switching between correction planes (i.e., a single-channel instrumentation). This is generally employed on balancing machines with stroboscopic angle indication and belt drive. In Figure 6-17B an indicating system is shown with two-channel instrumentation. Combined indication of amount of unbalance and its angular position is provided simultaneously for both correction planes on two vectormeters having illuminated targets projected on the back of translucent overlay scales. Displacement of a target from the central zero point provides a direct visual representation of the displacement of the principal inertia axis from the shaft axis. Concentric circles on the overlay scale indicate the amount of unbalance, and radial lines indicate its angular position.

Plane Separation

Consider the rotor in Figure 6-15 with only an unbalance mass on the left end of the rotor. This mass causes not only the left bearing to vibrate but, to a lesser degree, the right also. This influence is called correction plane interference or, for short, "cross effect." If a second mass is attached in the right plane of the rotor, the direct effect of the mass in the right plane combines with the cross effect of the mass in the left plane, resulting in a composite vibration of the right bearing. If the two unbalance masses are at the same angular position, the cross effect of one mass has the same angular position as the direct effect in the other rotor end plane; thus, their direct and cross effects are additive (Figure 6-18A). If the two unbalance masses are 180° out of phase, their direct and cross effects are subtractive (Figure 6-18B). In a hard-bearing balancing machine the additive or subtractive effects depend entirely on the ratios of distances between the axial positions of the correction planes and bearings. In a soft-

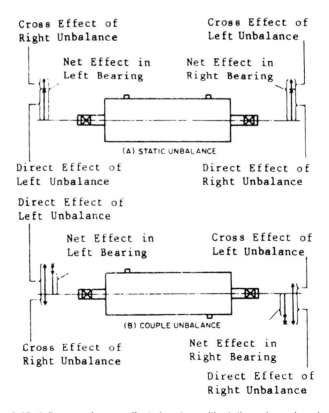

Figure 6-18. Influence of cross effects in rotors with static and couple unbalance.

bearing machine, the relationship is more complex because the masses and inertias of the rotor and its bearings must be taken into account.

If the two unbalance masses have an angular relationship other than 0 or 180°, the cross effect in the right bearing has a different phase angle than the direct effect from the right mass. Addition or subtraction of these effects is vectorial. The net bearing vibration is equal to the resultant of the two vectors, as shown in Figure 6-19. Phase angle indicated by the bearing vibration does not coincide with the angular position of either unbalance mass.

The unbalance illustrated in Figure 6-19 is the most common type, namely dynamic unbalance of unknown amount and angular position. Interaction of direct and cross effects will cause the balancing process to be a trial-and-error procedure. To avoid this, balancing machines incorporate a feature called "plane separation" which eliminates cross effect.

Before the advent of electrical networks, cross effect was eliminated by supporting the rotor in a cradle resting on a knife-edge and spring arrangement, as shown in Figure 6-20. Either the bearing-support members of the cradle or the knife edge pivot point are movable so that one unbalance correction plane always can be brought into the plane of the knife-edge.

Thus any unbalance in this plane will not cause the cradle to vibrate, whereas unbalance in all other planes will. The latter is measured and corrected in the other correction plane near the right end of the rotor body. Then the rotor is turned end for end, so that the knife-edge is in the plane of the first correction. Any vibration of the cradle is now due solely to unbalance present in the plane that was first over the knife-edge. Corrections are applied to this plane until the cradle ceases to vibrate. The

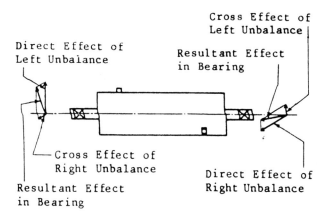

Figure 6-19. Influence of cross effects in rotors with dynamic unbalance. (All vectors seen from right side of rotor.)

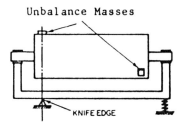

Figure 6-20. Plane separation by mechanical means.

rotor is now in balance. If it is again turned end for end, there will be no vibration.

Mechanical plane separation cradles restrict the rotor length, diameter, and location of correction planes. They also constitute a large parasitic mass which reduces sensitivity. Therefore, electric circuitry is used today to accomplish the function of plane separation. In principle, part of the output of each pickup is reversed in phase and fed against the output of the other pickup. Proper potentiometer adjustment of the counter voltage during calibration runs (with test masses attached to a balanced rotor) eliminates the cross effect.

Classification of Centrifugal Balancing Machines

Centrifugal balancing machines may be categorized by the type of unbalance a machine is capable of indicating (static or dynamic), the attitude of the journal axis of the workpiece (vertical or horizontal), or the type of rotor-bearing-support system employed (soft- or hard-bearing). In each category, one or more classes of machines are commercially built. The four classes are described in Table 6-1.

Class I: Trial-and-Error Balancing Machines. Machines in this class are of the soft-bearing type. They do not indicate unbalance directly in weight units (such as ounces or grams in the actual correction planes) but indicate only displacement and/or velocity of vibration at the bearings. The instrumentation does not indicate the amount of weight which must be added or removed in each of the correction planes. Balancing with this type of machine involves a lengthy trial-and-error procedure for each rotor, even if it is one of an identical series. The unbalance indication cannot be calibrated for specified correction planes because these machines do not have the feature of plane separation. Field balancing equipment usually falls into this class.

Table 6-1
Classification of Balancing Machines

Principle employed	Unbalance indicated	Attitude of shaft axis	Type of machine	Available classes
Gravity (nonrotating)	Static (single-plane)	Vertical	Pendulum	Not classified
		Horizontal	Knife-edges Roller sets	
Centrifugal (rotating)	Static (single-plane)	Vertical	Soft-bearing Hard-bearing	Not classified
		Horizontal	Not commercially available	
Centrifugal (rotating)	Dynamic (two-plane); also suitable for static (single-plane)	Vertical	Soft-bearing Hard-bearing	II, III III, IV
		Horizontal	Soft-bearing Hard-bearing	I, II, III IV

A programmable calculator or small computer with field balancing programs, either contained on magnetic strips or on a special plug-in ROM, will greatly reduce the trial-and-error procedure; however, calibration masses and three runs are still required to obtain magnitude and phase angle of unbalance on the first rotor. For subsequent rotors of the same kind, readings may be obtained in a single run but must be manually entered into the calculator and then suitably manipulated.

Class II: Calibratable Balancing Machines Requiring a Balanced Prototype. Machines in this class are of the soft-bearing type using instrumentation which permits plane separation and calibration for a given rotor type, if a balanced master or prototype rotor with calibration masses is available. However, the same trial-and-error procedure as for Class I machines is required for the first of a series of identical rotors.

Class III: Calibratable Balancing Machines Not Requiring a Balanced Prototype. Machines in this class are of the soft-bearing type using instrumentation which includes an integral electronic unbalance compensator. Any (unbalanced) rotor may be used in place of a balanced master rotor without the need for trial and error correction. Plane separation and calibration can be achieved in one or more runs with the help of calibration masses.

This class also includes soft-bearing machines with electrically driven shakers fitted to the vibratory part of their rotor supports.

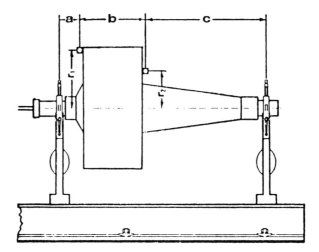

Figure 6-21. A permanently calibrated hard-bearing balancing machine, showing five rotor dimensions used in computing unbalance.

Class IV: Permanently Calibrated Balancing Machines. Machines in this class are of the hard-bearing type. They are permanently calibrated by the manufacturer for all rotors falling within the weight and speed range of a given machine size. Unlike the machines in other classes, these machines indicate unbalance in the first run without individual rotor calibration. This is accomplished by the incorporation of an analog or digital computer into the instrumentation associated with the machine. The following five rotor dimensions (see Figure 6-21) are fed into the computer: distance from left correction plane to left support (a); distance between correction planes (b); distance from right correction plane to right support (c); and r_1 and r_2, which are the radii of the correction masses in the left and right planes. The instrumentation then indicates the magnitude and angular position of the required correction mass for each of the two selected planes.

The compensation or "null-force" balancing machine falls into this class also. Although no longer manufactured, it is still widely used. It balances at the natural frequency or resonance of its suspension system including the rotor.

Maintenance and Production Balancing Machines

Balancing machines may also be categorized by their application in the following three groups:

1. Universal balancing machines.
2. Semi-automatic balancing machines.
3. Full automatic balancing machines with automatic transfer of work.

Each of these is available in both the nonrotating and rotating types, the latter for correction in either one or two planes.

Universal Balancing Machines

Universal balancing machines are adaptable for balancing a considerable variety of sizes and types of rotors. These machines commonly have a capacity for balancing rotors whose weight varies as much as 100 to 1 from maximum to minimum. The elements of these machines are adapted easily to new sizes and types of rotors. Amount and location of unbalance are observed on suitable instrumentation by the machine operator as the machine performs its measuring functions. This category of machine is suitable for maintenance or job-shop balancing as well as for many small and medium lot-size production applications.

Semi-Automatic Balancing Machines

Semi-automatic balancing machines are of many types. They vary from an almost universal machine to an almost fully automatic machine. Machines in this category may perform automatically any one or all of the following functions in sequence or simultaneously:

1. Retain the amount of unbalance indication for further reference.
2. Retain the angular location of unbalance indication for further reference.
3. Measure amount and position of unbalance.
4. Couple the balancing-machine drive to the rotor.
5. Initiate and stop rotation.
6. Set the depth of a correction tool depending on indication of amount of unbalance.
7. Index the rotor to a desired position depending on indication of unbalance location.
8. Apply correction of the proper magnitude at the indicated location.
9. Inspect the residual unbalance after correction.
10. Uncouple the balancing-machine drive.

Thus, the most complete semi-automatic balancing machine performs the entire balancing process and leaves only loading, unloading, and cycle initiation to the operator. Other semi-automatic balancing machines provide only means for retention of measurements to reduce operator fatigue and error. The features which are economically justifiable on a semi-automatic balancing machine may be determined only from a study of the rotor to be balanced and the production requirements.

Fully-Automatic Balancing Machines

Fully automatic balancing machines with automatic transfer of the rotor are also available. These machines may be either single- or multiple-station machines. In either case, the parts to be balanced are brought to the balancing machine by conveyor, and balanced parts are taken away from the balancing machine by conveyor. All the steps of the balancing process and the required handling of the rotor are performed without an operator. These machines also may include means for inspecting the residual unbalance as well as monitoring means to ensure that the balance inspection operation is performed satisfactorily.

In single-station automatic balancing machines, all functions of the balancing process (unbalance measurement, location, and correction) as well as inspection of the complete process are performed sequentially in a single station. In a multiple-station machine, the individual steps of the balancing process may be performed concurrently at two or more stations. Automatic transfer is provided between stations at which the amount and location of unbalance are determined; then the correction for unbalance is applied; finally, the rotor is inspected for residual unbalance. Such machines generally have shorter cycle times than single-station machines.

Establishing a Purchase Specification

A performance type purchase specification for a balancing machine should cover the following areas:

1. Description of the rotors to be balanced, including production rates, and balance tolerances.
2. Special rotor requirements, tooling, methods of unbalance correction, other desired features.
3. Acceptance test procedures.
4. Commercial matters such as installation, training, warranty, etc.

Rotor Description

To determine the correct machine size and features for a given application, it is first necessary to establish a precise description of the rotors to be balanced. To accumulate the necessary data. ISO 2953 suggests a suitable format. Refer to Appendix 6C.

Supporting the Rotor in the Balancing Machine

Means of Journal Support

A prime consideration in a balancing machine is the means for supporting the rotor. Various alternates are available, such as twin rollers, plain bearings, rolling element hearings (including slave bearings), V-roller bearings, nylon V-blocks, etc. (see also Appendix 6B, "Balancing Machine Nomenclature," and Appendix 6C.) The most frequently used and easiest to adapt are twin rollers. A rotor should generally be supported at its journals to assure that balancing is carried out around the same axis on which it rotates in service.

Rotors with More than Two Journals

Rotors which are normally supported at more than two journals may be balanced satisfactorily on only two journals provided that:

1. All journal surfaces are concentric with respect to the axis determined by the two journals used for support in the balancing machine.
2. The rotor is rigid at the balancing speed when supported on only two bearings.
3. The rotor has equal stiffness in all radial planes when supported on only two journals.

If the other journal surfaces are not concentric with respect to the axis determined by the two supporting journals, the shaft should be straightened. If the rotor is not a rigid body, or if it has unequal stiffness in different radial planes (e.g., crankshafts), the rotor should be supported in a (nonrotating) cradle at all journals during the balancing operation. This cradle should supply the stiffness usually supplied to the rotor by the rotor housing in which it is finally installed. The cradle should have minimum mass when used with a soft-bearing machine to permit maximum balancing sensitivity.

Rotors with Rolling Element Bearings

Rotors with stringent requirements for minimum residual unbalance and which run in rolling element bearings, should be balanced in their bearings, either in:

1. Special machines where the bearings are aligned and the outer races held in saddle bearing supports, rigidly connected by tie bars, or
2. In standard machines having supports equipped with V-roller carriages.

Frequently, practical considerations make it necessary to remove the bearings after balancing, to permit final assembly. If this cannot be avoided, the bearings should be match-marked to the rotor shaft and returned to the location used during balancing. Rolling element bearings with considerable radial play or bearings with a quality less than ABEC (Annular Bearing Engineers Committee) Standard grade 3 tend to cause erratic indications in the balancing machine. In some cases the outer race can be clamped tightly enough to remove excessive radial play. Only "fair" or lesser balance quality can be reached when rotors are supported on bearings of a grade lower than ABEC 3.

When maintenance requires antifriction bearings to be changed occasionally on a rotor, it is best to balance the rotor on the journals on which the inner races of the antifriction bearings fit. The unbalance introduced by displacement of the shaft axis due to eccentricity of the inner races can be minimized by use of high-quality bearings.

Driving the Rotor

If the rotor has its own journals, it may be driven in a horizontal balancing machine through:

1. A universal-joint or flexible-coupling drive from one end of the rotor.
2. A belt over the periphery of the rotor, or over a pulley attached to the rotor.
3. Air jets.
4. Other power means by which the rotor is normally driven in the final machine assembly.

The choice of end-drive can affect the residual unbalance substantially, even if the design considerations listed later in this text are carefully observed (see also "Balance Errors Due to Drive Elements" on page 328). Belt-drive has the advantage here, but it is somewhat limited in the amount of torque it can transmit to the rotor. Driving belts must be extremely flexible and of uniform thickness. Driving pulleys attached to the rotor should be used only when it is impossible to transmit sufficient driving torque by running the belt over the rotor. Pulleys must be as light as possible, must be dynamically balanced, and should be mounted on surfaces of the rotor which are square and concentric with the journal axis. The belt drive should not cause disturbances in the unbalance indication exceeding one-quarter of the permissible residual unbalance. Rotors driven by belt should not drive components of the balancing machine by means of any mechanical connection.

The use of electrical means or air for driving rotors may influence the unbalance readout. To avoid or minimize such influence, great care should be taken to bring in the power supply through very flexible leads, or have the airstream strike the rotor at right angles to the direction in which the balancing machine takes its readings.

If the electronic measuring system incorporates filters tuned to a specific frequency only, it is essential that means be available to control precisely the rotor speed to suit the filter setting.

Drive System Limitation

A given drive system has a certain rotor acceleration capability expressed in terms of the Wk^2n^2 value. This limiting value is generally part of the machine specification describing the drive, since it depends primarily on motor horsepower, motor type (squirrel-cage induction, wound-rotor, DC), and drive line strength.

The specified Wk^2n^2 value may be used to determine the maximum balancing speed (n) to which a rotor with a specific polar moment of inertia (Wk^2) can be accelerated; or conversely, to determine what maximum Wk^2 can be accelerated to a specified speed (n). (In each case the number of runs per hour must stay within the maximum number of cycles allowed.)

If a rotor is to be balanced which has a Wk^2n^2 value smaller than the maximum specified for a given drive, the stated cycles per hour may generally be exceeded in an inverse ratio.

On occasion it may happen that a large diameter rotor, although still within the weight capacity of the machine, cannot be accelerated to a given balancing speed. This may be due to the fact that the rotor's mass is located at a large radius, thus creating a large polar moment of inertia. As a result, a lower balancing speed may have to be selected.

Table 6-2
Factor C for Approximating Radius of Gyration k for Typical Rotors

Typical Rotor	C-Factor
Tube or Pipe	1
Solid Mass	0.7
Bladed Rotor	0.5–0.6
Propeller	0.4

A rotor's polar moment of inertia (Wk^2) is found by multiplying the rotor weight (W) in pounds by the square of the radius-of-gyration (k) in feet. The radius-of-gyration is the average of the radii from the shaft axis of each infinitesimal part of the rotor. It may be approximated by multiplying the outside radius of the rotor by a factor (C), shown in Table 6-2.

Example:

Wk^2 for a 2,500 lb solid steel flywheel, 3 ft diameter (1.5 ft outside radius).

$$Wk^2 = 2,500\,\text{lb}\ (1.5\,\text{ft} \times 0.7)^2 = 2,756\,\text{lb ft}^2$$

With the polar moment of inertia known, the maximum speed n (in rpm) to which the machine can accelerate this rotor may now be computed.

Example:

Machine specification limits Wk^2n^2 to $3,000 \cdot 10^6\,\text{lb ft}^2n^2$. The rotor has a Wk^2 of $2,750\,\text{lb ft}^2$.

$$n_{max} = \sqrt{\frac{Wk^2n^2}{Wk^2}} = \sqrt{\frac{3,000 \times 10^6}{2,750}} = 1,045\,\text{rpm}$$

To determine the maximum moment of inertia the machine can accelerate to a specific balancing speed, divide the limiting Wk^2n^2 value by the square of that speed (n^2).

Example:

Machine specification limits $Wk^2 n^2$ to $3,000 \cdot 10^6$ lb ft²n². The maximum rotor Wk^2 which can be accelerated to 900 rpm then is:

$$Wk^2 = \frac{Wk^2 n^2}{n^2} = \frac{3,000 \cdot 10^6}{900^2} = 3,700 \text{ lb ft}^2$$

If the moment of inertia of a given rotor is less than 3,700 lb ft², it may be balanced at 900 rpm.

Note: These calculations do not take air resistance and other frictional losses into account.

Weight-Speed Limitation (Wn²)

The weight-speed limitation stated by a balancing machine supplier for a given size machine serves (a) to prevent damage to the supports of soft-bearing machines, and (b) to prevent the hard-bearing machine support system from operating too closely to its natural frequency and giving false indications. The stated value of Wn^2 is based on the assumption that the rotors are approximately symmetrical in shape, rigid, and mounted between the supports.

Example:

Machine specification limits Wn^2 to $2,400 \cdot 10^6$ lb n².

A given symmetric rotor weighs 1,200 lb, and is to be balanced at 800 rpm. Its Wn^2 value is:

$$Wn^2 = 1,200 \cdot 800^2 = 768 \cdot 10^6$$

Therefore, the balancing speed of 800 rpm falls well within the capabilities of the machine.

For nonsymmetrical load distribution between the supports, and for outboard rotors, the following formula provides a fast approximation of (a) the maximum permissible balancing speed in a soft-bearing machine, and (b) the maximum balancing speed in a hard-bearing machine at which permanent calibration in the A-B-C mode is maintained.

$$W_e = W \left[\frac{(2s+1)^2}{D} + 1 \right]$$

Where: W_e = Weight equivalent to be used in Wn^2 formula, (lb).
   W = Weight of rotor, (lb).
    s = Distance from the rotor CG to the nearest support. (If the CG is outboard of the supports, s is positive; if the CG is inboard, s is negative.)
   D = Distance between the supports.

Determining the Right Balancing Speed

The question is often asked whether a given rotor such as a crankshaft, fan, roll or other rotating component should be balanced at its respective service speed. The answer, in most cases, is no. The next question, usually, is why not? Doesn't unbalance increase with the square of the rotational speed? The answer, again, is no. Only the centrifugal force that a given unbalance creates increases proportionately to the square of the speed, but the actual unbalance remains the same. In other words, an ounce-inch of unbalance represents a one ounce unbalance mass with its center-of-gravity located at a one inch radius from the shaft axis, no matter whether the part is at rest or rotating (see also earlier in this chapter on "Units of Unbalance").

What balancing speed should be used then? To answer that question, consider the following requirements:

1. The balancing speed should be as low as possible to decrease cycle time, horsepower requirement, wind, noise, and danger to the operator.
2. It should be high enough so that the balancing machine has sufficient sensitivity to achieve the required balance tolerance with ease.

However, there is one other important consideration to be made before deciding upon a balancing speed substantially lower than the rotor's service speed; namely, is the part (or assembly) rigid?

Is the Rotor "Rigid"?

Theoretically it is not, since no workpiece is infinitely rigid. However, for balancing purposes there is another way of looking at it (see definition of "Rigid Rotor" in Appendix 6A).

Any rotor satisfying this definition can be balanced on standard balancing machines at a speed which is normally well below the service speed. When selecting the balancing speed, consider the following guidelines:

1. Determine the proper balance tolerance by consulting Table 6-5 and subsequent nomograms.
2. Select the lowest available speed at which the balancing machine provides at least $\frac{1}{4}$ in. amount-of-unbalance indicator deflection or 5 digital units of indication for the required balance tolerance. It is usually of no advantage to select a higher speed for achieving greater sensitivity, since the repeatability of a good quality balancing machine is well in line with today's exacting balance tolerances.

Whether a given rotor can be termed "rigid" as defined in Appendix 6A depends on numerous factors that should be carefully evaluated. For instance:

1. Rotor configuration and service speed. Technical literature provides reference tables which permit approximating the critical speed of the first flexural mode from the significant geometric rotor parameters (see Appendix 6D). In most cases it can be assumed that a rotor can be balanced successfully at low speed if its service speed is less than 50 percent of the computed first flexural critical speed.
2. Rotor design and manufacturing procedures. Rotors which are known to be flexible or unstable may, nevertheless be balanced satisfactorily at low speed if certain precautions are taken. Rotors of this type are classified as "quasi-rigid rotors."

Examples:

- A gas turbine compressor assembly, consisting of a series of bladed disks which can all be balanced individually prior to rotor assembly. Considerable effort has been made by the turbine designers to provide for accurate component balancing so that standard (low speed) balancing machines can be employed in production and overhaul of these sophisticated rotor assemblies.
- A turbine rotor with flexible or unstable mass components, such as governors or loose blades. To obtain, at low balancing speed, a position of governor or blades which most nearly approximates their position at the much higher service speed, it may be necessary to block the governor or "stake" the blades.
- A large diesel crankshaft normally rotating in five or even seven journals. When running such a shaft on only two journals in a balancing

machine, the shaft may bend from centrifugal forces caused by large counterweights and thus register a large (erroneous) unbalance. To avoid these difficulties, the balancing speed must be extremely low and/or the shaft must be supported in the balancing machine on a rigid cradle with three, five, or even seven precisely aligned bearings.

- Rotors which can not be satisfactorily balanced at low speed, require special high-speed or "modal" balancing techniques, since they must be corrected in several planes at or near their critical speed(s)[2].

Flexibility Test

This test serves to determine if a rotor may be considered rigid for balancing purposes, or if it must be treated as a flexible rotor. The test is carried out at service speed either in the rotor service bearings or in a high-speed, hard-bearing balancing machine.

The rotor should first be balanced fairly well at low speed. Then one test mass is added at the same angular position in each end plane of the rotor near its journals. During a subsequent test run, vibration is measured on both bearings. Next, the rotor is stopped and the test masses are moved to the center of the rotor, or where they are expected to cause the largest rotor distortion. In a second run the vibration is again measured at the bearings. If the total of the first readings is designated A, and the total of the second readings B, then the ratio of (B-A)/A should not exceed 0.2. Experience has shown that, if the ratio stays below 0.2, the rotor can be satisfactorily corrected at low speed by applying correction masses in two or three planes. Should the ratio exceed 0.2, the rotor will generally have to be balanced at or near its service speed.

Direction of Rotation

The direction of rotation in which the rotor runs while being balanced is usually unimportant with the exception of bladed rotors. On these (or others that create windage) it is recommended to run in the direction that creates the least turbulence and thus, uses the least drive power. Certain fans need close shrouding to reduce drive power requirements to an acceptable level. Turbine rotors with loose blades should be run backward (opposite to operational direction) to approximate the blade position in service, while compressor rotors should run forward (the same as under service conditions).

End-Drive Adapters

Design Considerations

End-drive adapters used on horizontal balancing machines to drive workpieces need to be carefully balanced so as not to introduce a balance error into the workpiece.

Considerations should be given to the following details when designing an end-drive adapter:

1. Make the adapters as light in weight as possible, consistent with capability to transmit the required driving torque. This will reduce balance errors due to fit tolerances which allow the adapter to locate eccentrically, i.e., offset from the shaft axis of the workpiece.
2. Maintain close tolerances on fit dimensions between end-drive adapter and workpiece, and between adapter and balancing machine drive. Loose fits cause shifting of the adapter and consequent changes in adapter balance. Multiply the weight of the adapter in grams by one half of the maximum radial runout possible due to a loose fit to obtain the maximum balance error in gram-inches that may result.
3. Design adapters so that they may be indexed 180° relative to the workpiece. This will allow checking and correcting the end-drive adapter balance on the balancing machine.
4. Harden and grind adapters to be used in production runs to reduce wear and consequent increase in fit clearances.

Balancing Keyed End-Drive Adapters

An adapter for a keyed rotor shaft should be provided with two 180° opposed keyways. The correct procedure for balancing the adapter depends entirely on which of the two methods was used to take care of the mating keyway when balancing the component which, on final assembly, mounts to the keyed shaft end of the workpiece being balanced.

Half-Key Method

This is the method most commonly used in North American industry. Shafts with keyways, as well as the mating components are individually balanced with half-keys fitted to fill the void the keys will occupy upon final assembly of the unit (see Figure 6-22A). To balance the end-drive adapter using this method, proceed as follows:

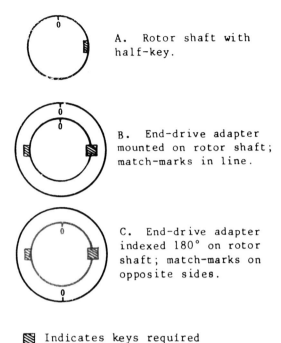

A. Rotor shaft with half-key.

B. End-drive adapter mounted on rotor shaft; match-marks in line.

C. End-drive adapter indexed 180° on rotor shaft; match-marks on opposite sides.

▨ Indicates keys required

Figure 6-22. Half-key method.

1. Mount the adapter to the workpiece shaft using a full key in the shaft keyway and fill the half-key void in the opposite side of the adapter with a half-key (see Figure 6-22B). Balance the assembly by adding balancing clay to the workpiece.
2. Index the adapter 180° on rotor shaft (see Figure 6-22C). If the adapter is out of balance, it will register on the balancing machine instrumentation. Note the gram-inch unbalance value in the plane closest to the adapter. Eliminate half of the indicated unbalance by adding clay to the adapter, the other half by adding clay to the workpiece.
3. Index the adapter 180° once again, back to the position shown in Figure 6-22, and check unbalance indication. Repeat correction method outlined above. Then replace clay on adapter with permanent unbalance correction, such as drilling, grinding, etc., on opposite side.

If it is not possible to reduce the unbalance in the adapter to a satisfactory level by this method, it is an indication that the tolerances on fit dimensions are not adequate.

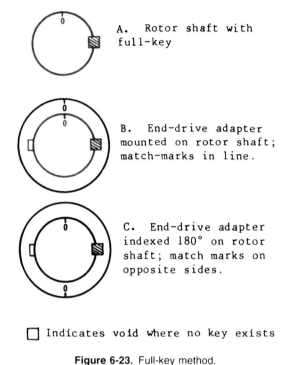

A. Rotor shaft with full-key

B. End-drive adapter mounted on rotor shaft; match-marks in line.

C. End-drive adapter indexed 180° on rotor shaft; match marks on opposite sides.

☐ Indicates void where no key exists

Figure 6-23. Full-key method.

This is the method most commonly used in European industry. Shafts are balanced with full keys and mating components without a key. To balance the end-drive adapter using this method, proceed as follows:

1. Place a full key into the keyway of the workpiece shaft (see Figure 6-23 A). Mount adapter to the workpiece shaft, leaving the opposite half-key void in the adapter empty (see Figure 6-23B). Balance the assembly using balancing clay.
2. Follow the index balancing procedure outlined in paragraphs 2 and 3 of the half-key method.

Balancing Arbors

Definition

A balancing arbor (or mandrel) generally is an accurately machined piece of shafting on which rotors that do not have journals are mounted prior to balancing. Flywheels, clutches, pulleys and other disc-shaped parts fall into this category. Arbors are employed on horizontal as well as vertical balancing machines. Particularly when used on the latter, they are also referred to as "adapters," "fixtures," or "tooling."

Since an arbor becomes part of the rotating mass being balanced, several criteria must be carefully observed during its design, manufacture, and use.

Basic Design Criteria

As is the case with most balancing machine tooling, an arbor should be as light as possible to have minimum effect on machine sensitivity. This is particularly important when using a soft-bearing machine. At the same time, the arbor must be rigid enough not to flex or bend at balancing speed.

For ease of set-up in a horizontal machine, the arbor should be designed so that the rotor can be mounted near the center (Figure 6-24). Where this is not possible, perhaps because the rotor has a blind or very small bore, the rotor may be mounted in an outboard position (Figure 6-25). If the center-of-gravity of the combined rotor and arbor falls outboard of the machine supports, a negative load bearing is required on the opposite support to absorb the uplift.

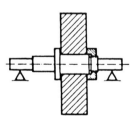

Figure 6-24. Rotor in center of arbor.

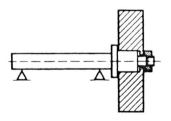

Figure 6-25. Rotor mounted outboard.

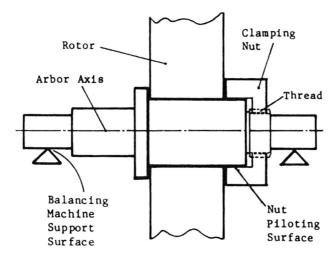

Figure 6-26. Rotor held on arbor with clamping nut.

A light push fit between arbor and rotor will facilitate assembly and dis-assembly, but may allow the rotor to slip during acceleration or decelera-tion. To prevent this, a hydraulically or mechanically expanding arbor is ideal. If none is available, a set screw may do. A small, flat area should be provided on the shaft for set screw seating.

If the rotor has a keyway, the arbor should be provided with a mating key of the same length as the final assembly key. If the arbor has no keyway, the void of the rotor keyway should be filled with a half-key having the same length as the final assembly key, even if it differs from the length of the keyway.

Threads are not a good locating or piloting surface. Sometimes a nut is used to hold the rotor on the arbor (see Figure 6-26). The nut should be balanced in itself and have a piloting surface to keep it concentric with the arbor axis.

Error Analysis

The tighter the balance tolerance, the more important it is to keep all working surfaces of the arbor as square and concentric as possible. Any eccentricity of the rotor mounting surface to the arbor axis and/or loose-ness in the fit of the rotor on the arbor causes balance errors.

To determine the balance error U (i.e., unbalance) caused by eccentricity e of the rotor mounting surface (and by rotor clearance), use the following formula:

$$U(g \cdot in) = W(g) \cdot e(inches)$$

Where:

W = Weight of rotor (grams)
e = eccentricity (inches) (= $\frac{1}{2}$ Total indicator runout [TIR] of rotor mounting surface relative to arbor axis, plus $\frac{1}{2}$ clearance between rotor and arbor)
$U_{(1-4)}$ = Unbalance (gram · inches) caused by eccentric rotor mounting surface and/or rotor/arbor fit clearance.

Example:

W = 1,000 grams
e = $\frac{1}{2}$ TIR (mounting surface to shaft axis), say $\frac{1}{2}$ of 0.004 in.
= 0.002 in. + $\frac{1}{2}$ the total clearance between in. rotor and arbor, say $\frac{1}{2}$ of 0.002 = 0.001 in. = 0.003 in.

1. U_1 = 1,000 g · 0.003 in. = 3 gram · inches
   To this may have to be added:
2. Unbalance caused by eccentricity and thread clearance of the clamping nut, assume:

   W = 100 grams
   e = 0.001 in.
   U_2 = 100 g · 0.001 in. = 0.1 g · in. (For simplification, the residual unbalance of the nut is ignored)

3. Residual unbalance U_3 of the arbor, assume 0.1 g · in.
4. Eccentricity plus $\frac{1}{2}$ fit clearance in mounting surfaces of the final rotor installation. Assuming that similar tolerances prevail as were used in making the arbor, the same unbalance will result, or:

   $$U_4 = U_1 = 3g \cdot in.$$

Total unbalance caused by arbor eccentricity and fit clearance U_1, nut eccentricity U_2, arbor residual unbalance, and installation error therefore may add up to a maximum of:

$$U_1 + U_2 + U_3 + U_4 = 3 + 0.1 + 0.1 + 3$$
$$= 6.2 \ (g \cdot in.)$$

Statistical Evaluation of Errors

One can readily see that if the rotor balance tolerance is, say, $10\,g \cdot in.$, 62 percent of it ($6.2\,g \cdot in.$) is already used up by tooling and mounting errors. Thus, the balancing machine operator is forced to balance each part to $10 - 6.2 = 3.8$ ($g \cdot in.$) or better, to be sure that the maximum permissible residual unbalance of $10\,g \cdot in.$ will be attained in the final assembly. This may be rather time consuming and, therefore, costly. To allow a larger working tolerance, the various tooling errors could be reduced by a more precisely machined arbor and final shaft. However, this too may be costly or impractical.

A solution may be found in a statistical approach. Since the various unbalance errors are vectors and may have different angular directions, they add to each other vectorially, not arithmetically. If certain errors have opposite angular directions, they actually subtract, thus resulting in a smaller total error than assumed above.

To determine the probable maximum error, the root of the sum of the squares (RSS) method should be used. This statistical method requires that the individual errors (U_1 to U_4) each be squared, then added and the square root drawn of the total. In the aforementioned example, the computation would look as follows:

$$U_{Total} = \sqrt{U_1^2 + U_2^2 + U_3^2 + U_4^2}$$
$$= \sqrt{3^2 + 0.1^2 + 0.1^2 + 3^2}$$
$$= 4.25 \ (g \cdot in.)$$

Now the operator is allowed a working tolerance of $10 - 4.25 = 5.75\,g \cdot in.$, or 50 percent more than when the unbalance errors were added arithmetically. If this still presents a problem, a more sensitive machine may be needed, or the rotor may have to be trim- or field-balanced after assembly. Under certain conditions "biasing" of the arbor may help. This method is described in the third subheading down.

Balancing the Arbor

Since residual unbalance in the arbor itself is one of the factors in the error analysis, every arbor should be carefully balanced and periodically checked. If the arbor has a keyway, it should be of the same length as the final assembly key and be filled completely during balancing with a half-key (split lengthwise) for rotors of North American origin, with a full key for rotors of European origin (see Figures 6-22 and 6-23). If the arbor has a nut, the arbor should be balanced first without it. Then the nut should be added and any residual unbalance corrected in the nut. The nut should be checked in several angular positions to make sure it stays in balance. If it does not, its locating surface must be corrected.

Special Design Features

If the arbor is to be used on a horizontal balancing machine with end-drive, one arbor face must be provided with a pilot and bolt hole circle to interface with the drive flange of the universal-joint shaft that transmits the driving torque from the balancing machine headstock.

If a horizontal machine with belt-drive is to be used, and if the rotor has no surface over which the drive belt may be placed, the arbor must be provided with a belt pulley, unless the belt can run over the arbor itself. In either case, balancing speed and drive power requirements must be taken into consideration. On machines with fixed drive motor speeds, the ratio between drive pulley diameter and driven rotor (or arbor pulley) diameter determines the desired balancing speed.

If arbors are to be used often, for instance for production balancing, they should be hardened and ground. Special care must be taken during storage to prevent corrosion and damage to locating and running surfaces.

Biasing an Arbor

This method is helpful whenever the runout (primarily radial runout) of the arbor surface which locates the rotor represents a significant factor in the error analysis. Biasing means the addition of artificial unbalance(s) to the (otherwise balanced) arbor. The bias masses are intended to compensate for the unbalance error caused by rotor displacement from the arbor's axis of rotation; rotor displacement being caused, for instance, by radial runout of the arbor surface which locates the rotor and/or, on vertical machines, runout of the machine spindle pilot.

Since the attachment of masses to a (horizontal machine) arbor may prevent it from being inserted in the rotor bore, biasing is often accomplished by grinding or drilling two light spots into the arbor, equidistant to the left and right of the rotor. The light spots must have the same angular location as the high spot of the arbor surface which locates the rotor radially.

The combined approximate unbalance value (g·in.) of the two high spots may be calculated by multiplying the rotor weight (g) by $\frac{1}{2}$ of the TJR (in.). On vertical machines the addition of bias masses to the arbor is often the simpler method. Whether the proper bias has been reached can be tested by balancing a rotor to the machine's minimum achievable residual unbalance, and then indexing it 180° on the arbor. One half of the unbalance which shows up after indexing is corrected in the rotor, the other half in the arbor. This indexing procedure is repeated until no further residual unbalance is detectable. The total correction made in the arbor is now considered its bias correction compensating for its runout, but only for the particular type of rotor used. If the rotor weight changes, the bias will have to be corrected again. Bias correction requires a good rotor fit. It will not overcome locating errors caused by loose fits.

To eliminate the need for physically biasing an arbor, balancing machine instrumentation can be furnished with a "double compensator." This feature permits biasing of the machine indication by means of suitable electrical circuits.

The Double Compensator

As its name indicates, the double compensator has a two-fold purpose: to eliminate errors in unbalance caused by tooling (thereby biasing the tooling or arbor), and to compensate for initial workpiece unbalance during machine setup.

Used in conjunction with 180° indexing, the compensator allows the machine to indicate only the rotor's true unbalance. Typically, this works as follows (see Figure 6-27).

1. Mount first workpiece on adapter. Start machine, on Schenck Trebel equipment depress compensator switch "$U + K_1$," and observe initial indication, I_1. This represents the combination (vectorial addition) of workpiece unbalance, U_1, and tooling error E (adapter eccentricity e and/or adapter-spindle unbalance), both of which are of unknown amount and angle.
2. Adjust compensator until indicator I_1 becomes zero. Compensator voltage, K_1, has now compensated for U_1 and E.
3. Index workpiece 180° in reference to the adapter. This does not change the magnitude nor the angle of the tooling error E. The initial workpiece unbalance, however, moves 180° with the workpiece.

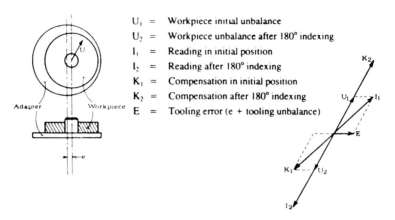

U$_1$ = Workpiece initial unbalance
U$_2$ = Workpiece unbalance after 180° indexing
I$_1$ = Reading in initial position
I$_2$ = Reading after 180° indexing
K$_1$ = Compensation in initial position
K$_2$ = Compensation after 180° indexing
E = Tooling error (e + tooling unbalance)

Figure 6-27. Schematic representation of double compensator.

Since the U$_1$ component of K$_1$ now adds to the reversed workpiece unbalance U$_2$, indication I$_2$ will be opposite U$_1$ and twice its magnitude.

4. Depress switch "U + K$_2$" and adjust compensation voltage K$_2$ until I$_2$ is zero.
5. Depress switch "U + ½K$_2$." This divides compensation voltage K$_2$ in half. The remaining indication is U$_1$, or the true initial unbalance in the workpiece. The tooling error E remains compensated by K$_1$ and thus has no more influence on this reading or on readings taken on subsequent workpieces of the same type. If the workpiece type changes, the double compensator procedure described above must be repeated for a new setup.

Just as the compensator is used to correct for unwanted errors, it can also be used to bias tooling, thereby producing a specified unbalance in a part. A typical example would be a crankshaft for a single or dual piston pump which might call for a given amount of compensating unbalance in the counterweights.

Before using a compensator for this purpose, the required accuracy for the bias must be evaluated. For large biases with tight tolerances, it may be necessary to add precisely made (and located) bias masses to the tooling. An error analysis and statistical evaluation (see earlier chapters) may then be required to take into account all error sources such as weight of bias mass, its CG uncertainty due to unbalance and mounting fit tolerance, distance of bias mass to the shaft axis of the arbor, angular location, etc.

Unbalance Correction Methods

Corrections for rotor unbalance are made either by the addition of mass to the rotor, by the removal of material, or in some cases, by relocating the shaft axis ("mass centering"). The selected correction method should ensure that there is sufficient capacity to allow correction of the maximum unbalance which may occur. The ideal correction method permits reduction of the maximum initial unbalance to less than balance tolerance in a single correction step. However, this is often difficult to achieve. The more common methods described below, e.g., drilling, usually permit a reduction of 10:1 in unbalance if carried out carefully. The addition of mass may achieve a reduction ratio as large as 20:1 or higher, provided the mass and its position are closely controlled. If the method selected for reduction of maximum initial unbalance cannot be expected to bring the rotor within the permissible residual unbalance in a single correction step, a preliminary correction is made. Then a second correction method is selected to reduce the remaining unbalance to its permissible value.

Addition of Mass

1. Addition of solder or two-component epoxy. It is difficult to apply the material so that its center-of-gravity is precisely at the desired correction location. Variations in location introduce errors in correction. Also, this method requires a fair amount of time.
2. Addition of bolted or riveted washers. This method is used only where moderate balance quality is required.
3. Addition of cast iron, lead, or lead masses. Such masses, in incremental sizes, are used for unbalance correction.
4. Addition of masses by resistance-welding them to a suitable rotor surface. This method provides a means of attaching a wide variety of correction masses at any desired angular locations. Care must be taken that welding heat does not distort the rotor.

Removal of Mass

1. Drilling. Material is removed from the rotor by a drill which penetrates the rotor to a measured depth, thereby removing the intended mass of material with a high degree of accuracy. A depth gage or limit switch can be provided on the drill spindle to ensure that the hole is drilled to the desired depth. This is probably the most effective method of unbalance correction.

2. Milling, shaping, or fly cutting. This method permits accurate removal of mass when the rotor surfaces, from which the depth of cut is measured, are machined surfaces, and when means are provided for accurate measurement of cut with respect to those surfaces; used where relatively large corrections are required.
3. Grinding. In general, grinding is used as a trial-and-error method of correction. It is difficult to evaluate the actual mass of the material which is removed. This method is usually used only where the rotor design does not permit a more economical type of correction.

Mass Centering

For the definition of mass centering see Appendix 6A. Such a procedure is used, for instance, to reduce initial unbalance in crankshaft forgings. The shaft is mounted in a balanced cage or cradle which, in turn, is rotated in a balancing machine. The shaft is adjusted radially with respect to the cage, until the unbalance indication for the combined shaft and cradle assembly is within a given tolerance. At this point the principal inertia axis of the shaft essentially coincides with the shaft axis of the balanced cage. Center-drills, guided along the axis of the cage, drill the shaft centers and thereby provide an axis in the crankshaft about which it is in balance. The subsequent machining of the crankshaft is carried out between these centers.

Because material removal is uneven at different parts of the shaft, the machining operation will introduce some new unbalance. A final balancing operation is therefore still required. It is generally accomplished by drilling into the crankshaft counterweights. However, final unbalance corrections are small and balancing time is significantly shortened. Furthermore, final correction by drilling does not exceed the material available for it, nor does it reduce the mass of the counterweights to a level where they no longer perform their proper function, namely to compensate for the opposed masses of the crankshaft.

Testing Balancing Machines

Total verification of all purchase specification requirements may be possible for a production machine, but usually not for a general purpose machine, such as a machine in a motor repair shop, because a rotor of the maximum specified weight or polar moment of inertia may not be available at the time of acceptance tests. Nevertheless, essential conformance with the specification may be ascertained by a complete physical inspection and performance tests with typical workpieces and/or a "proving rotor." Physical inspection needs to take into account all specified dimensions, features, instrumentation, tooling, and accessories that are listed in the purchase specification and/or the seller's proposal. Performance tests are somewhat more involved and should be witnessed by a representative of the buyer who is well acquainted with balancing machines and the particular specification applying to the machine to be tested.

Tests for Production Machines

A production machine is usually purchased for balancing a given part or parts in large quantities. Acceptance tests, therefore, are generally performed by running samples of such parts, so that total compliance with specified indicating accuracy and cycle time can be ascertained under simulated production conditions. At the same time, tooling is checked for locating accuracy and balance. Additional tests, as described in the following paragraphs, may then be confined to just the first part (U_{mar} Test), since compliance with the specified cycle time may already be considered sufficient proof that the machine achieves a satisfactory "Unbalance Reduction Ratio." This, however, is only the case if the initial unbalance of the sample rotors is representative of the whole range of initial unbalances that will be encountered in actual production parts.

Basic Test Concepts

From time to time over the last 30 or 40 years, the devising of procedures for testing balancing machines, particularly dynamic balancing machines, has occupied many experts and various committees of engineering societies. The chief problem usually has been the interaction of errors in amount indication, angle indication, and plane separation. A requirement for a given accuracy of amount indication becomes meaningless if the machine's indicating system has poor plane separation or lacks accuracy of angle indication; or the best plane separation is useless if the amount and angle indication are inaccurate.

As an example of interdependence between amount and angle indication, Figure 6-28 illustrates how an angle error of 10° results in an amount indication error of 17.4 percent. The initial unbalance of 100 g was corrected 10° away from where the correction mass should have been attached. The residual unbalance indicated in the next run is 17.4 g at 85°, nearly at a right angle to the initial unbalance.

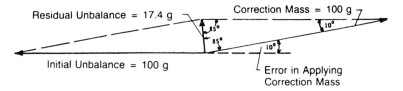

Figure 6-28. Residual unbalance due to angle error.

Table 6-3
Interdependence of Angle and Amount Indication

Angle Error	Amount Error*
1 degree	1.7%
2 degrees	3.5%
3 degrees	5.2%
4 degrees	7.0%
5 degrees	8.7%
6 degrees	10.5%
8 degrees	14.0%
10 degrees	17.4%
12 degrees	20.9%
15 degrees	26.1%

Percent of initial unbalance.

Listed in Table 6-3 are residual unbalances expressed in percent of initial unbalances which result from applying unbalance correction of proper amount but at various incorrect angular positions.

Eventually it was recognized that most balancing machine users are really not so much interested in how accurately the individual parameter is indicated, but rather, in the accuracy of the combination of all three. In other words, the user wants to reduce the initial unbalance to the specified permissible residual unbalance in a minimum number of steps. Acceptance of this line of reasoning resulted in the concept of the "Unbalance Reduction Ratio," URR for short (see definition in Appendix 6A). It expresses the percentage of initial unbalance that one correction step will eliminate. For instance, a URR of 95 percent means that an initial unbalance of 100 units may be reduced to a residual unbalance of 5 units in one measuring and correction cycle—provided the correction itself is applied without error. A procedure was then developed to verify whether a machine will meet a specified URR. This test is called the Unbalance

Reduction Test, or UR Test. It tests a machine for combined accuracy of amount indication, angle indication, and plane separation, and should be part of every balancing machine acceptance test.

Note: On single-plane machines, the UR test only checks combined accuracy of amount and angle indication.

Aside from the UR test, acceptance test procedures should also include a check whether the machine can indicate the smallest unbalance specified. For this purpose, a test for "Minimum Achievable Residual Unbalance" was developed, called, "U_{mar} Test" or "Traverse Test," for short. Both U_{mar} and UR tests are described in subsequent chapters. They should be repeated periodically; for instance, once a month if the machine is used daily, to assure that it is still in proper operating condition.

Table 6-4 lists various current standards for testing balancing machines (see also Appendix 6C).

Inboard Proving Rotors for Horizontal Machines

For general purpose machines, and in the absence of a proving rotor supplied by the balancing machine manufacturer, any rigid rotor such as an armature, roll, flywheel, etc, may be made into a proving rotor. Ideally, its weight and shape should approximate the actual rotors to be balanced. Since these usually vary all over the capacity range of a general purpose machine, ISO 2953 suggests one rotor to be near the minimum weight limit, a second rotor near the maximum.

Particularly for soft-bearing machines, it is important to make the U_{mar} test with a small rotor since that is where parasitic mass of the vibratory system (carriages, bridge, springs, etc.) has its maximum effect on the sensitivity of unbalance indication. As a general rule, it would probably be sufficient if the rotor fell within the bottom 20 percent of the machine weight range. For hard-bearing machines, it is not as important to test the lower end of the weight range, since parasitic mass has little effect on the readout sensitivity of such machines.

Testing both soft- or hard-bearing machines in the upper 20 percent of their weight range will verify their weight carrying and drive capability, but add little additional knowledge concerning the measuring system. On machines with weight ranges larger than 10,000 lbs it may be impractical to call for a test near the upper weight limit before shipment, since a balancing machine manufacturer rarely has such heavy rotors on hand. A final test after installation with an actual rotor may then be the better choice. In any case, it will generally suffice to include one small, or on hard-bearing machines, one small to medium size proving rotor, in the purchase of a machine. Rotors weighing several thousand pounds might possibly

Table 6-4
Standards for Testing Balancing Machines

Application	Title	Issuer	Document no.
General industrial balancing machines	Balancing Machines— Description and Evaluation	International Standards Organization (ISO)	DIS 2953 1983*
Jet engine rotor balancing machines (for two-plane correction)	Balancing Equipment for Jet Engine Components, Compressor and Turbine, Rotating Type, for Measuring Unbalance in One or More Than One Transverse Plane	Society of Automotive Engineers, Inc. (SAE)	ARP 587 A
Jet engine rotor balancing machines (for single-plane correction)	Balancing Equipment for Jet Engine Components Compressor and Turbine, Rotating Type, for Measuring Unbalance in One Transverse Plane	Society of Automotive Engineers, Inc. (SAE)	ARP 588 A
Gyroscope rotor balancing machines	Balancing Machine— Gyroscope Rotor	Defense General Supply Center, Richmond, Va.	FSN 6635- 450-2208 NT
Field balancing equipment	Field Balancing Equipment— Description and Evaluation	International Standards Organization (ISO)	ISO 2371

The 1983 version contains important revisions in the test procedure.

be furnished temporarily by the balancing machine manufacturer for the acceptance test.

For all sizes of proving rotors, a symmetrical shape is preferred to which test masses can be attached at precisely defined positions in 2 transverse planes. Two typical kinds of proving rotors are shown in Figure 6-29.

ISO 2953 suggests the solid roll-type rotors, with the largest one weighing 1,100 lb. For larger rotors (or even at the 1,100 lb level) a dumbbell-type rotor may be more economical. This also depends on available material and manufacturing facilities.

Critical are the roundness of the journals, their surface quality, radial runout of the test mass mounting surfaces, and the axial and angular loca-

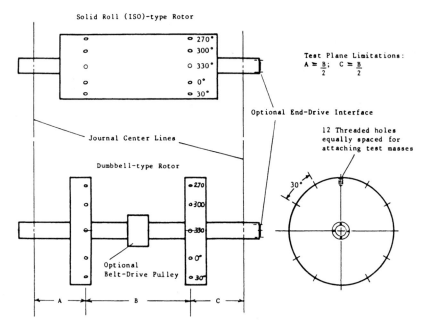

Figure 6-29. Typical proving rotors for horizontal machines.

tion of the threaded holes which hold the test masses. For guidance in determining machining tolerances, refer to the section on Test Masses.

Before using a proving rotor, it will have to be balanced as closely to zero unbalance as possible. This can generally be done on the machine to be tested, even if its calibration is in question. The first test (U_{mar} Test) will reveal if the machine has the capability to reach the specified minimum achievable residual unbalance, the second test (UR Test) will prove (or disprove) its calibration.

Whenever the rotor is reused at some future time, it should be checked again for balance. Minor correction can be made by attaching balancing clay or wax, since the rotor will probably change again due to aging, temperature distortion or other factors. The magnitude of such changes generally falls in the range of a few microinches displacement of CG, and is not unusual.

Test Masses

Test masses are attached to a balanced proving rotor to provide a known quantity of unbalance at a precisely defined location. The rotor is then run in the balancing machine at a given speed and the unbalance indication is observed. It should equal the unbalance value of the test mass within a permissible plus/minus deviation.

Since the rotor with test masses functions as a gage in assessing the accuracy of the machine indication, residual unbalance and location errors in the test masses should be as small as possible. The test procedure makes allowance for the residual unbalance in the proving rotor but not for test mass errors. Therefore, the following parameters must be carefully controlled to minimize errors.

1. Weight of test mass.
2. Distance of test mass mounting surface to proving rotor shaft axis.
3. Distance of test mass center of gravity (CG) to mounting surface.
4. Angular position of test mass.
5. Axial position of test mass.

Since all errors are vector quantities, they should be treated as was done in the error analysis in the section on balancing arbors, i.e., adjusted by the RSS method. The resulting probable maximum error should ideally not use up more than one tenth of the reciprocal of the specified Unbalance Reduction Ratio factor. For example, if a URR of 95 percent is to be proven, the total test mass error from parameters 1 to 5 should not exceed $0.1 \cdot 5$ percent $= 0.5$ percent of the test mass weight.

Often test masses need to be so small that they become difficult to handle. It is then quite common to work with differential test masses, i.e., two masses 180° opposite each other in the same transverse plane. The effective test mass is the difference between the two masses, called the "differential unbalance." For instance, if one mass weighs 10 grams and the other 9, the difference of 1 gram represents the differential unbalance.

When working with differential test masses, the errors of the two comparatively large masses affect the accuracy of the differential unbalance in an exaggerated way. In the example used above, each differential test mass would have to be accurate within approximately 0.025 percent of its own value to keep the maximum possible effect on the differential unbalance to within $2 \cdot 0.25$ percent $= 0.5$ percent. In other words, if the opposed masses are about ten times as large as their difference, each mass must be ten times more accurate than the accuracy required for the difference.

Test Procedures

To test the performance of a balancing machine, ISO 2953 prescribes two separate tests, the U_{mar} Test and the Unbalance Reduction Test. The origin and philosophy behind these tests and their purpose were explained. Here are the actual test procedures:

U_{mar} (or Traverse) Test

1. Perform the mechanical adjustment, calibration and/or setting of the machine for the particular proving rotor being used for the test, ensuring that the unbalance in the rotor is smaller than five times the claimed minimum achievable residual unbalance for the machine.
2. Put 10 to 20 times the claimed minimum achievable residual unbalance on the rotor by adding two unbalance masses (such as balancing clay). These masses shall not be:
   - in the same transverse plane
   - in a test plane
   - at the same angle
   - displaced by 180°
3. Balance the rotor, following the standard procedure for the machine, by applying corrections in two planes other than test planes or those used for the unbalance masses in a maximum of four runs at the balancing speed selected for the U_{mar} Test.
4. In the case of horizontal machines, after performing the actions described in 1 to 3, change the angular reference system of the machine by 60 or 90°, e.g., turn the end-drive shaft with respect to the rotor, turn black and white markings, etc.
5. For horizontal or vertical two-plane machines, attach in each of the two prepared test planes a test mass equal to ten times the claimed minimum achievable residual unbalance.

   For example, if the ISO proving rotor No. 5 weighing 110 lbs (50,000 g) is used, the weight of each test mass is calculated as follows:

   The claimed minimum achievable residual specific unbalance is, say

   $$1 \, e_{mar} = 0.000020 \text{ in.}$$

   The claimed minimum achievable residual unbalance per test plane, i.e., for half the rotor weight, is therefore:

   $$1 \, U_{mar} \text{(per plane)}$$

   $$= \frac{50{,}000 \text{ g}}{20} \cdot 0.000020 \text{ in.}$$

   $$= 0.5 \text{ g} \cdot \text{in.}$$

The desired 10 U_{mar} test mass per plane is therefore equivalent to:

10 U_{mar} (per plane)

= $10 \cdot 0.5$ g · in.

= 0.5 g · in.

If the test mass is attached so that its center of gravity is at a radius of four in. (effective test mass radius), the actual weight of each test mass will be:

$$m = \frac{5 \text{ g} \cdot \text{in.}}{4 \text{ in.}} = 1.25 \text{ g}$$

When two of these test masses are attached to the rotor (one in each test plane as shown in Figure 6-30), they create a combined static unbalance in the entire rotor of 10 U_{mar} (or specific unbalance of 10 e_{mar}), since each test mass had been calculated for only one half of the rotor weight.

Note 1: If a proving rotor with asymmetric CG and/or test planes is used, the test masses should be apportioned between the two test planes in such a way that an essentially parallel displacement of the principal inertia axis from the shaft axis results.

Note 2: U_{mar} Tests are usually run on inboard rotors only. However, if special requirements exist for balancing outboard rotors, a U_{mar} Test may be advisable which simulates those requirements.

6. Attach the test masses in phase with one another in all 12 equally spaced holes in the test planes, using an arbitrary sequence. Record amount-of-unbalance readings in each plane for each position of the masses in a log shown in Figure 6-31. For the older style 8-hole rotors, a log with 45° test mass spacing must be used.

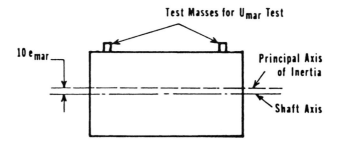

Figure 6-30. Proving rotor with test masses for "Umar" test.

Angle of Test Mass		0°	30°	60°	90°	120°	150°	180°	210°	240°	270°	300°	330°
Amount of Unbalance Readings	Left Plane (Lower Plane*)												
	Right Plane (Upper Plane*)												

* Lower and upper plane designations apply to vertical machines.

Figure 6-31. Log for "Umar" test.

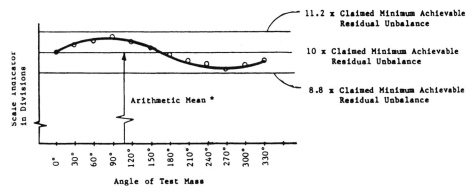

Figure 6-32. Diagram showing residual unbalance.

7. Plot the logged results as shown in Figure 6-32 in two diagrams, one for the left and one for the right plane (or upper and lower planes on vertical machines). For 8-hole rotors, use a diagram with 45° spacing.

Connect the points in each diagram by an averaging curve. It should be of sinusoidal shape and include all test points.

If the rotor has been balanced (as in 3) to less than $\frac{1}{2}$ U_{mar}, the plotted test readings may scatter closely around the 10 U_{mar} line and not produce a sinusoidal averaging curve. In that case add $\frac{1}{2}$ U_{mar} residual unbalance to the appropriate test plane and repeat the test. Draw a horizontal line representing the arithmetic mean of the scale reading into each diagram and add two further lines representing ±12 percent of the arithmetic mean for each curve, which accounts for 1 U_{mar} plus 20 percent for the effects of variation in the position of the masses and scatter of the test data.

If all the plotted points are within the range given by those two latter lines for each curve, the claimed minimum achievable residual unbalance has been reached.

If the amount-of-unbalance indication is unstable, read and plot the maximum and minimum values for all angular positions of the test mass. Again, all points must be within the range given.

Note: If different U_{mar} values are specified for different speeds, the test should be repeated for each.

8. On horizontal and vertical single-plane balancing machines designed to indicate static unbalance only, proceed in the same way as described in 1 and 7 but use only one test mass in the left (or lower) plane of the proving rotor. This test mass must be calculated using the *total* weight of the proving rotor.

9. On vertical machines, the spindle balance should be checked. Remove the proving rotor and run the machine. The amount of unbalance now indicated should be less than the claimed minimum achievable residual unbalance.

Unbalance Reduction Test

This test is intended to check the combined accuracy of amount-of-unbalance indication, angle indication, and plane separation. Experience gained with running the test in accordance with the procedure described in ISO 2953 (1973 version) showed that the operator could influence the test results because he knew in advance what the next reading should be. For instance, if a reading fluctuated somewhat, he could wait until the indicator showed the desired value and at that moment actuate the readout retention switch.

To avoid such operator influence, a somewhat modified procedure has been developed similar to that used in ARP 587 (see Appendix 6C). In the new procedure (ISO 2953—second edition) a stationary mass is attached to the rotor in the same plane in which the test mass is traversed. The unbalance resulting from the combination of two test masses, whose angular relationship changes with every run, is nearly impossible to predict.

To have a simultaneous check on plane separation capability of the machine, a stationary and a traversing (or "traveling") test mass are also attached in the other plane. Readings are taken in both planes during each run.

Unbalance readings for successive runs are logged on the upper "log" portion of a test sheet, and subsequently plotted on the lower portion containing a series of URR limit circles. All plotted points except one per plane must fall within their respective URR limit circles to have the machine pass the test. A similar procedure has been used by the SAE for more than ten years and has proven itself to be practical and foolproof.

The new Unbalance Reduction Test is divided into an inboard and an outboard test. The inboard test should be conducted for all machines; in addition, the outboard test should be conducted for all horizontal two-plane machines on which outboard rotors are to be balanced.

Each test consists of two sets of 11 runs, called "low level" and "high level" tests. When using the older style proving rotor with eight holes per plane, only seven runs are possible. The low level tests are run with a set of small test masses, the high level tests with a larger set to test the machine at different levels of unbalance. Test mass requirements and procedures are described in detail in Figure 6-33.

Balance Tolerances

Every manufacturer and maintenance person who balances part of his product, be it textile spindles or paper machinery rolls, electric motors or gas turbines, satellites or re-entry vehicles, is interested in a better way to determine an economical yet adequate balance tolerance. As a result, much effort has been spent by individual manufacturers to find the solution to their specific problem, but rarely have their research data and conclusions been made available to others.

In the 1950s, a small group of experts, active in the balancing field, started to discuss the problem. A little later they joined the Technical Committee 108 on Shock and Vibration of the International Standards Organization and became Working Group 6, later changed to Subcommittee 1 on Balancing and Balancing Machines (ISO TC-108/Sc1). Interested people from other countries joined, so that the international group now has representatives from most major industrialized nations. National meetings are held in member countries under the auspices of national standards organizations, with balancing machine users, manufacturers and others interested in the field of dynamic balancing participating. The national committees then elect a delegation to represent them at the annual international meeting.

One of the first tasks undertaken by the committee was an evaluation of data collected from all over the world on required balance tolerances for millions of rotors. Several years of study resulted in an ISO Standard No. 1940 on "Balance Quality of Rotating Rigid Bodies" which, in the meantime, has also been adopted as S2.19-1975 by the American National Standards Institute ("ANSI," formerly USASI and ASA). The principal points of this standard are summarized below. Balance tolerance

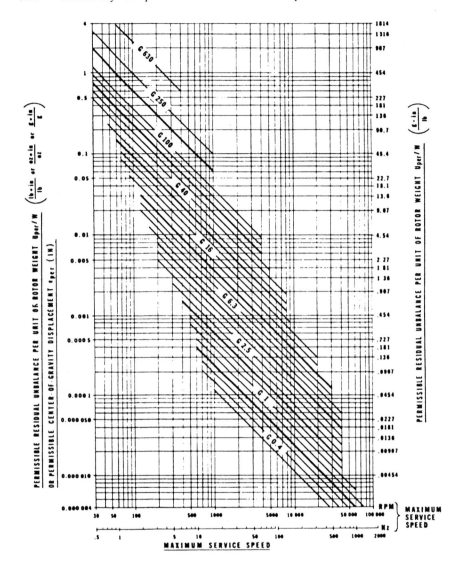

Figure 6-33. Maximum permissible residual specific unbalance corresponding to various balancing quality grades "G," in accordance with ISO 1940.

NOTES:

(1) In general, for rigid rotors with two correction planes, one half of the recommended residual unbalance is allowed for each plane; these values apply usually for any two arbitrarily chosen planes. (See paragraph on Applying Tolerances to Two-plane Rotors). For disc-shaped rotors the full recommended value applies to one plane.

(2) To obtain acceptable residual unbalance in oz-in per lb of rotor weight multiply lb-in value by 16.

(3) To avoid need for multiplication use appropriate Balance Tolerance Nomogram on following pages.

nomograms, developed by the staff of Schenck Trebel Corporation from the composite ISO metric table, have been added to provide a simple-to-use guide for ascertaining recommended balance tolerances (see Figures 6-34 and 6-35).

Balance Quality Grades

We have already explained the detrimental effects of unbalance and the purpose of balancing. Neither balancing cost considerations, nor various rotor limitations such as journal concentricity, bearing clearances or fit, thermal stability, etc., permit balancing every rotor to as near zero unbalance as might theoretically be thought possible. A tolerance must be set to allow a certain amount of residual unbalance, just as tolerances are set for various other machine shop operations. The question usually is, how much residual unbalance can be permitted while still holding detrimental effects to an insignificant or acceptable level?

The recommendations given in ISO 1940 will usually produce satisfactory results. The heart of the Standard is a listing of various rotor types, grouped according to "quality grades" (see Table 6-5). Anyone trying to determine a reasonable balance tolerance can locate his rotor type in the table and next to it find the assigned quality grade number. Then the graph in Figure 6-33 or the nomograms in Figures 6-34 and 6-35 are used to establish the gram·inch value of the applicable balance tolerance (i.e., "permissible residual unbalance" or U_{per}).

Except for the upper or lower extremes of the graph in Figure 6-33, every grade incorporates 4 bands. For lack of a better delineation, the bands might be considered (from top to bottom in each grade) substandard, fair, good, and precision. Thus, the graph permits some adjustment to individual circumstances within each grade, whereas the nomograms list only the median values (centerline in each grade). The difference in permissible residual unbalance between the bottom and top edge of each grade is a factor of 2.5. For particularly critical applications it is, of course, also possible to select the next better grade.

CAUTION: The tolerances recommended here apply *only* to *rigid rotors*. Recommendations for flexible rotor tolerances are contained in ISO 5343 (see Appendix 6C) or in Reference 2.

Special Conditions to Achieve Quality Grades G1 and G0.4

To balance rotors falling into Grades 1 or 0.4 usually requires that the following special conditions be met:

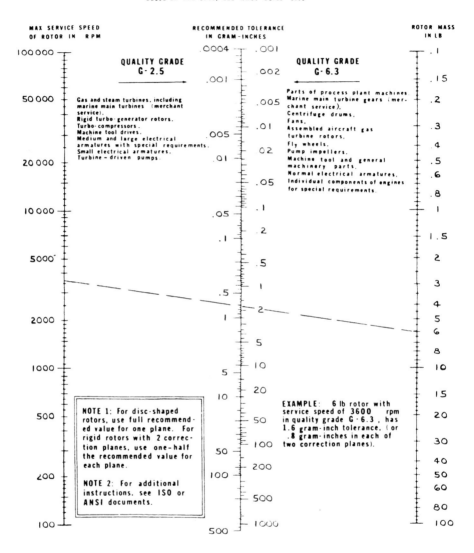

Based on ISO 1940, and ANSI S2.19 - 1975

1 g·in = .0353 oz·in
1 oz·in = 28.35 g·in

Figure 6-34. Balance tolerance nomogram for G-2.5 and G-6.3, small rotors.

Based on ISO 1940, and ANSI S2.19 - 1975

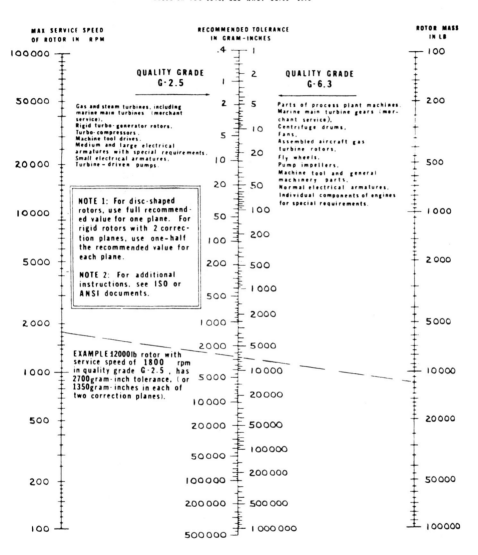

1 g·in = .0353 oz·in
1 oz·in = 28.35 g·in

Figure 6-35. Balance tolerance nomogram for G-2.5 and G-6.3, large rotors.

Table 6-5
Balance Quality Grades for Various Groups of Representative Rigid Rotors in Accordance with ISO 1940 and ANSI S2.19-1 975

Balance Quality Grade G	Rotor Types—General Examples
G 4000	Crankshaft-drives (2) of rigidly mounted slow marine diesel engines with uneven number of cylinders (3).
G 1600	Crankshaft-drives of rigidly mounted large two-cycle engines.
G 630	Crankshaft-drives of rigidly mounted large four-cycle engines. Crankshaft-drives of elastically mounted marine diesel engines.
G 250	Crankshaft-drives of rigidly mounted fast four-cylinder diesel engines (3).
G 100	Crankshaft-drives of fast diesel engines with six and more cylinders (3). Complete engines (gasoline or diesel) for cars, trucks and locomotives (4).
G 40	Car wheel (5), wheel rims, wheel sets, drive shafts. Crankshaft-drives of elastically mounted fast four-cycle engines (gasoline or diesel) with six and more cylinders (3). Crankshaft-drives for engines of cars, trucks and locomotives.
C 16	Drive shafts (propeller shafts, cardan shafts) with special requirements. Parts of crushing machinery. Parts of agricultural machinery. Individual components of engines (gasoline or diesel) for cars, trucks and locomotives. Crank-shaft-drives of engines with six or more cylinders under special requirements.
G 6.3	Parts of process plant machines. Marine main turbine gears (merchant service). Centrifuge drums. Fans. Assembled aircraft gas turbine rotors. Flywheels. Pump impellers. Machine-tool and general machinery parts. Medium and large electric armatures (of electric motors having at least 80 mm shaft height) without special requirements. Small electric armatures, often mass produced, in vibration insensitive applications and/ or with vibration damping mountings. Individual components of engines under special requirements.
G 2.5	Gas and steam turbines, including marine main turbines (merchant service). Rigid turbogenerator rotors. Rotors. Turbo-compressors. Machine-tool drives. Medium and large electrical armatures with special requirements. Small electric armatures not qualifying for one or both of the conditions stated in G6.3 for such. Turbine-driven pumps.
G 1	Tape recorder and phonograph drives. Grinding-machine drives. Small electrical armatures with special requirements.
G 0.4	Spindles, discs, and armatures of precision grinders. Gyroscopes.

NOTES:
1. *The quality grade number represents the maximum permissible circular velocity of the center of gravity in mm/sec.*
2. *A crankshaft drive is an assembly which includes the crankshaft, a flywheel, clutch, pulley, vibration damper, rotating portion of connecting rod, etc.*
3. *For the purposes of this recommendation, slow diesel engines arc those with a piston velocity of less than 30 ft. per sec., fast diesel engines are those with a piston velocity of greater than 30 ft per sec.*
4. *In complete engines, the rotor mass comprises the sum of all masses belonging to the crankshaft-drive.*
5. *G 16 is advisable for off-the-car balancing due to clearance or runout in central pilots or bolt hole circles.*

For Quality Grade 1:

- Rotor mounted in its own service bearings
- *No* end-drive

For Quality Grade 0.4:

- Rotor mounted in its own housing and bearings
- Rotor running under service conditions (bearing preload, temperature)
- Self-drive

Only the highest quality balancing equipment is suitable for this work.

Applying Tolerances to Single-Plane Rotors

A single-plane rotor is generally disc-shaped and, therefore, has only a single correction plane. This may indeed be sufficient if the distance between bearings is large in comparison to the width of the disc, and provided the disc has little axial runout. The entire tolerance determined from such graphs as shown in Figures 6-34 and 6-35 may be allowed for the single plane.

To verify that single-plane correction is satisfactory, a representative number of rotors that have been corrected in a single plane should be checked for residual couple unbalance. One component of the largest residual couple (referred to the two-bearing planes) should not be larger than one half the total rotor tolerance. If it is larger, moving the correction plane to the other side of the disc (or to some optimal location between the disc faces) may help. If it does not, a second correction plane will have to be provided and a two-plane balancing operation performed.

Applying Tolerances to Two-Plane Rotors

In general, one half of the permissible residual unbalance is applied to each of the two correction planes, provided the distance between (inboard) rotor CG and either bearing is not less than $\frac{1}{3}$ of the total bearing distance, and provided the correction planes are approximately equidistant from the CG, having a ratio no greater than $3:2$.

If this ratio is exceeded, the total permissible residual unbalance (U_{per}) should be apportioned to the ratio of the plane distances to the CG. In other words, the larger portion of the tolerance is allotted to the correction plane closest to the CG; however, the ratio of the two tolerance portions should never exceed 7:3, even though the plane distance ratio may be higher.

For rotors with correction plane distance (b) larger than the bearing span (d), the total tolerance should be reduced by the factor d/b before any apportioning takes place.

For rotors with correction plane distance smaller than $\frac{1}{3}$ of the bearing span and for rotors with two correction planes outboard of one bearing, it is often advisable to measure unbalance and state the tolerance in terms of (quasi-) static and couple unbalance. Satisfactory results can generally be expected if the static residual unbalance is held within the limits of

$$U_{static} \leq U_{per}/3$$

and the couple residual unbalance within

$$U_{couple} \leq \frac{U_{per}}{3} \cdot d$$

(where d = bearing span)

If separate indication of static and couple unbalance is not desired or possible, the distribution of the permissible residual unbalance must be specially investigated, taking into account, for instance, the permissible bearing loads[4]. It may also be necessary to state a family of tolerances, depending on the angular relationship between the residual unbalances in the two correction planes.

For all rotors with narrowly spaced (inboard or outboard) correction planes, the following balancing procedure may prove advantageous if U_{per} is specified in terms of residual unbalance per correction plane.

1. Calibrate respectively the balancing machine to indicate unbalance in the two chosen correction planes I and II (see Figure 6-36).
2. Measure and correct unbalance in plane I only.
3. Recalibrate or set the balancing machine to indicate unbalance near bearing plane A and in plane II.
4. Measure and correct unbalance in plane II only.
5. Check residual unbalance with machine calibrated or set as in 3. Allow residual unbalance portions for the inboard rotor as discussed above (inversely proportional to the correction plane distances from the CG), considering A and II as the correction planes; for the outboard rotor allow no more than 70 percent of U_{per} in plane II, and no less than 30 percent in plane A.

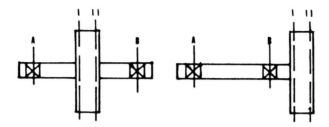

Figure 6-36. Inboard and outboard rotors with narrowly spaced correction planes.

Experimental Determination of Tolerances

For reasons of rotor type, economy, service life, environment or others, the recommended tolerances may not apply. A suitable tolerance may then be determined by experimental methods. For instance, a sample rotor is balanced to the smallest achievable residual unbalance. Test masses of increasing magnitude are then successively applied, with the rotor undergoing a test run under service conditions before each test mass is applied. The procedure is repeated until the test mass has a noticeable influence on the vibration, noise level, or performance of the machine. In the case of a two-plane rotor, the effects of applying test masses as static or couple unbalance must also be investigated. From the observations made, a permissible residual unbalance can then be specified, making sure it allows for differences between rotors of the same type, and for changes that may come about during sustained service.

Applying Tolerances to Rotor Assembly Components

If individual components of a rotor assembly are to be pre-balanced (on arbors for instance), the tolerance for the entire assembly is usually distributed among the components on the basis of the weight that each component contributes to the total assembly weight. However, allowance must be made for additional unbalance being caused by fit tolerances and mounting surface runouts. To take all these into account, an error analysis should be made.

Testing a Rotor for Tolerance Compliance

If the characteristics of the available balancing equipment do not permit an unbalance equivalent to the specified balance tolerance to be measured with sufficient accuracy (ideally within ± 10 percent of value), the U_{mar} test described earlier may be used to determine whether the specified tolerance has been reached. The test should be carried out separately for each correction plane, and a test mass equivalent to 10 times the tolerance should be used for each plane.

Balance Errors Due to Drive Elements

During balancing in general, and during the check on tolerance compliance in particular, significant errors can be caused by the driving elements (for example, driving adapter and universal-joint drive shaft).

In Figure 6-37 seven sources of balance errors are illustrated:

1. Unbalance from universal-joint shaft.
2. Unbalance-like effect from excessive looseness or tightness in universal joints.
3. Loose fit of adapter in universal-joint flange.
4. Offset between adapter pilot (on left) and adapter bore (on right).
5. Unbalance of adapter.
6. Loose fit of adapter on rotor shaft.
7. Eccentricity of shaft extension (on which adapter is mounted) in reference to journals.

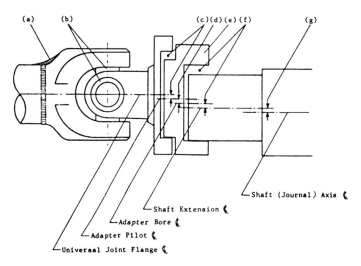

Figure 6-37. Error sources in end-drive elements.

The effects of errors 1, 3, 4, and 5 may be demonstrated by indexing the rotor against the adapter. These errors can then be jointly compensated by an alternating index-balancing procedure described below. Error 2 will generally cause reading fluctuation in case of excessive tightness, nonrepeating readings in case of excessive looseness. Error 6 may be handled like 1, 3, 4, and 5 if the looseness is eliminated by a set screw (or similar) in the same direction after each indexing and retightening cycle. If not, it will cause nonrepeating readings. Error 7 will not be discovered until the rotor is checked without the end-drive adapter, presumably under service conditions with field balancing equipment. The only (partial) remedy is to reduce the runout in the shaft extension and the weight of the end-drive elements to a minimum.

Balance errors from belt-drive pulleys attached to the rotor are considerably fewer in number than those caused by end-drive adapters. Only the pulley unbalance, its fit on the shaft, and the shaft runout at the pulley mounting surface must be considered. Such errors are avoided altogether if the belt runs directly over the part. Certain belt-drive criteria should be followed.

Air- and self-drive generally introduce minimal errors if the cautionary notes mentioned previously are observed.

Balance Errors Due to Rotor Support Elements

Various methods of supporting a rotor in a balancing machine may cause balance errors unless certain precautions are taken. For instance, when supporting a rotor journal on roller carriages, the roller diameter should differ from the journal diameter by at least 10 percent, and the roller speed should never differ less than 60 rpm from the journal speed. If this margin is not maintained, unbalance indication becomes erratic.

A rotor with mounted rolling element bearings should be supported in V-roller carriages (see Nomenclature, Appendix 6B). Their inclined rollers permit the bearing outer races to align themselves to the inner races and shaft axis, letting the rolling elements run in their normal tracks.

Rotors with rolling element bearings may also be supported in sleeve or saddle bearings; however, the carriages or carriage suspension systems must then have "vertical axis freedom" (see Terminology, Appendix 6A). Without this feature, the machine's plane separation capability will be severely impaired because the support bridges (being connected via the rotor) can only move in unison toward the front and rear of the machine; thus only static unbalance will be measured.

Vertical axis freedom is also required when the support bridges or carriages are connected by tiebars, cradles, or stators. Only then can couple unbalance be measured without misaligning the bearings in each (out-of-phase) back-and-forth movement of the support bridges. This also holds true for hard-bearing machines, even though bridge movement is microscopically small.

Index-Balancing Procedure

A procedure of repetitively balancing and indexing (by 180°) one component against another leads to diminishing residual unbalance in both, until eventually one component can be indexed against the other without a significant change in residual unbalance. Index-balancing may be used to eliminate the unbalance errors in an end-drive adapter, for biasing an arbor or for improving the residual unbalance in a rotor mounted on an arbor.

If the procedure is used for a single-plane application (e.g., an end-drive adapter), one half of the residual unbalance (after the first indexing) is corrected in the adapter, the other half in the rotor. The cycle may have to be repeated once or twice until a satisfactory residual unbalance is reached.

Care should be taken that after each indexing step, set screws are tightened with the same torque and in the same sequence. The procedure does not work well unless the position of the indexed component is precisely repeatable.

If the above iterative process becomes too tedious, a graphic solution may be used. It is described below for a two-plane rotor mounted on an arbor, with Figure 6-38 showing a typical plot for one plane.

1. Balance arbor by itself to minimum achievable residual unbalance.
2. Mount rotor on arbor, observing prior cautionary notes concerning keyways, set screws, and fits.
3. Take unbalance readings for both planes and plot points P on separate graphs for each plane. (Only one plot for one plane is shown in Figure 6-38. The reading for this, say the left plane, is assumed to be 35 units at an angular position of 60°.)
4. Index rotor on arbor by 180°.
5. Take unbalance readings for both planes and plot them as points P′. (Reading for left plane assumed to be 31 units at an angle of 225°.)
6. Find midpoint R on line connecting points P and P′.
7. Draw line SS′ parallel to PP′ and passing through O.
8. Determine angle of OS (52° for left plane) and distance RP (32.5 units for left plane). Add correction mass of 32.5 units at 52° to rotor in left correction plane.

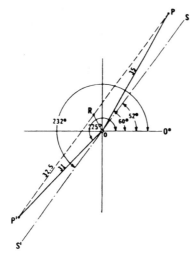

Figure 6.38. 180° indexing plot.

9. Steps 6–8 must also be performed for the right plane. The rotor is now balanced and the residual unbalance (OR in Figure 6-38) remaining in the arbor/rotor assembly is due to arbor unbalance and run-out. If this residual unbalance is corrected by adding a correction mass to the arbor equal hut opposite to OR, the arbor is corrected *and* biased for subsequent rotors of the same weight and configuration.

Additional indexing and unbalance measurement may uncover a much reduced residual unbalance, which can again be plotted and pursued through steps 4–9. This will probably refine both the balance of the rotor and the bias of the arbor. If additional indexing produces inconsistent data, the minimum level of repeatability in locating the rotor on the arbor has been reached. Further improvement in rotor balance is not possible with this arbor.

Recommended Margins Between Balance and Inspection Tolerances

Quite often the residual unbalance changes (or appears to have done so) between the time when:

1. The rotor was balanced originally.
2. It is checked for balance by an inspector.

Table 6-6
Recommended Margins Between Balance and
Inspection Tolerances

Quality Grade	Adjustment to Recommended Tolerances	
	For Balancing	For Inspection
G4000-G40	−10 percent	+10 percent
G16-G2.5	−15 percent	+15 percent
G1	−20 percent	+20 percent
G0.4	−30 percent	+30 percent

The following factors may contribute to these changes:

- Calibration differences between balancing and inspection machines
- Tooling and/or drive errors
- Bearing or journal changes
- Environmental differences (heat, humidity)
- Shipping or handling damage
- Aging or stress relieving of components

To provide a margin of safety for such changes, it is recommended that the tolerance allowed the balancing machine operator be set below, and the tolerance allowed at time of inspection be set above the values given in graphs such as Figures 6-33 and 6-34. Percentages for these margins vary between quality grades as shown in Table 6-6.

Computer-Aided Balancing

In recent years, the practice of balancing has entered into a new stage: computerization. While analog computers have been in use on hard-bearing machines ever since such machines came on the market, it is the application of digital computers to balancing that is relatively new. At first, desk top digital computers were used in large, high-speed balancing and overspeed spin test facilities for multi-plane balancing of flexible rotors. As computer hardware prices dropped, their application to more common balancing tasks became feasible. The constant demand by industry for a simpler balancing operation performed under precisely controlled conditions with complete documentation led to the marriage of the small, dedicated, table-top computer to the hard-bearing balancing machine as

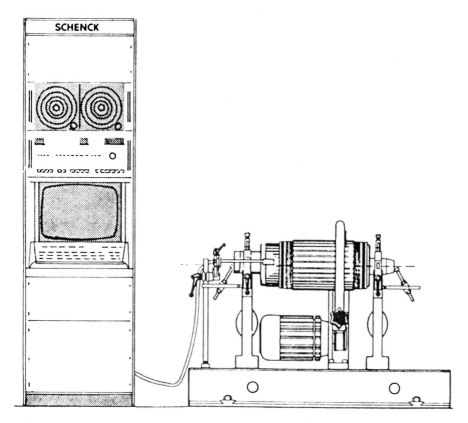

Figure 6-39. Hard-bearing balancing machine controlled by a combination of desk top computer and vectometer instrumentation.

shown in Figure 6-39. And a happy marriage it is indeed, because it proved cost effective right away in many production applications.

Features

The advantages that a computerized balancing system provides versus the customary manual system are the many standard and not-so-standard functions a computer performs and records with the greatest of ease and speed. Here is a list of basic program features and optional subroutines:

- Simplification of setup and operation
- Reduction of operator errors through programmed procedures with prompting

- More precise definition of required unbalance corrections in terms of different practical correction units
- Direct indication of unbalance in drill depth for selectable drill diameters and materials
- Averaging of ten successive readings for increased accuracy
- Automatic storage of readings taken in several runs with subsequent calculation of the mean reading (used to average the effect of blade scatter on turbine rotors with loose blades)
- Readout in any desired components (in certain workpieces, polar correction at an exact angular location is not possible. Instead, correction may only be applied at specific intervals, e.g., 30°, 45°, 90°, etc. The computer then calculates the exact correction mass to be applied at two adjacent components for any desired angle between components)
- Optimized distribution of fixed-weight correction masses to available locations
- Automatic comparison of initial unbalance with maximum permissible correction, and machine shutdown if the initial unbalance is too large
- Automatic comparison of residual unbalance with predetermined tolerances or with angle-dependent family of tolerances
- Translation of unbalance readings from one plane to another without requiring a new run (for crankshafts, for instance, where correction may only be possible at certain angular locations in certain planes)
- Permanent record of the balancing operation, including initial and residual unbalance, rotor identification, etc. (for quality control)
- Machine operation as a sorter
- Statistical analysis of unbalance
- Connection to an X-Y plotter for graphic display of unbalance
- Operator identification
- Inspector identification
- Monitoring of the number of runs and correction steps
- Time and/or date record

Prompting Guides, Storage, and Retrieval

Prompting displays on a computer screen guide the operator every step of the way through the program (Figure 6-40). Rotor data are stored electronically and can be recalled at a later date for balancing the same type of rotor. Thus ABC, R1, and R2 rotor dimensions need to be entered into the computer only once.

Balancing Program No. 002
Rotor Data

ROTOR NAME		= SAMPLE
ROTOR BALANCING SPEED		= 750 RPM
INSTRUMENT SPEED RANGE		= 390-1000 RPM
BALANCING MODE		= DYNAMIC
LEFT PLANE TO LEFT SUPPORT DISTANCE	'A'	= 4.10 INCHES
LEFT PLANE TO RIGHT PLANE DISTANCE	'B'	= 10.00 INCHES
RIGHT PLANE TO RIGHT SUPPORT DISTANCE	'C'	= 4.10 INCHES
CORRECTION RADIUS, LEFT PLANE	'R1'	= 3.50 INCHES
CORRECTION RADIUS, RIGHT PLANE	'R2'	= 3.50 INCHES
BALANCING TOLERANCE, LEFT PLANE		= 20.00 GRAM-INCHES
BALANCING TOLERANCE, RIGHT PLANE		= 20.00 GRAM-INCHES

        ROTOR S/N = 123ABC

Unbalance Data

RUN	LEFT PLANE		RIGHT PLANE	
NUMBER	AMOUNT	ANGLE	AMOUNT	ANGLE
INIT.UNBAL.	****76.332 G	278 DEGREES	****58.983 G	179 DEGREES
CHECK RUN 1	****19.936 G	314 DEGREES	******.118 G	*82 DEGREES
--------->	****19.936 G	314 DEGREES	RIGHT PLANE IN TOLERANCE	
CHECK RUN 2	******.852 G	186 DEGREES	******.478 G	277 DEGREES
--------->	LEFT PLANE IN TOLERANCE		RIGHT PLANE IN TOLERANCE	

```
ROTOR S/N 123ABC IS BALANCED WITHIN TOLERANCES
RESIDUAL UNBALANCE DATA:
PLANE 1 AMOUNT = 2.9819  GRAM-INCHES
PLANE 1 ANGLE  = 186 DEGREES
PLANE 2 AMOUNT = 1.6717  GRAM-INCHES
PLANE 2 ANGLE  = 277 DEGREES
```

Figure 6-40. Printout of unbalance data.

Multiple Machine Control and Programs

Different computers may be used depending on the application. They may be mounted in the balancing machine's instrumentation console, or in a central electronic data processing room. A computer may control one balancing machine, or a series of machines.

Basic computer programs are available for single and two plane balancing, field balancing, and flexible rotor balancing. Software libraries for optional subroutines are continually growing.

Of course, the user may modify the available programs to his particular requirements, write his own programs, or have them written for him. Thus, the potential applications of the computerized hard-bearing balancing machines are unlimited.

Field Balancing Overview

Once a balanced rotor has been mounted in its housing and installed in the field, it will not necessarily stay in balance forever. Corrosion, temperature changes, build-up of process material and other factors may cause it to go out of balance again and, thus, start to vibrate. However, unbalance is not the only reason for vibration. Bearing wear, belt problems, misalignment, and a host of other detrimental conditions will also cause it. In fact, experience has shown that vibration is an important indication of a machine's mechanical condition. During normal operation, properly functioning fans, blowers, motors, pumps, compressors, etc., emit a specific vibration signal, or "signature." If the signature changes, something is wrong.

Excessive vibration has a destructive effect on piping, tanks, walls, foundations, and other structures near the vibrating equipment. Operating personnel may be influenced too. High noise levels from vibration may exceed legal limitations and cause permanent hearing damage. Workers may also experience loss of balance, blurred vision, fatigue, and other discomfort when exposed to excessive vibration.

Methods of vibration detection, analysis, diagnostics, and prognosis have been described by the authors previously in detail[5]. A quick review of the hardware required to perform field balancing should therefore suffice.

Field Balancing Equipment

Many types of vibration indicators and measuring devices are available for field balancing. Although these devices are sometimes called "portable balancing machines," they never provide direct readout of amount and location of unbalance.

Basically, field balancing equipment consists of a combination of a suitable transducer and meter which provides an indication proportional to the vibration magnitude. The vibration magnitude indicated may be displacement, velocity, or acceleration, depending on the type of transducer and readout system used. The transducer can be held by an operator, or attached to the machine housing by a magnet or clamp, or permanently mounted. A probe thus held against the vibrating machine is presumed to cause the transducer output to be proportional to the vibration of the machine.

At frequencies below approximately 15 cps, it is almost impossible to hold the transducer sufficiently still by hand to give stable readings. Frequently, the results obtained depend upon the technique of the operator;

this can be shown by obtaining measurements of vibration magnitude on a machine with the transducer held with varying degrees of firmness. Transducers of this type have internal seismic mountings and should not be used where the frequency of the vibration being measured is less than three times the natural frequency of the transducer.

A transducer responds to all vibration to which it is subjected, within the useful frequency range of the transducer and associated instruments. The vibration detected on a machine may come through the floor from adjacent machines, may be caused by reciprocating forces or torques inherent in normal operation of the machine, or may be due to unbalances in different shafts or rotors in the machine. A simple vibration indicator cannot discriminate between the various vibrations unless the magnitude at one frequency is considerably greater than the magnitude at other frequencies.

The approximate location of unbalance may be determined by measuring the phase of the vibration; for instance, with a stroboscopic lamp that flashes each time the output of an electrical transducer changes polarity in a given direction. Phase also may be determined by use of a phase meter or by use of a wattmeter. Vibration measurements in one end of a machine are usually affected by unbalance vibration from the other end. To determine more accurately the size and phase angle of a needed correction mass in a given (accessible) rotor plane, three runs are required. One is the "as is" condition, the second with a test mass in one plane, the third with a test mass in the other correction plane. All data are entered into a hand-held computer and, with a few calculation steps, transformed into amount and phase angle of the necessary correction masses with two selected planes. To simplify the calculation process even further, software has recently become available which greatly facilitates single plane or multiplane field balancing.

Field Balancing Examples

As we saw, two methods are available for the systematic balancing of rotors:

- Balancing on a balancing machine
- Field balancing in the assembled state

Both methods have specific fields of application. The balancing machine is the correct answer from the technical and economic point of view for balancing problems in production. Field balancing, on the other hand, provides a practical method for the balancing of completely assembled machines during test running, assembly, and maintenance.

It is the purpose of this section to illustrate the possibilities of field balancing on the basis of three typical balancing problems. The machines chosen for these problems represent examples only and could at any time be exchanged for machines with similar rotor systems. The solutions to the problems indicated, therefore, apply equally to other machines and types of rotor not specifically mentioned here. The following classification will simplify the allocation of different types of machines to the three problem solutions:

1st Problem Solution: Machines with narrow disc-shaped rotors such as blowers, fans, grinding wheels, belt pulleys, flywheels, couplings, chucks, gear wheels, impellers, atomizer discs, etc.
2nd Problem Solution: Machines with long roll-shaped rotors such as centrifuges, paper rolls, electric motors and generators, beater shafts, machine tool spindles, grinding rolls, internal combustion engines, etc.
3rd Problem Solution: Machines having multiple bearing coupled rotors such as standing machines, twisting machines, motor generator sets, turbo generators, cardan shafts, etc.

Field balancing of very low speed rotating assemblies (cooling tower fans, etc.) may require special techniques which are not covered here. The reader should discuss special requirements with the machinery manufacturer.

First Problem: Unbalance Vibration in Blowers

Build-up on blades, corrosion, wear, and thermal loading regularly lead to unbalance in blowers. The presence of such unbalance shows itself externally in the form of mechanical vibration generated by the blower rotor and transmitted via the bearings and the frame into the foundations and finally into the environment. If the danger of this unbalance vibration is not recognized, after a very short operating period costly damage may be caused. This may frequently result in the destruction of the bearings, cracks in the bearing housing and in the air channels, damage to the foundation, and cracks in the building.

Solution: Field Balancing in One Plane

For economic reasons rebalancing of a blower should always be carried out in the assembled state. This does away with the need to disassemble the whole plant, to make available a balancing machine and to transport the blower rotor to the balancing machine. Only the electronic balancing instrument needs to be brought to where the blower is installed. The operation of the instrument and the determination of the masses required to correct the unbalance is carried out by trained personnel. Not only does this result in considerable cost savings but the method also provides a significant saving in time compared to rebalancing on a balancing machine. With careful preparation, field balancing of a blower need not take more than 30–60 minutes.

It has been shown in practice that in over 95 percent of all blowers field balancing in one plane is sufficient to reduce unbalance vibrations to permissible and safe values. The method is as follows:

1. The balancing instrument is used to measure the unbalance vibration at the bearing positions nearest to the blower rotor (Figures 6-41 and 6-42). The instrument suppresses with a high degree of separation any extraneous vibration and shows the amount and the angular position of the rotational frequency vibrations. Both measured values—also designated as "initial unbalance"—are entered on a vector diagram (Figure 6-43).
2. After bringing the rotor to a standstill, a known unbalance (calibrating mass) is applied in the vicinity of the center of gravity plane of

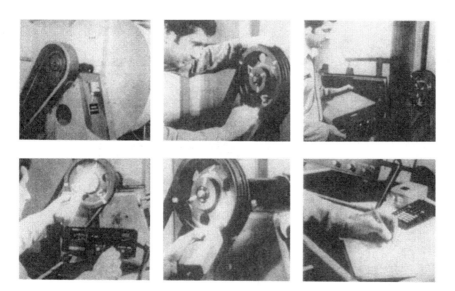

Figure 6-41. Individual stages in the field balancing of a blower using balancing and vibration analyzer "VIBROTEST."

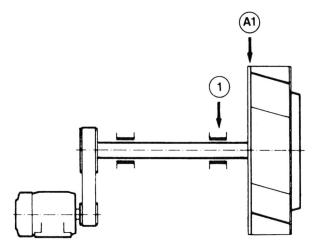

Figure 6-42. Unbalance vibration of blower is measured at bearing position (1)—"measuring point." The unbalance determined this way is corrected in the center of gravity plane A1—"correction plane."

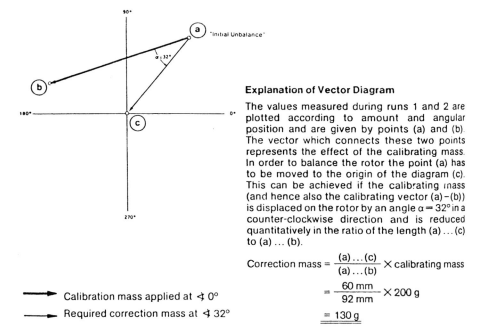

Explanation of Vector Diagram

The values measured during runs 1 and 2 are plotted according to amount and angular position and are given by points (a) and (b). The vector which connects these two points represents the effect of the calibrating mass. In order to balance the rotor the point (a) has to be moved to the origin of the diagram (c). This can be achieved if the calibrating mass (and hence also the calibrating vector (a)−(b)) is displaced on the rotor by an angle $\alpha = 32°$ in a counter-clockwise direction and is reduced quantitatively in the ratio of the length (a)...(c) to (a)...(b).

Calibration mass applied at ∢ 0°

Required correction mass at ∢ 32°

$$\text{Correction mass} = \frac{(a)...(c)}{(a)...(b)} \times \text{calibrating mass}$$

$$= \frac{60 \text{ mm}}{92 \text{ mm}} \times 200 \text{ g}$$

$$= \underline{\underline{130 \text{ g}}}$$

Figure 6-43. Vector diagram of field balancing in one plane.

the blower. When the blower has again reached its operational speed the unbalance vibration is measured again and the results are also entered in the vector diagram (Figure 6-43).

3. The graphic evaluation of this vector diagram provides amount and angular position of the correction masses required for balancing.
4. The calculated correction mass is welded to the blower rotor and a check measurement of the residual vibration is carried out. The balancing process is completed as soon as residual vibration lies within the permissible tolerance. As far as tolerance values are concerned, reference should be made to the standards of Figure 6-34.

Second Problem: Unbalance Vibration in Centrifuges

Centrifuges are high-speed machines. High rotational speeds demand a high balance quality of the rotating parts, mainly the centrifuge drum, the worm, the belt pulley, etc. Balancing of the individual rotors on a balancing machine does not always suffice to achieve the required residual unbalance. Tolerances and fits of the components, errors in the roller bearings, variations in wall thickness of the drum, etc., may mean that the unbalance vibration of the completely assembled centrifuge exceeds the permissible values. The need to correct this may arise when test running a new centrifuge and after repair and overhaul of older installations.

Solution: Field Balancing in Two Planes

Disassembly, additional machining, excess costs, and user complaints may be avoided by rebalancing on the test stand (Figure 6-44) or at the final point of installation. Because of the geometry of the centrifuge drum, field balancing in two planes is almost always necessary in order to improve the unbalance condition effectively. For this purpose the unbalance vibration is measured at two bearing positions as shown in Figure 6-45 and the unbalance determined in this way is corrected in two radial planes A1 and A2.

Measurement is carried out with a portable electronic balancing instrument that indicates the amount and angular position of the unbalance vibration for both measuring positions with frequency selectivity. For the evaluation of measured results, graphical methods have been used almost exclusively up to the present. They require experience, accuracy and time (approximately 30 minutes). The appearance of relatively inexpensive pro-

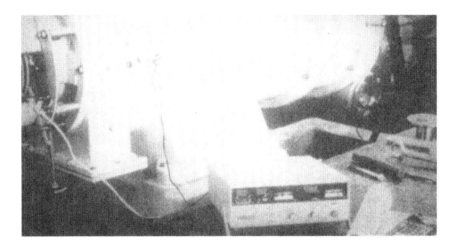

Figure 6-44. A centrifuge is rebalanced on the test stand in two planes.

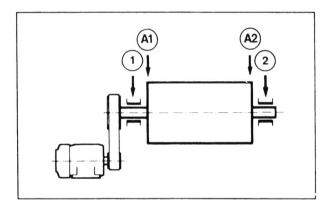

Figure 6-45. Sketch of a centrifuge. The two bearing locations (1) and (2) are chosen as measuring points. Unbalance correction is made in the end planes A1 and A2 by applying or removing mass.

grammable pocket calculators (Figure 6-46) in the late 1970s made it possible to replace these methods with more accurate numerical methods. The determination of amount and angular position of the correction masses for both correction planes could be carried out in approximately two–three minutes even by untrained personnel. In detail, the following method could be followed:

Figure 6-46. Programmable pocket calculators with balancing module make it easier and faster to determine the correction masses when field balancing.

1. Using the balancing instrument, the angular position and amount of the unbalance vibration is measured at bearing positions 1 and 2 and the values are entered into the schedule "initial unbalance run" (Figure 6-47).
2. The centrifuge is brought to a standstill and a known calibrating mass is applied in correction plane A1. After again reaching the operational speed the unbalance vibration is measured again and entered "test run 1".
3. The calibrating mass is removed from plane A1 and applied to plane A2. The resulting measured values are again noted down "test run 2."
4. The evaluation of the measurement results listed in Figure 6-47 using the pocket calculator gives the correction masses that must be applied at the calculated angular positions. A subsequent check run of the centrifuge will determine the correctness of the balancing measures and will show whether an additional correction process is required.

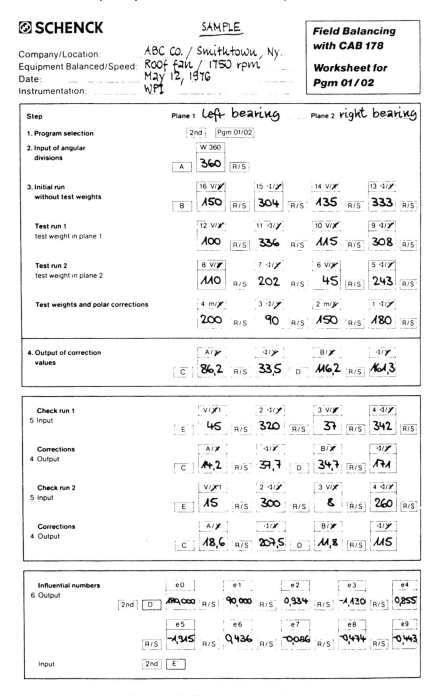

Figure 6-47. Field balancing worksheet.

Explanation of Schedule and of Calculator Program

The results of the initial unbalance run, of both test runs, and the magnitude of the calibrating masses used are entered in appropriately numbered spaces. After inserting balancing module, the measured values are keyed into the pocket calculator. By calling up the stored data, the pocket calculator immediately indicates the required masses for unbalance correction either in polar form or in the form of 90° components.

If the residual unbalance of the rotor exceeds the allowable tolerance, it is possible to calculate the correction masses for further correction by using the measured values of the check run but without any need for new test runs. The influence coefficients which may on demand be indicated and noted make it possible to rebalance a rotor without test runs even after a long time interval.

Third Problem: Unbalance Vibration in Twisting and Stranding Machine

Machines for the production of wire rope, cable and flex operate with multi-bearing rotor systems which consist of two or more part rotors coupled together with angular rigidity. A type which frequently occurs in practice is shown schematically in Figure 6-48. Rotor systems of this type are difficult to balance in their completely assembled form on balancing machines but are better balanced divided into their individual rotors. After assembling the balanced component rotors, new unbalances can occur due to fits and tolerances, alignment errors and centrifugal force loading. This is also the case when replacing rotating wear parts such as, for example, the wire guide tube in stranding machines.

Any excessively large residual unbalance leads to considerable mechanical vibration and to the excitation of mounting and machine resonances. Both of these factors can lead to damage of the machine and physical and psychological strain on the operating personnel. Frequently the only first aid measure available is a reduction in the production rate by reducing the operational speed. This, however, is only tolerable over a limited time span. The economics of the process require a longer term solution that can only be found in field balancing the complete rotor system.

Solution: Field Balancing in Several Planes

The unbalance and vibrational behavior of multiple bearing rotor systems may be improved in a systematic manner by multi-plane balanc-

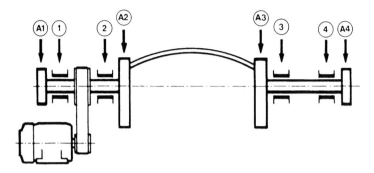

Figure 6-48. Sketch of the rotating parts of a stranding machine. Each of the measuring positions (1) to (4) is related to a corresponding correction plane A1 to A4.

ing in the assembled state (Figure 6-48). For unbalance correction, there must be a correction plane for every bearing position on the rotor. For example, the four-bearing rotor shown in Figure 6-48 requires four correction planes.

The electronic balancing instrument for this task is identical to the instruments used for single and two plane balancing. It is resting on top of the bunching machine being shown, during field balancing, in Figure 6-49. The balancing process only differs in the number of calibrating runs and the way the measured results are evaluated:

1. Using the portable balancing instrument the phase position and amount of the unbalance vibrations is measured at the four bearing positions of the stranding machine. These "initial unbalance values" are entered on four vector diagrams similar to that in Figure 6-50.
2. The stranding machine is switched off and a known calibrating mass is applied in the correction plane A1 (Figure 6-48). After running up to the operational speed the unbalance values are again measured at the four bearing positions. These values are also entered in the vector diagram.
3. The procedure described under 2 above is repeated with calibrating masses in the correction planes A2, A3, and A4.
4. The graphical evaluation of the vector diagram gives the amount and angular position of the correction masses which must be applied in the four chosen correction planes.

Figure 6-49. Field balancing of a bunching machine. Using the balancing and vibration analyzer "VIBROTEST," the unbalance vibration in the four bearing planes is measured successively in terms of amount and angular position.

5. After unbalance correction a check measurement is carried out and the residual vibrations determined by this are compared with the permissible tolerance values.

The Vector Diagram

The results from the five measuring runs are entered in four vector diagrams. Figure 6-50 shows the vector diagram for plane 1. The vector (a) . . . (b) represents the effect of the calibration mass. The other three vectors show the effect which the calibration masses applied in planes A1, A3, and A4 exert in plane 1.

The graphic evaluation of the vector diagram is carried out by an approximation method. By means of a systematic trial and error process, combinations of masses are determined which result in a reduction of unbalance in all four planes. For this purpose the effects and influences of the correction masses in all planes must be taken into account and entered into the vector diagram.

Extract from the Measuring Schedule

Run No	Observations	Unbalance Vibrations							
		Plane 1		Plane 2		Plane 3		Plane 4	
		Amount	∡	Amount	∡	Amount	∡	Amount	∡
1	Initial unbalance	380	105°	255	126°	210	243°	340	276°
2	Calibrating mass m = 350g at ∡ 0° applied in plane A1	330	31°	160	103	235	220	310	270°
3	Calibrating mass m = 350g at 0° applied in plane A2	415	72°	340	47°	190	262°	330	292°

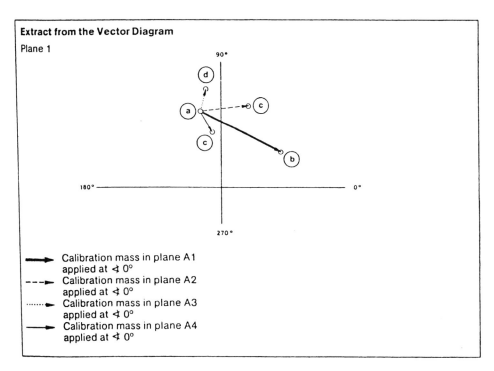

Extract from the Vector Diagram

Plane 1

→ Calibration mass in plane A1
applied at ∡ 0°
---→ Calibration mass in plane A2
applied at ∡ 0°
······→ Calibration mass in plane A3
applied at ∡ 0°
——→ Calibration mass in plane A4
applied at ∡ 0°

Figure 6-50. Field balancing in four planes. Example of graphic evaluation of measured results.

References

1. Schenck Trebel Corporation, *Fundamentals of Balancing*, Schenck Trebel Corporation, Deer Park, New York, second edition, 1983, 115 pages.
2. Schenck Trebel Corporation, *Aspects of Flexible Rotor Balancing*, Schenck Trebel Corporation, Deer Park, New York, third edition, 1980, Pages 1–34.
3. Reference 1, Pages 35–42.
4. Schneider, H., *Balancing Technology*, Schenck Trebel Corporation, Deer Park, New York.
5. Bloch, H. P. and Geitner, F. K., *Machinery Failure Analysis and Troubleshooting*, Gulf Publishing Co., Houston, Texas, third edition, 1997, Pages 351–433.

Bibliography

1. ISO 1925: *Balancing—Vocabulary* (1981).
2. ISO 1940: *Balance Quality of Rigid Rotors* (1973) and DIS 2953 1982 with revised test procedure.
3. ISO 2953: *Balancing Machines—Description and Evaluation* (1973).
4. Society of Automotive Engineers, Inc. ARP 587A: *Balancing Equipment for Jet Engine Components, Compressors and Turbines, Rotating Type, for Measuring Unbalance in One or More Than One Transverse Planes* (1972).
5. Society of Automotive Engineers, Inc. ARP 588A: *Static Balancing Equipment for Jet Engine Components, Compressor and Turbine, Rotating Type, for Measuring Unbalance in One Transverse Plane* (1972).
6. Society of Automotive Engineers, Inc. ARP 1382: *Design Criteria for Balancing Machine Tooling* (1977).
7. Federn, Klaus: *Auswuchttechnik* Volume 1, Springer Verlag, Berlin (1977).
8. Luhrs, Margret H.: *Computers in the Balancing Industry* (December, 1979).
9. McQueary, Dennis: *Understanding Balancing Machines*, American Machinist (June, 1973).
10. Meinhold, Ted F.: *Measuring and Analyzing Vibration*, Plant Engineering (October 4, 1979).
11. Muster, Douglas and Stadelbauer, Douglas G.: *Balancing of Rotating Machinery*, Shock and Vibration Handbook, second edition, McGraw-Hill, New York (1976).

12. Rieger, Neville F. and Crofoot, James F.: *Vibrations of Rotating Machinery*, The Vibration Institute, Clarendon Hills, Illinois (1977).
13. Schenck Trebel Corporation: *Aspects of Flexible Rotor Balancing*, Company Publication (1976).
14. Schenck Trebel Corporation: *Theory of Balancing*, Company Publication (1973).
15. Schneider, Hatto: *Balancing Technology*, VDI Publication T29, second edition (1977), distributed by Schenck Trebel Corporation.
16. Stadelbauer, D. G.: *Balancing of Fans and Blowers*, Vibration and Acoustic Measurement Handbook, Spartan Books (1972).
17. Stadelbauer, D. G.: *Balancing Machines Reviewed*, Shock and Vibration Digest, Volume 10, No. 9 (September, 1978).
18. ISO DIS 7475: *Enclosures and Other Safety Measures for Balancing Machines.*

Appendix 6-A

Balancing Terminology

NOTE: The definitions followed by a number are taken from ISO 1925 2nd Edition.

Amount of Unbalance (3.3): The quantitative measure of unbalance in a rotor (referred to a plane), without referring to its angular position. It is obtained by taking the product of the unbalance mass and the distance of its center of gravity from the shaft axis.

Angle of Unbalance (3.4): Given a polar coordinate system fixed in a plane perpendicular to the shaft axis and rotating with the rotor, the polar angle at which an unbalance mass is located with reference to the given co-ordinate system.

Angle Indicator (5.18): The device used to indicate the angle of unbalance.

Angle Reference Generator (5.19): In balancing, a device used to generate a signal which defines the angular position of the rotor.

Angle Reference Marks (5.20): Marks placed on a rotor to denote an angle reference system fixed in the rotor; they may be optical, magnetic, mechanical, or radioactive.

Balancing (4.1): A procedure by which the mass distribution of a rotor is checked and, if necessary, adjusted in order to ensure that the vibration of the journals and/or forces on the bearings at a frequency corresponding to service speed are within specified limits.

Balancing Machine (5.1): A machine that provides a measure of the unbalance in rotor which can be used for adjusting the mass distribution of that rotor mounted on it so that once per revolution vibratory motion of the journals or force on the bearings can be reduced if necessary.

Balancing Machine Accuracy (5.24): The limits within which a given amount and angle of unbalance can be measured under specified conditions.

351

Balancing Machine Minimum Response (5.23): The measure of the machine's ability to sense and indicate a minimum amount of unbalance in terms of selected components of the unbalance vector.

Balancing Machine Sensitivity (5.28): Of a balancing machine under specified conditions, the increment in unbalance indication expressed as indicator movement or digital reading per unit increment in the amount of unbalance.

Balancing Run (5.43): On a balancing machine: A run consisting of one measure run and the associated correction process.

Balancing Speed (2.18): The rotational speed at which a rotor is balanced.

Calibration (5.36): The process of adjusting a machine so that the unbalance indicator(s) read(s) in terms of selected correction units in specified correction planes for a given rotor and other essentially identical rotors; it may include adjustment for angular location if required.

Centrifugal (Rotational) Balancing Machine (5.3): A balancing machine that provides for the support and rotation of a rotor and for the measurement of once per revolution vibratory forces or motions due to unbalance in the rotor.

Compensating (Null Force) Balancing Machine (5.9): A balancing machine with a built-in calibrated force system which counteracts the unbalanced forces in the rotor.

Center of Gravity (1.1): The point in a body through which passes the resultant of the weights of its component particles for all orientations of the body with respect to a uniform gravitational field.

Component Measuring Device (5.22): A device for measuring and displaying the amount and angle of unbalance in terms of selected components of the unbalance vector.

Correction Plane Interference (Cross-Effect) (5.25): The change of balancing machine indication at one correction plane of a given rotor, which is observed for a certain change of unbalance in the other correction plane.

Correction Plane interference Ratios (5.26): The interference ratios (I_{AB}, I_{BA}) of two correction planes A and B of a given rotor are defined by the following relationships:

$$I_{AB} = \frac{U_{AB}}{U_{BB}}$$

Calibration Mass: A precisely defined mass used (a) in conjunction with a proving rotor to calibrate a balancing machine, and (b) on the first rotor of a kind to calibrate a soft-bearing balancing machine for that particular rotor and subsequent identical rotors.

Calibration (Master) Rotor (5.35): A rotor (usually the first of a series) used for the calibration of a balancing machine.

Correction Mass: A mass attached to a rotor in a given correction plane for the purpose of reducing the unbalance to the desired level.

Correction Mass Set: A number of precisely apportioned correction masses used for correcting a given unbalance in (a) a single plane or (b) more than one plane. For flexible rotor balancing the number of correction masses in a set is usually related to the flexural mode they are intended to correct.

Counterweight (5.16): A weight added to a body so as to reduce a calculated unbalance at a desired place.

Claimed Minimum Achievable Residual Unbalance (5.41): A value of minimum achievable residual unbalance stated by the manufacturer for his machine, and measured in accordance with the procedure specified in ISO 2953.

Couple Unbalance (3.8): That condition of unbalance for which the central principal axis intersects the shaft axis at the center of gravity.

Component Correction: Correction of unbalance in a correction plane by mass addition or subtraction at two or more of a predetermined number of angular locations.

Correction (Balancing) Plane (4.6): A plane perpendicular to the shaft axis of a rotor in which correction for unbalance is made.

Cycle Rate: Cycle rate of a balancing machine for a given rotor having a specified polar moment of inertia and for a given balancing speed is the number of starts and stops that the machine can perform per hour without damage to the machine when balancing the rotor.

Differential Test Masses: Two masses representing different amounts of unbalance added to a rotor in the same transverse plane at diametrically opposed positions.

NOTE

1. Differential test masses are used, for example, in cases where a single test mass is impractical.
2. In practice, the threaded portion and height of the head of the test mass are kept the same. The diameter of the head is varied to achieve the difference in test mass.
3. The smaller one of the two differential test masses is sometimes called tare mass, the larger one tare-delta mass.

Differential Unbalance: The difference in unbalance between two differential test masses.

Dynamic Unbalance (3.9): That condition in which the central principal axis is not coincident with the shaft axis.

NOTE

The quantitative measure of dynamic unbalance can be given by two complementary unbalance vectors in two specified planes (perpendicular to the shaft axis) which completely represent the total unbalance of the rotor.

Dummy Rotor: A balancing fixture of the same mass and shape as the actual rotor it replaces.

NOTE

A dummy rotor is used when balancing one or more rotors of an assembly of rotors, simulating the mass of the missing rotor.

Field Balancing (4.11): The process of balancing a rotor in its own bearings and supporting structure rather than in a balancing machine.

NOTE

Under such conditions, the information required to perform balancing is derived from measurements of vibratory forces or motions of the supporting structure and/or measurements of other responses to rotor unbalance.

Index Balancing: A procedure whereby a component is repetitively balanced and then indexed by 180° on an arbor or rotor shaft.

NOTE

After each indexing, half of the residual unbalance is corrected in the arbor (or rotor shaft), the other half in the indexed component.

Indexing: Incremental rotation of a rotor about its journal axis for the purpose of bringing it to a desired position.
Mass Centering: The process of determining the mass axis of a rotor and then machining journals, centers or other reference surfaces to bring the axis of rotation defined by these surfaces into close proximity with the mass axis.
Measuring Plane (4.7): A plane perpendicular to the shaft axis in which the unbalance vector is determined.
Method of Correction (4.5): A procedure whereby the mass distribution of a rotor is adjusted to reduce unbalance, or vibration due to unbalance, to an acceptable value. Corrections are usually made by adding material to, or removing it from, the rotor.

Multi-Plane Balancing (4.4): As applied to the balancing of flexible rotors, any balancing procedure that requires unbalance correction in more than two correction planes.

Outboard Rotor (2.12): A two-journal rotor which has its center of gravity located other than between the journals.

Overhung Rotor: A two-journal rotor with inboard CG but with significant masses and at least one correction plane located other than between the journals.

Perfectly Balanced Rotor (2.10): A rotor the mass distribution of which is such that it transmits no vibratory force or motion to its bearings as a result of centrifugal forces.

Plane Translation: The process of converting a given amount and angle of unbalance in two measuring (or correction) planes into the equivalent unbalance in two other planes.

Polar Correction: Correction of unbalance in a correction plane by mass addition or subtraction at a single angular location.

Parasitic Mass (5.31): Of a balancing machine, any mass, other than that of the rotor being balanced, that is moved by the unbalance force(s) developed in the rotor.

Permanent Calibration (5.33): The property of a hard-bearing balancing machine that permits the machine to be calibrated once and for all, so that it remains calibrated for any rotor within the capacity and speed range of the machine.

Plane Separation (5.27): Of a balancing machine, the operation of reducing the correction plane interference ratio for a particular rotor.

Plane Separation (Nodal) Network (5.30): An electrical circuit, interposed between the motion transducers and the unbalance indicators, that performs the plane-separation function electrically without requiring particular locations for the motion transducers.

Practical Correction Unit (5.15): A unit corresponding to a unit value of the amount of unbalance indicated on a balancing machine. For convenience, it is associated with a specific radius and correction plane; and is commonly expressed as units of an arbitrarily chosen quantity such as drill depths of given diameter, weight, lengths of wire solder, plugs, wedges, etc.

Production Rate (5.45): The reciprocal of floor-to-floor time.

NOTE

The time is normally expressed in pieces per hour.

Proving (Test) Rotor (5.32): A rigid rotor of suitable mass designed for testing balancing machines and balanced sufficiently to permit the introduction of exact unbalance by means of additional masses with high reproducibility of the magnitude and angular position.

Quasi-Rigid Rotor (2.17): A flexible rotor that can be satisfactorily balanced below a speed where significant flexure of the rotor occurs.

Rigid Rotor (2.2): A rotor is considered rigid when it can be corrected in any two (arbitrarily selected) planes and, after that correction, its unbalance does not significantly exceed the balance tolerances (relative to the shaft axis) at any speed up to maximum service speed and when running under conditions which approximate closely to those of the final supporting system.

Rotor (2.1): A body, capable of rotation, generally with journals which are supported by bearings.

Resonance Balancing Machine (5.7): A machine having a balancing speed corresponding to the natural frequency of the suspension-and-rotor system.

Setting (5.37): Of a hard-bearing balancing machine, the operation of entering into the machine information concerning the location of the correction planes, the location of the bearings, the radii of correction, and the speed if applicable.

Single-Plane (Static) Balancing Machine (5.4): A gravitational or centrifugal balancing machine that provides information for accomplishing single-plane balancing.

Soft-Bearing (Above Resonance) Balancing Machine (5.8): A machine having a balancing speed above the natural frequency of the suspension-and-rotor system.

Swing Diameter (5.11): The maximum workpiece diameter that can be accommodated by a balancing machine.

Specific Unbalance "e" (Mass Eccentricity) (3.17): The amount of static unbalance (U) divided by mass of the rotor (M); it is equivalent to the displacement of the center of gravity of the rotor from the shaft axis.

Static Unbalance (3.6): That condition of unbalance for which the central principal axis is displaced only parallel to the shaft axis.

Service Speed (2.19): The rotational speed at which a rotor operates in its final installation or environment.

Shaft Axis (2.7): The straight line joining the journal centers.

Slow Speed Runout. The total indicated runout measured at a low speed (i.e., a speed where no significant rotor flexure occurs due to unbalance) on a rotor surface on which subsequent measurements are to be made at a higher speed where rotor flexure is expected.

Two-Plane (Dynamic) Balancing (4.3): A procedure by which the mass distribution of a rigid rotor is adjusted in order to ensure that the residual dynamic unbalance is within specified limits.

Test Plane: A plane perpendicular to the shaft axis of a rotor in which test masses may be attached.

Trial Mass: A mass selected arbitrarily and attached to a rotor to determine rotor response.

NOTE

A trial mass is usually used in trial and error balancing or field balancing where conditions cannot be precisely controlled and/or precise measuring equipment is not available.

Test Mass: A precisely defined mass used in conjunction with a proving rotor to test a balancing machine.

NOTES

1. The use of the term "test weight" is deprecated; the term "test mass" is accepted in international usage.
2. The specification for a precisely defined Test Mass shall include its mass and its center of gravity location; the aggregate effect of the errors in these values shall not have a significant effect on the test results.

Traverse (U_{mar}) Test (5.46): A test by which the residual unbalances of a rotor can be found (see ISO 1940 or ISO 2953), or with which a balancing machine may be tested for conformance with the claimed minimum achievable unbalance (U_{mar}, see ISO 2953).

Turn-Around Error: Unbalance indicated after indexing two components of a balanced rotor assembly in relation to each other; usually caused by individual component unbalance, run-out of mounting (locating) surfaces, and/or loose fits. (See also index balancing).

Unbalance (3.1): That condition which exists in a rotor when vibratory force or motion is imparted to its bearings as a result of centrifugal forces.

NOTES

1. The term "unbalance" is sometimes used as a synonym for "amount of unbalance," or "unbalance vector."
2. Unbalance will in general be distributed throughout the rotor but can be reduced to
   a. static unbalance and couple unbalance described by three unbalance vectors in three specified planes, or
   b. dynamic unbalance described by two unbalance vectors in two specified planes.

Unbalance Mass (3.5): That mass which is considered to be located at a particular radius such that the product of this mass and its centripetal acceleration is equal to the unbalance force.

NOTE

The centripetal acceleration is the product of the distance between the shaft axis and the unbalance mass and the square of the angular velocity of the rotor, in radians per second.

Unbalance Reduction Ratio (URR) (5.34): The ratio of the reduction in the unbalance by a single balancing correction to the initial unbalance.

$$\text{URR} = \frac{U_1 - U_2}{U_1} = 1 - \frac{U_2}{U_1}$$

where

U_1 is the amount of initial unbalance;

U_2 is the amount of unbalance remaining after one balancing correction.

Vector Measuring Device (5.21): A device for measuring and displaying the amount and angle in terms of an unbalance vector, usually by means of a point of line.

Vertical Axis Freedom (5.47): The freedom of a horizontal balancing machine bearing carriage or housing to rotate by a few degrees about the vertical axis through the center of the support.

NOTE

This feature is required when dynamic or couple unbalance is to be measured in a rotor supported on sleeve bearings, cylindrical roller bearings, flat twin rollers, or in cradles, stators or tiebars.

Appendix 6-B

Balancing Machine Nomenclature

V-type Headstock With End-Drive And Bed

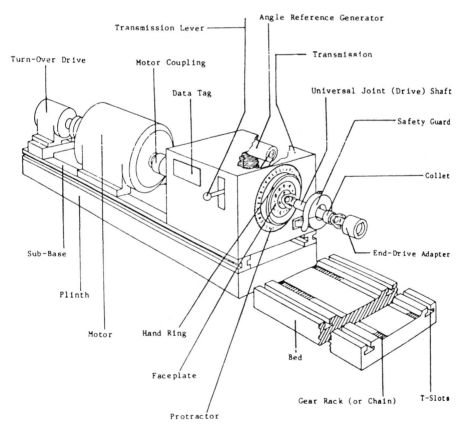

6-B-1

Support

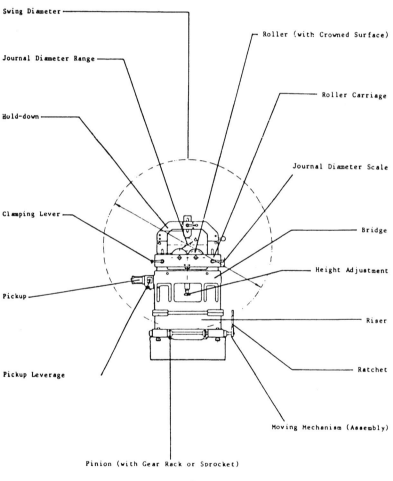

Swing Diameter

Roller (with Crowned Surface)

Journal Diameter Range

Roller Carriage

Hold-down

Journal Diameter Scale

Clamping Lever

Bridge

Height Adjustment

Pickup

Riser

Pickup Leverage

Ratchet

Moving Mechanism (Assembly)

Pinion (with Gear Rack or Sprocket)

6-B-2

Accessories

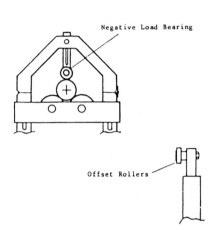

Negative Load Bearing

Offset Rollers

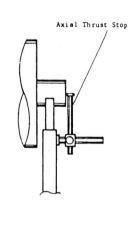

Axial Thrust Stop

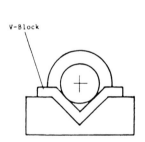

V-Block

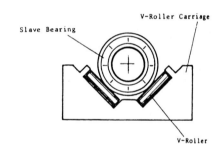

Slave Bearing

V-Roller Carriage

V-Roller

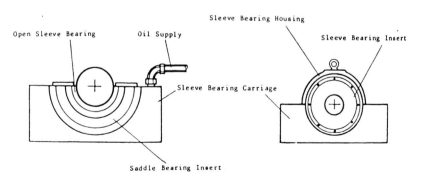

Sleeve Bearing Housing

Open Sleeve Bearing

Oil Supply

Sleeve Bearing Insert

Sleeve Bearing Carriage

Saddle Bearing Insert

6-B-3

Tooling

Tiebar Frame

Tiebar

Tiebar Arm

Shroud (Clam-Shell Type)

Cradle

Stator

Mandrel (Balancing Arbor)

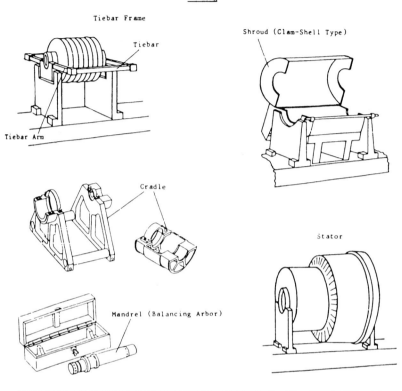

Belt and Other Drives

—Underslung Belt-Drive—   —Tangential Belt-Drive—   —Friction Roller Drive—

Belt-Driven Surface

Tensioning Pulley

Belt-Drive Bracket

Belt

Idler Pulley

Belt Tension Adjustment

Rotor

Motor Pulley

Friction Roller Pressure Adjustment

Friction Drive Arm

Friction Roller

Friction Drive Bracket

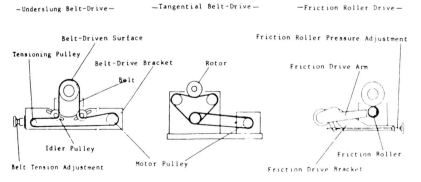

Appendix 6-C

Balancing and Vibration Standards

Balancing Standards

ISO 1925 *Balancing Vocabulary.* Contains definitions of most balancing and related terms. (Same as ANSI S2.7-1982.)

ISO 1940 *Balance Quality of Rotating Rigid Bodies.* Classifies all rigid rotors and recommends balance tolerances for them. (Same as ANSI S2.19-1975.)

ISO 3080 *The Mechanical Balancing of Marine Steam Turbine Machinery for Merchant Service.* Furnishes guidance in applying ISO 1940 to this type of rotor.

ISO 2371 *Field Balancing Equipment—Description and Evaluation.* Recommends to the equipment manufacturer how to describe his systems, and to the user how to evaluate them. (Same as ANSI S2.38-1982.)

ISO 2953 *Balancing Machines—Description and Evaluation.* Tells a prospective balancing machine user how to describe his requirements to a balancing machine manufacturer, then enumerates the points that a proposal should cover, and finally explains how to test a machine to assure compliance with the specification.

ISO 5406 *The Mechanical Balancing of Flexible Rotors.* Classifies rotors into groups in accordance with their balancing requirements and gives guidance on balancing procedures for flexible rotors. (Same as ANSI S2.42.)

ISO 5343 *Criteria for Evaluating Flexible Rotor Unbalance.* Recommends balance tolerances for flexible rotors. Must be read in conjunction with ISO 1940 and ISO 5406.

ISO 3719*¹   *Balancing Machines—Symbols for Front Panels.* Establishes symbols for control panels of balancing machines.

DIS 7475*²   *Enclosures and Other Safety Measures for Balancing Machines.* Identifies hazards associated with spinning rotors in balancing machines, classifies enclosures, and specifies protection requirements.

Vibration Standards

ISO 2041   *Vibration and Shock Vocabulary.* Contains definitions of most vibration and shock related terms.

ISO 2372   *Mechanical Vibration of Machines with Operating Speeds from 10 to 200 Rev/s.* Basis for specifying evaluation standards.

ISO 2373   *Mechanical Vibration of Certain Rotating Electrical Machinery with Shaft Heights Between 80 and 400 mm.* Measurement and evaluation of vibration severity.

ISO 2954   *Mechanical Vibration of Rotating and Reciprocating Machinery.* Requirements for instruments for measuring vibration severity.

ISO 3945   *Mechanical Vibration of Large Rotating Machines with Speed Range from 10 to 200 RPS.* Measurement and evaluation of vibration severity in situ.

US National Standards

ANSI S2.l9-1975:*³   Balance Quality of Rotating Rigid Bodies (Identical to ISO 1940-1973)

ANSI S2.l7-1980:   Techniques of Machinery Vibration Measurement

ANSI S2.7-1982:   Balancing Terminology (Identical to ISO 1925-1981)

ANSI S2.38-1982:   Field Balancing Equipment—Description and Evaluation (Identical to ISO 2371-1982)

ANSI S2.42-1982:   Procedure for Balancing Flexible Rotors (Identical to ISO 5406-1980)

All standards (ISO as well as ANSI) may be ordered from:
American National Standards Institute, Inc.
1430 Broadway
New York, NY 10018

---

*¹ISO = International Standard Organization, Geneva, Switzerland.
*²Draft International Standard. to be released 1983.
*³ANSI = American National Standards Institute.

For Jet Engine Balancing:

May be ordered from:	Society of Automotive Engineers, Inc. (SAE) 400 Commonwealth Drive Warrendale, PA 15096 Tel. (412) 776-4841
SAE ARP 587A	Balancing Equipment for Jet Engine Components, Compressors and Turbines, Rotating Type, for Measuring Unbalance in One or More Than One Transverse Planes. Contains machine and proving rotor parameters and the all important SAE acceptance test for horizontal machines.
SAE ARP 588A	Static Balancing Equipment for Jet Engine Components, Compressor and Turbine, Rotating Type, for Measuring Unbalance in One Transverse Plane. Contains machine and proving rotor parameters and acceptance test for vertical machines.
SAE ARP 1340	Periodic Surveillance Procedures for Horizontal Dynamic Balancing Machines. An abbreviated test that may be run periodically to assure proper machine function.
SAE ARP 1342	Periodic Surveillance Procedures for Vertical, Static Balancing Machines. An abbreviated test that may be run periodically to assure proper machine function.
SAE ARP 1382	Design Criteria for Balancing Machine Tooling. Describes rotor supports, cradles, arbors, shrouds and other typical accessories for horizontal and vertical balancing machines. Also useful for general balancing work.
SAE ARP 1202	Bell Type Slave Bearings for Rotor Support in Dynamic Balancing Machines. Specifies dimensions and tolerances for special balancing bearings.
SAE ARP 1134	Adapter Interface—Turbine Engine Blade Moment Weighing Scale. Standardizes adapter tooling interface for blade moment weighing scales.
SAE ARP 1136	Balance Classification of Turbine Rotor Blades. Standardizes blade data and markings for classifying moment weight.

Appendix 6-D

Critical Speeds of Solid and Hollow Shafts

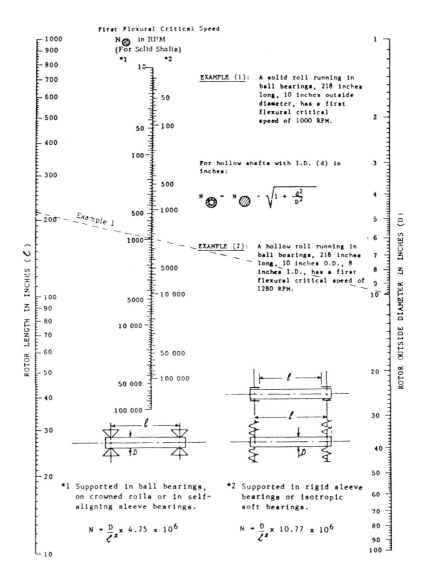

First Flexural Critical Speed

N in RPM
(For Solid Shafts)

EXAMPLE (1): A solid roll running in ball bearings, 218 inches long, 10 inches outside diameter, has a first flexural critical speed of 1000 RPM.

For hollow shafts with I.D. (d) in inches:

$$N_{\text{hollow}} = N_{\text{solid}} \cdot \sqrt{1 + \frac{d^2}{D^2}}$$

EXAMPLE (2): A hollow roll running in ball bearings, 218 inches long, 10 inches O.D., 8 inches I.D., has a first flexural critical speed of 1280 RPM.

ROTOR LENGTH IN INCHES (ℓ)

ROTOR OUTSIDE DIAMETER IN INCHES (D)

Example 1

*1 Supported in ball bearings, on crowned rolls or in self-aligning sleeve bearings.

$$N = \frac{D}{\ell^2} \times 4.75 \times 10^6$$

*2 Supported in rigid sleeve bearings or isotropic soft bearings.

$$N = \frac{D}{\ell^2} \times 10.77 \times 10^6$$

366

Part III

Maintenance and Repair of Machinery Components

Chapter 7

Ball Bearing Maintenance and Replacement

The fundamental purpose of a bearing is to reduce friction and wear between rotating parts that are in contact with one another in any mechanism. The length of time a machine will retain its original operating efficiency and accuracy will depend upon the proper selection of bearings, the care used while installing them, proper lubrication, and proper maintenance provided during actual operation.

The manufacturer of the machine is responsible for selecting the correct type and size of bearings and properly applying the bearings in the equipment. However, maintenance of the machine is the responsibility of the user. A well-planned and systematic maintenance procedure will assure extended operation of the machine. Failure to take the necessary precautions will generally lead to machine downtime. It must also be remembered that factors outside of the machine shaft may cause problems.

Engineering and Interchangeability Data

Rings and Balls—The standard material used in ball bearing rings and balls is a vacuum processed high chromium steel identified as SAE 52100 or AISI-52100. Material quality for balls and bearing rings is maintained by multiple inspections at the steel mill and upon receipt at the bearing manufacturing plants. The 52100 bearing steel with standard heat treatment can be operated satisfactorily at temperatures as high as 250°F

(121°C). For higher operating temperatures, a special heat treatment is required in order to give dimensional stability to the bearing parts.

Seals—Standard materials used in bearing seals are generally nitrile rubber. The material is bonded to a pressed steel core or shield. Nitrile rubber is unaffected by any type of lubricant commonly used in anti-friction bearings. These closures have a useful temperature range of −70° to +225°F (−56° to 107°C). For higher operating temperatures, special seals of high temperature materials can be supplied.

Ball Cages—Ball cages are pressed from low carbon steel of SAE 1010 steel. This same material is used for bearing shields. Molded nylon cages are now available for many bearing sizes. The machined cages ordinarily supplied in super-precision ball bearings are made from laminated cotton fabric impregnated with a phenolic resin. This type of cage material has an upper temperature limit of 225°F (107°C) with grease and 250°F (121°C) with oil for extended service. For periods of short exposure, higher temperatures can be tolerated.

Lubricant—Prelubricated bearings are packed with an initial quantity of high quality grease which is capable of lubricating the bearing for years under certain operating conditions. As a general rule, standard greases will yield satisfactory performance at temperatures up to 175°F (79°C), as long as proper lubrication intervals and lube quantities are observed. Special greases are available for service at much higher temperatures. Estimation of grease life at elevated temperatures involves a complex relationship of grease type, bearing size, speed, and load. Volume 4 of this series can provide some guidance, although special problems are best referred to the product engineering department of major bearing manufacturers.

Standardization

Bearing envelope dimensions and tolerances shown in this chapter are based on data obtained from MRC/TRW Bearing Division. They comply with standards established in the United States by the Annular Bearing Engineers' Committee (ABEC) of the Anti-Friction Bearing Manufacturers Association (AFBMA). These standards have also been approved by the American Standards Association (ASA) and the International Standards Organization (ISO). This assures the bearing user of all the advantages of dimensional standardization. However, dimensional interchangeability is not necessarily an indication of functional interchangeability. Cage type, lubricant grade, internal fitting practice, and many other details are necessary to establish complete functional interchangeability.

Ball Bearing Variations

Special purpose bearings are generally one of the types shown in Table 7-1 but with special features as noted. For ease of reference we are including Table 7-2, "Commonly Used MRC Bearing Symbols," and Table 7-3, "Ball Bearing Interchange Table."

(Text continued on page 376)

Table 7-1
Special Purpose Bearings

Bearing Type	Description	Load Capability	Relative Capacity
Type S Conrad Deep Groove	Ordinarily supplied with loose internal clearance. Other degrees of internal clearance may be necessary for special conditions. Outer and inner races have full depth on each side. Ball assembly is made by eccentric displacement of rings.	Has equal load-carrying capacities in either direction. Recommended for moderately heavy radial loads, thrust loads in either direction, or combination loads.	
Type M Maximum-Capacity Filling Notch	Ordinarily supplied with standard internal clearance. Other degrees of internal clearance may be necessary for special conditions. This type has a filling notch on one side of the outer and inner rings to insert full quota of balls.	Can carry heavier radial loads than Type S and heavier combined radial and thrust loads. Recommended for a majority of heavily loaded bearing positions — heavy industrial machinery, farm equipment, tractors and trucks.	
Type R Angular-Contact Counterbored Outer Ring	Supplied with loose internal clearance for normal applications. Other degrees of internal clearance may be necessary for special conditions. Counterbored outer ring. To assemble full quota of balls, rings are expanded by heating. Non-separable type	Higher radial load-carrying capacity than Type S. Heavy thrust load in one direction only, or combined loads where radial load predominates and thrust is always against heavy shoulder of outer ring	
7000 Series Angular-Contact Counterbored Outer Ring	Counterbored outer ring, non-separable type with initial contact angle of 29°. Two-piece steel cage for normal use or one-piece non-metallic or solid bronze cage for high speed, high operating temperature or severe vibration applications.	Very high thrust load in one direction, combined radial and thrust load where thrust load predominates.	
7000-P Series Angular-Contact Heavy Duty Type	Similar in design to 7000 Series Bearings but 40° contact angle. Ball complement and race groove depth designed for increased thrust capacity which varies with individual sizes depending on change in ball complement.	Capacity is 1.18 to 1.40 times that of 7000 Series, varies with individual sizes. Restricted to primarily thrust loads. Should not be used for radial loads only or combined radial and thrust loads where radial load is predominant	
9000 Series Angular-Contact Split Inner Ring	Designed with solid one-piece outer ring and two-piece inner ring, maximum ball complement and one-piece machined cage.	Construction allows bearing to carry greater thrust in either direction than Type S. May be used where there is substantial radial load providing there is always sufficient thrust load present.	
5000 Series Double-Row C Type Conrad M Type Maximum-Capacity	C Type assembled by elastic displacement of rings. Does not have ball filling notches. M Type filling notches on one side only for inserting full quota of balls. Supplied with standard clearance but tighter or looser internal clearance can be supplied to suit special service conditions.	C Type will support very heavy radial loads, and equal heavy thrust loads in either direction, or heavy combined radial and thrust loads. M Type has very heavy radial capacity. Also has heavy thrust capacity in one direction and can take light thrust load in reversing direction	
	Sizes 5200-5203, 5300-5303 (SB suffix) made in conrad type with inverted contact angle. 5400 Series, a non-rigid type, has filling notches on both sides. Contact angles converge inwardly. 5219 thru 5222 and 5330 thru 5322 have similar construction.		

Table 7-2
Commonly Used MRC Bearing Symbols

Prefix	Suffix	Description	Prefix	Suffix	Description
	-B	Outwardly convergent contact angle when used with 5000 series (see SB, SBK)	R-		Cylindrical roller bearing
CONV.		Conveyor roll bearing	R-		Extra small, inch size bearing (was S-S)
	-D	Bearing with controlled relationship of ring faces, used in duplex sets		-R	Single-row, maximum-capacity, counterbore outer, angular-contact type bearing
	-DB	Duplex bearing, back-to-back mounting		-RJ	Angular-contact type bearing, inner and outer rings counterbored, non-separable
	-DD	Two glass fabric reinforced PTFE (polytetrafluoroethylene) seals.		-S	Single-row, deep-groove, conrad type bearing (1900, 200, 300)
	-DF	Duplex bearing, face-to-face mounting		-SB	Double-row, conrad type with outwardly convergent contact angles
	-DS	Duplex bearing, universal, preload requirement		-SBK	Double-row, conrad type bearing with outwardly convergent contact angles, standard width
	-DT	Duplex bearing, tandem mounting		-SFFC	Cartridge width bearing, conrad type with two shields
	-DU	Duplex bearing, universal, free running, no end play		-SFFCG	Cartridge width bearing, conrad type with two shields, snap ring
	-F	One shield		-ST	Stainless Steel
FB-		Flanged bearing, tapered O.D.		-SWI	Single-row, deep-groove, conrad type bearing with wide inner ring
	-FF	Two shields		-SZZC	Cartridge width bearing, conrad type with two Synthe-Seals
	-FFM	Two metal shields		-SZZCG	Cartridge width bearing, conrad type with two Synthe-Seals, snap ring
	-FFP	Two rubber-beaded shields		-U	Split inner ring bearing (9000 series)
	-FFS	Two felt seals		-UH	Split inner and split outer ring bearing (9000 series)
	-FFSG	Two felt seals and snap ring		-UK	Split inner ring bearing (9100 series)
	-FM	One metal shield		-UP	Split inner ring bearing (9000 series)
	-FP	One rubber-beaded shield		-WI	Single-row, maximum-capacity, filling notch type bearing, wide inner ring
	-FS	One felt seal			
	-FSF	One felt seal and one shield		-X	Tapered bore ball bearing
	-FSFG	One felt seal and one shield, snap ring			
F-	-WI	Shield on flush side of wide inner	XLS-		Single-row, maximum-capacity, counterbore outer, radial type bearing, inch dimensions
	-G	Snap ring			
G-	-MF	Snap ring opposite side from standard shield (also G-SF)	XO-	-RBDS	Excello replacement bearings
	-H	Split outer ring bearing (9000 series)		-Y	Adapter sleeve
	-K	Standard width, double-row bearing		-Z	One synthetic rubber seal
	-KR	Single-row, maximum-capacity, counterbore outer, radial type bearing (100-K and 7100-K series)		-ZZ	Two synthetic rubber seals
	-KS	Single-row, deep-groove, conrad type bearing (100-K series)		-ZZC	Cartridge width bearing, extra-small type, two Synthe-Seals
	-KSB	Conrad type bearing with counterbored separable inner ring			
	-M	Single-row, maximum-capacity, filling notch type bearing			
	-P	Heavy duty, angular-contact bearing			

Table 7-3
Ball Bearing Interchange Table

MRC Ball Bearing Series	B C A	Fafnir	FAG	Federal	Hoover NSK	New Departure-Hyatt	SKF
R-2	....	33-K-3	R-2	R2	R2	R-2	EE-0
R-2-F	....	33-KD-3	R-2-Z	R2F	7R2	7-R-2	EE-O/Z
R-2-FF	....	33-KDD-3	R-2-2Z	R2FF	77R2	77-R-2	EE-O/2Z
R-2-Z	....	....	R-2-RS	R2R	9R2	9-R-2	EE-O/RS
R-2-ZZ	....	....	R-2-2RS	R2RR	99R2	99-R-2	EE-O/2RS
1R-2-A	....	33-K-4	R-2-A	R2A	R2A	R-2-A	....
1R-3	....	33-K-5	R-3	R3	R3	R-3	EE-1
1R-4	....	S-1-K-7	R-4	R4	R4	R-4	EEA-2
1R-4-A-4	....	S-1-K	R-4-A	R4A	R4A	R-4-A	EE-2
1R-6	....	S-3-K	R-6	R6	R6	R-6	EE-3
1R-8	....	S-5-K	R-8	R8	R8	R-8	EE-4
1R-10	....	S-7-K	R-10	R10	R10	R-10	EE-5
1R-12	....	S-8-K	R-12	R12	R12	R-12	EE-6
1R-14	....	S-9-K	R-14	R14	R14	R-14	EE-8
1R-16	....	S-10-K	R-16	R16	R16	R-16	EE-9
1R-18	....	S-11-K	R-18	R18	R18	R-18	EE-10
1R-20	....	S-12-K	R-20	R20	R20	R-20	EE-11
1R-22	....	....	R-22	R22	....	R-22	....
1R-24	....	....	R-24	R24	R24	R-24	....
XLS	XLS§	XLS§	XLS§	XLS§	....	....	....
30	30	30-K	600	30	30	30	R-4
30-F	30-S	30-KD	600-Z	30-F	7030	7030	R-4-Z
30-FF	30-SS	30-KDD	600-2Z	30-FF	77030	77030	R-4-2Z
30-FFS	....	30-KLL	....	FS88030	....	88030	....
30-FS	....	30-KL	....	FS8030	....	8030	....
30-FSF	....	30-KLD	....	FS87030	....	87030	....
30-FZ	....	....	600-RSZ	....	....	97030	....
30-Z	30-D	30-P	600-RS	30-R	9030	9030	R-4-RS
30-ZZ	30-DD	30-PP	600-2RS	30-RR	99030	99030	R-4-2RS
100-KR	7100-N	9100-WI	B-7000-C	....	....	0-L-00	7000-C
100-KRD	D7100-N	9100-WI-DU	(2) B-7000-CUO	....	....	0-L-00-D	(2) 7000-CG
100-KS	100	9100-K	6000	9100	3L00	3-L-00	6000
100-KSF	100-S	9100-KD	6000-Z	9100-F	73L00	73-L-00	6000-Z
100-KSFF	100-SS	9100-KDD	6000-2Z	9100-FF	773L00	773-L-00	6000-2Z
100-KSFFG	100-SSL	9100-KDDG	6000-2ZNR	....	773L00-G	4773-L-00	6000-2ZNR
100-KSFG	100-SL	9100-KDG	6000-ZNR	9100-GF	73L00-G	473-L-00	6000-ZNR
100-KSFZ	....	9100-NPD	6000-RSZ	....	973L00	973-L-00	6000-RSZ
100-KSFZG	....	....	6000-RSZNR	....	973L00-G	4973-L-00	6000-RSZNR
100-KSG	100-L	9100-KG	6000-NR	9100-CG	3L00 G	43-L-00	6000-NR
100-KSZ	100-D	9100-P	6000-RS	9100-R	93L00	93-L-00	6000-RS
100-KSZG	100-DL	9100-PG	6000-RSNR	9100-GR	93L00-G	493-L-00	6000-RSNR
100-KSZZ	100-DD	9100-PP	6000-2RS	9100-RR	993L00	993-L-00	6000-2RS
100-KSZZG	100-DDL	9100-PPG	6000-2RSNR	9100-GRR	993L00-G	4993-L-00	6000 2RSNR
200-FFS	88500	200-KLL	....	FS-88500	....	88500	....
200-FFSG	....	200-KLLG	....	....	....	488500	....
200-FS	8500	200-KL	....	FS-8500	8500	8500	....
200-FSF	87500	200-KLD	....	FS-87500	87500	87500	....
200-FSFG	....	200-KLDG	....	....	87500-G	487500	....
200-M	1200	200-W	200	1200-M	M-200	1200	200
200-MF	1200-S	200-WD	200-Z	1200-MF	M-7200	7200	200-Z
200-MFF	1200-SS	200-WDD	200-2Z	1200-MFF	M-77200	77200	200-2Z
200-MFFG	1200-SSL	200-WDDG	200-2ZNR	1200-MGFF	M-77200-G	477200	200-2ZNR
200-MFG	1200-SL	200-WDG	200-ZNR	1200-MGF	M-7200-G	47200	200-ZNR
200-MFZ	....	....	....	....	M-97200	....	....
200-MG	1200-L	200-WG	200-NR	1200-MG	M-200 G	41200	200-NR
200-MZ	....	....	....	....	M-9200	9200	....
200-R	7200-N	7200-W	B-7200-C	7200	7200-C	20200	7200-C
200-RD	D7200-N	7200-W-DU	(2) B-7200-CUO	....	7200-C-D	20200-D	(2) 7200-CG
200-S	200	200-K	6200	1200	200	3200	6200
200-SF	200-S	200-KD	6200-Z	1200-F	7200	7500	6200-Z
200-SFF	200-SS	200-KDD	6200-2Z	1200-FF	77200	77500	6200-2Z
200-SFFC	W-200-CC	W-200-KLL	S-3500	W1200-FF	S-3500	....	462200
200-SFFCG	....	....	S-3500 NR	W1200-GFF	S-3500-G	....	....
200-SFFG	200-SSL	200-KDDG	6200-2ZNR	1200-GFF	77200-G	477500	6200-2ZNR
200-SFG	200-SL	200-KDG	6200-ZNR	1200 GF	7200-G	47500	6200-ZNR
200-SFZ	....	200-PD	6200-RSZ	1200-FR	97200	97500	6200-RSZ
200-SFZG	....	....	6200-RSZNR	....	97200-G	497500	....
200-SG	200-L	200-KG	6200-NR	1200-CG	200-G	43200	6200-NR
200-SX	....	5500N	....	1200-X	....	....	....
200-SXY	....	5500-N-A	....	1200-XY	....	....	....
200-SZ	200-D	200-P	6200-RS	1200-R	9200	9500	6200-RS
200-SZG	200-DL	200-PG	6200-RSNR	1200-GR	9200-G	49500	6200-RSNR
200-SZZ	200-DD	200-PP	6200-2RS	1200-RR	99200	99500	6200-2RS
200-SZZC	....	W-200-KLL	S-3500-2RS	W1200 KK	....	....	462200
200-SZZG	200-DDL	200-PPG	6200-2RSNR	1200-GRR	99200-G	499500	6200-2RSNR
300-M	1300	300-W	300	1300-M	M-300	1300	300
300-MF	1300-S	300-WD	300-Z	1300-MF	M-7300	7300	300-Z

Shielded or sealed types—to determine interchange see MRC R-2, R-2-F, R-2-FF, R-2-Z, R-2-ZZ. Add to competitive bearing number their prefix or suffix as shown opposite MRC R-2 Series. For Fafnir sealed bearing numbers, substitute P for single seal, PP for double seal in place of D or DD.

* This table indicates methods of numbering. Some sizes not in production.

§ MRC Bearing has counterbored outer ring.

Table 7-3
Ball Bearing Interchange Table—cont'd

BALL BEARING INTERCHANGE TABLE* — BASIC SERIES

MRC Ball Bearing Series	B C A	Fafnir	FAG	Federal	Hoover NSK	New Departure-Hyatt	SKF
300-MFF 300-MFG 300-MG	1300-SS 1300-SL 1300-L	300-WDD 300-WDG 300-WG	300-2Z 300-ZNR 300-NR	1300-MFF 1300-MGF 1300-MG	M-77300 M-7300-G M-300-G	77300 47300 41300	300-2Z 300-ZNR 300-NR
300-MZ 300-R 300-RD	1300-D 7300-N D7300-N	7300-W 7300-W-DU	B-7300-C B-7300-C-UO	7300	M-9300 7300-C 7300-C-D	20300 20300-D	
300-S 300-SF 300-SS	300 300-S 300-SS	300-K 300-KD 300-KDD	6300 6300-Z 6300-2Z	1300 1300FF 1300FF	300 7300 77300	3300 7600 77600	6300 6300-Z 6300-2Z
300-SFFC 300-SFFCG 300-SFFG	W-300-CC 300-SSL	W-300-KLL 300-KDDG	S-3600 S-3600-NR 6300-2ZNR	W1300FF W1300GFF 1300GFF	S-3600 S-3600-G 77300-G	S-3600 477600	462300 6300-2ZNR
300-SFG 300-SFZ 300-SFZG	300-SL	300-KDG 300-PD	6300-ZNR 6300-RSZ 6300-RSZNR	1300GF 1300FR 1300GFR	7300-G 97300 97300-G	47600 97600 497600	6300-ZNR 6300-RSZ 6300-RSZNR
300-SG 300-SWI	300-L	300-KG WIR-300-K	6300-NR	1300CG	300-G	43300 4600	6300-NR
300-SX 300-SXY 300-SZ	300-D	6600N 6600-N-A 300-P	6300-RS	1300X 1300XY 1300R	9300	9600	6300-RS
300-SZG 300-SZZ 300-SZZC 300-SZZG	300-DL 300-DD 300-DDL	300-PP W-300-KLL 300-PPG	6300-RSNR 6300-2RS S-3600-2RS 6300-2RSNR	1300GR 1300RR W1300KK 1300GRR	9300-G 99300 S-3600 99300-G	49600 99600 499600	6300-RSNR 6300-2RS 462300 6300-2RSNR
300-WI 400-M 400-MF	1400 1400-S	WIR-300-W 400-W		1300MWI 1400-M 1400-MF	M-400 M-7400	4300 1400 7400	
400-MFF 400-MG 400-R	1400-SS 1400-L			1400-MFF 1400-MG 7400	M-77400 M-400-G	77400 41400 20400	
400-S 400-SF 400-SG	400 400-S 400-L	400-K	6400 6400-Z	1400 1400-F 1400-CG	400 7400 400-G	3400 7700 43400	6400
400-SZ 400-SZZ 1900-R	400-D 400-DD			1400-L 1400-LL			
1900-S 5200 •5200-F		9300-K	61900	1900 5200N 5200F§		3LL00	6900
•5200-FG •5200-K •5200-KF	5200 5200-S	5200-W† 5200-WD†§	3200† 3200-Z†	5200 FS5500	5200† 5200-F†	5200-W 5500-W	5200†
•5200-KFF •5200-KFG •5200-KG	5200-SS 5200-SL 5200-L	5200-SS 5200-WG†	3200-2Z†	FS55500 FS5500CG 5200CG	5200-FF† 5200-FG† 5200-G†	55500-W 45500-W 45200-W	
•5200-SB •5200-SBF •5200-SBK		5200-K 5200-KD 5200-K	3200 3200	A5200 A5200	C-5200 C-5200	5200 5200	5200 5200
•5200-SBKF •5200-SBKFF •5200-SBKFG		5200-KD§ 5200-KDD§ 5200-KDG§	3200-Z 3200-2Z	FSA5500 FSA55500 FSA5500CG	C-5200-F C-5200-FF C-5200-FG	5500 55500 45500	
•5200-SBKG •5200-SBKZ •5200-SBKZZ		5200-KG		A5200CG	C-5200-G	45200 95200 995200	
•5300 •5300-F •5300-FG	5300 5300-S 5300-SL	5300-W† 5300-WD†§	3300†	5300 5300F	5300†	5300-W	5300†
•5300-G •5300-KF •5300-KFG	5300-L 5300-S 5300-SL	5300-WG†	3300-Z†	5300CG FS5600 FS5600CG	5300-G† 5300-F† 5300-FG†	45300-W 5600-W 45600-W	
•5300-SB •5300-SBF •5300-SBG		5300-K 5300-KG	3300	A5300 A5300CG	C-5300 C-5300-F C-5300-G	5300 45300	5300
•5300-SBKF •5300-SBKFF •5300-SBKFG			3300-Z 3300-2Z	FSA5600 FSA55600 FSA5600-CG	C-5300-F C-5300-FF C-5300-FG	5600 55600 45600	
5400 5400-F 5400-G	5400 5400-L	5400-W† 5400-WD† 5400-WG†				5400 5700	5400†
7100-KR 7100-KRD 7200	7100-A D7100-A 7200-A	7200-W	B-7000-E (2) B-7000-EUO B-7200-E	7200	7200-A	H0-L-00 H20200	7000-C (2) 7000-CG 7200
7200-D 7200-P 7200-PD	D7200-A 7200-J D7200-J	7200-W-DU 7200-PW 7200-PW-DU	(2) B-7200-EUO 7200-B (2) 7200-BUO	DU7200	7200-A-D 7200-B 7200-B-D	H20200-D 30200 30200-D	(2) 7200-G 7200-B (2) 7200-BG
7300 7300-D 7300-P	7300-A 7300-A D7300-J	7300-W 7300-W-DU 7300-PW	B-7300-E (2) B-7300-EUO 7300-B	7300 DU7300	7100-A 7100-A-D 7100-B	H-20300 H-20300-D 30300	7300 (2) 7300-G 7300-B
7300-PD 7400 7400-D		7300-PW-DU 7400 7400-D	(2) 7300-BUO	7400 DU7400	7300-B-D	30300-D H-20400 H-20400-D	(2) 7300-BG 7400 (2) 7400-G

* This table indicates methods of numbering. Some sizes not in production.
† MRC bearing contact angles converge inwardly.
§ On certain sizes, width may be greater than corresponding MRC bearing.
• These double-row bearing types have been superseded by new C and M types. Interchange to new types and with other makes shown on page 50.

Table 7-3
Ball Bearing Interchange Table—cont'd

5000 SERIES DOUBLE-ROW BALL BEARINGS

Interchangeability of MRC 5000 Series
Bearings with various other makes.
This table indicates methods of marking. Some competitive sizes in the
series are not in production

MRC Ball Bearing Series	BCA	Fafnir	FAG	Federal	Hoover NSK	New Departure-Hyatt	SKF
5200-C	....	5200K	....	....	5200	5200	5200H
5200-CF	....	5200KD	....	....	....	5500	5200Z
5200-CFF	....	5200KDD	....	....	5200ZZ	55500	5200-2Z
5200-CFFG	....		....		....	455500	....
5200-CFG	....	5200KDG	....		....	45500	....
5200-CFZ	....	....	....		....	....	....
5200-CFZG	....	....					
5200-CG	....	....	....	....	5200NR	45200	5200NRH
5200-CZ	....	....	....	....	....	....	5200RS
5200-CZG	....	....	....	....	....	....	
5200-CZZ	....	....	....	....	....	....	5200-2RS
5200-CZZG	....	....	....	....	....	....	....
5200-M	5200W	5200W	3200	A5200	3200	5200	....
5200-MF	5200WS	5200WD	....	FSA5500	....	5500	....
5200-MFF	5200WSS	....	....	FSA55500	3200ZZ	55500	....
5200-MFFG		....	....		....	455507	....
5200-MFG	5200WSL	....	....	FSA5500CG	....	45500	....
5200-MFZ	....	....	....	....	....	....	....
5200-MFZG	....	....					
5200-MG	5200WL	5200WG	....	FSA5200CG	3200NR	45200	....
5200-MZ	....	....	....	....	....	....	....
5200-MZG	....						
5200-SB	....	5200K	....	....	5200	5200	5200
5200-SBKF	....	5200KD	....	....	....	....	5200Z
5200-SBKFF	....		....	....	5200ZZ	....	5200-2Z
5200-SBKFG	....	5200KDG	....	....	....	....	....
5200-SBKZ	....		....	....	....	....	5200RD
5200-SBKZZ	....		....	....	....	....	5200-2RS
5300-C	....	5300K	3300	....	5300	5300	5300H
5300-CF	....	....	....	....	....	5600	....
5300-CFF	....	....	....	....	....	55600	....
5300-CFFG	....	....	....	....	....	455600	....
5300-CFG	....	....	....	....	....	45600	....
5300-CFZ	....	....	....	....	....	....	....
5300-CFZG	....	....	....	....	....	....	....
5300-CG	....	5300KG	....	....	5300NR	45300	5300NRH
5300-CZ	....	....	....	....	....	....	....
5300-CZG	....	....	....	....	....	....	....
5300-CZZ	....	....	....	....	....	....	....
5200-CZZG	....		....	....	....	....	....
5300-M	5300W†	5300W	3300	A5300	3300	5300	....
5300-MF	5300WS†	....	....	FSA5600	....	5600	....
5300-MFF		....	....	FSA55600	3300ZZ	55600	....
5300-MFFG	....	....	....		....	455600	....
5300-MFG	5300WSL†	....	....	FSA5600CG	....	45600	....
5300-MFZ	....	....	....	....	....	....	....
5300-MFZG	....	....	....	....	....	....	....
5300-MG	5300WL†	5300WG	....	FSA5300CG	3300NR	45300	....
5300-MZ	....	....	....	....	....	....	....
5300-MZG	....	....	....	....	....	....	....
5300-SB	....	5300K	....	....	5300	5300	....
5400	5400	5400W•	....	....	....	5400•	5400
5400F		....	....	....	....	....	....
5400G	5400L	5400WG•	....	....	....	....	....

† BCA bearing contact angles converge inwardly
• MRC bearing contact angles converge inwardly

(Text continued from page 371)
"Special" bearings include:

Adapter Type—Conrad type with a tapered adapter sleeve.
Aircraft Bearings—A category by themselves. Not related to other types listed here.
Cartridge Type—Conrad type with both rings same width as a double-row bearing.
Conveyor Roll—Conrad type. Special construction, wider than standard built-in seals.
Felt Seal—Conrad type, unequal width rings.

Cleanliness and Working Conditions in Assembly Area

Many ball bearing difficulties are due to contaminants that have found their way into the bearing after the machine has been placed in operation. Contaminants generally include miscellaneous particles which, when trapped inside the bearing, will permanently indent the balls and raceways under the tremendous pressures generated by the operating load (Figure 7-1).

Average contact area stresses of 250,000 lbs per square in. are not uncommon in bearings. Due to the relatively small area of contact between

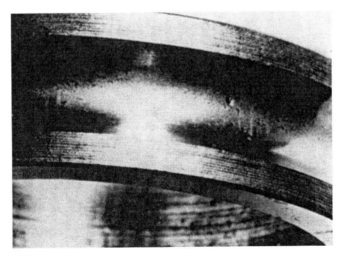

Figure 7-1. Hard, coarse foreign matter causes small, round-edged depressions of various sizes.

the ball and raceway, contact area pressures are very high even for lightly loaded bearings. When rolling elements roll over contaminants, the contact areas are greatly reduced and the pressure becomes extremely high.

When abrasive material contaminates the lubricant, it is frequently crushed to finer particles that cause wear to the ball and race surfaces. The wear alters the geometry of the balls and races, increases the internal looseness of the bearing, and roughens the load-carrying surfaces (Figure 7-2). Therefore, it is highly important to maintain a clean environment when working on all bearing applications during servicing operations.

The assembly area should be isolated from all possible sources of contamination. Filtered air will help eliminate contamination and a pressurized and humidity-controlled area is advantageous to avoid moist and/or corrosive atmospheres. Work benches, tools, clothing, and hands should be free from dirt, lint, dust, and other contaminants detrimental to bearings.

Surfaces of the work bench should be of splinter-free wood, phenolic composition, or rubber-covered to avoid possible nicking of spindle parts that could result from too hard a bench top. To maintain cleanliness, it is suggested that the work area be covered with clean poly-coated kraft paper, plastic, or other suitable material (Figure 7-3) which, when soiled, can be easily and economically replaced.

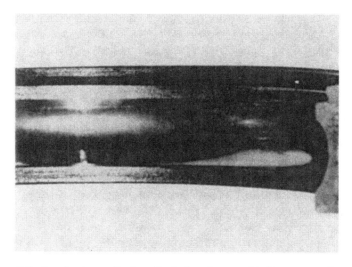

Figure 7-2. Fine foreign matter laps the ball surfaces and ball races, causing wear.

Figure 7-3. Cover workbench with clean, lint-free paper, plastic, or similar material. Also, isolate work area from contamination sources.

Removal of Shaft and Bearings from Housing

The first step in dismantling a spindle or shaft is to remove the shaft assembly from the housing. To do this, it is generally necessary to take off the housing covers from each end.

Most machine tool spindle and API pump housings are constructed with bearing seats as an integral part of the housing. This contributes to the rigidity of the spindle. However, it makes disassembly more difficult and extreme care must be taken to avoid bearing damage. Also, it is not generally possible to remove bearings from the shaft unless the shaft assembly is first removed from the housing.

On most spindle assemblies this can be done by first placing the entire spindle in an arbor press and in alignment with the press ram. Next, carefully apply pressure to the end of the shaft making sure that there is clearance for the expulsion of the shaft assembly on the press table. As pressure is applied, the shaft is forced from the housing along with the bearing mounted on the opposite end of the shaft.

The bearing on the end where pressure is applied remains in the housing. It is removed from the housing either with hand pressure or by carefully pushing it out of the housing from the opposite side with rod tubing having a diameter slightly smaller than the housing bore. The tubing should contact the bearing outer ring and should push it from the housing with little or no pressure on the balls and inner ring. Following this procedure will help avoid brinelling of the raceways due to excessive pressure on the rolling elements and races.

Electric motor shafts are generally constructed to permit removal of one end bell, leaving the shaft and bearings exposed. The rotor or shaft assembly is then free to be removed by drawing it through the stator.

Bearing Removal from Shaft

Removal of bearings from spindle shafts is a highly important part of the maintenance and service operation. In most cases, it is far more difficult to remove a bearing from the shaft than to put it on. For this reason, a bearing can be damaged unnecessarily in the process. Every precaution must be taken to avoid damage to any of the parts including the bearings. If the bearings are damaged during removal, the damage often is not noticed and may not become known until the spindle is completely reassembled.

Bearing damage during removal from the shaft can occur in many ways, of which these are the most common:

- *The smooth, highly-polished surface of the ball raceways may be brinelled, i.e., indented, by the balls* (Figure 7-4). Brinell marks on

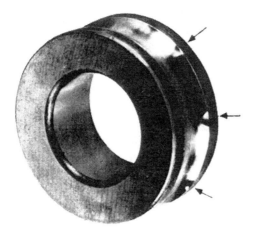

Figure 7-4. Brinell marks or nicks, indicated by arrows, are the most common result of improper bearing removal.

the surface of the races are usually caused when a bearing is forced off the shaft by applying excessive or uneven pressure through the rolling element complement. Any shock load, such as hammer blows on the inner or outer rings, is apt to cause brinelling. Major brinelling can sometimes be discovered on the job by applying a thrust load from each direction while rotating the inner or outer ring slowly. As the ring is turned through the brinelled area on either of the race shoulders, it can often be felt as a catch or rough spot. *A brinelled bearing is unfit for further use. Never put it back into service.*

- *Ball raceways may be roughened due to dirt particles or metal chips working into the bearing.* As soon as the shaft has been removed from the housing, it should be placed in a clean work area and suitably covered so that no contaminant can become lodged in the bearing prior to removal from the shaft. If contaminants enter the housing and the bearing is subsequently rotated, it is possible that they will roughen and damage the raceways.
- *The ball cage* may *be damaged if the bearing puller is used incorrectly.* Use of improper tools such as a hammer or chisel to pound or pry the bearing off the shaft may result in damage to the bearing in addition to the hazard of contaminating the bearing.

Removal From Shaft

Because of operating conditions or location of the shaft, bearings are often tight and resist easy removal. This holds true even though they were originally mounted with a "push" fit, usual in most machine tool spindle applications. A "push" fit means ability to press the bearing on the shaft with hand pressure.

If these conditions occur, mechanical means such as a bearing puller (Figure 7-5) or the use of an arbor press (Figure 7-6) should be employed to effect bearing removal. The hammer and drift tube method, sometimes used to pound the bearing from the shaft, generally is not recommended, especially on machine tool spindle bearings. There is always the chance that the hammer shocks conducted through the tube will cause brinelling.

For some types of bearings, electrical means of removal are possible as well. These removal methods will be described later.

Bearings are mounted on shafts or spindles in several ways so that dismounting must be accomplished by different means. Here are the most common conditions:

- The bearing is free of grease and/or other parts. Place the shaft in an arbor press in line with the ram and with the inner ring of the bearing

Figure 7-5. Bearing puller with two claws.

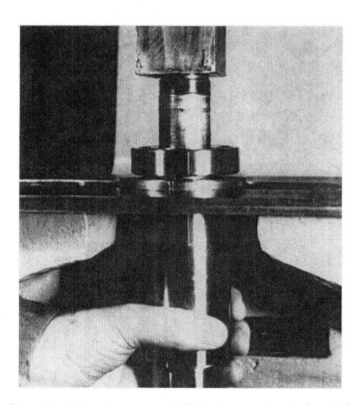

Figure 7-6. Using arbor press and split ring to remove bearing from shaft.

supported by a split ring having a bore slightly larger than the shaft (Figure 7-7). Press the shaft from the bearing with an even pressure, making sure it does not drop free and become damaged. If the split ring is not available, two flat bars of equal height could support the bearing (Figure 7-8).

Another means of removing a bearing from the shaft is by use of a bearing puller, several of which are shown in Figures 7-13 to 7-15.

- The bearing mounted with gears and/or other parts abutting it (Figure 7-9). In most cases, a bearing in this location can only be removed by a bearing puller which applies pressure on the outer ring (Figure 7-10). Extreme care must be exercised when applying pressure to

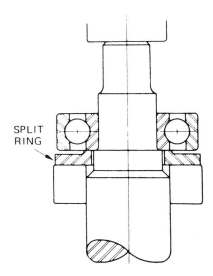

Figure 7-7. Split ring supports inner ring of bearing.

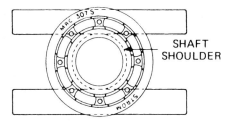

Figure 7-8. Equal height bars spaced to support both inner and outer rings.

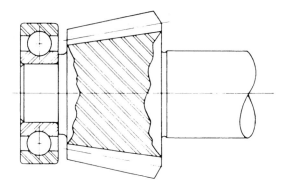

Figure 7-9. Bearing mounted with other parts abutting it.

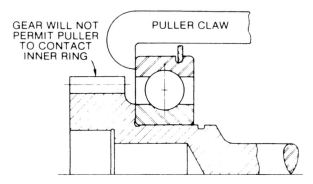

GEAR WILL NOT
PERMIT PULLER
TO CONTACT
INNER RING

PULLER CLAW

Figure 7-10. Where shaft parts obstruct inner ring accessibility, apply pressure with bearing puller on outer ring as evenly and squarely as possible. On bearings with one high and one low shoulder, pressure should be applied against the deep shoulder only.

make sure that the pull is steady and equal all around the outer ring. If the gears or other parts are removable, it may be possible to apply pressure through them to force the bearing off the shaft. An arbor press may be employed to do the job if the bearing or gear can be adequately supported while pressure is applied.

Applying Pressure with Bearing Puller

Whenever possible, bearings always should be moved from the shaft by square and steady pressure against the *tight* ring. Thus with a tight fit on the shaft, pressure should be against the inner ring; with a tight fit in the housing, pressure should be against the outer ring. If it is impractical to

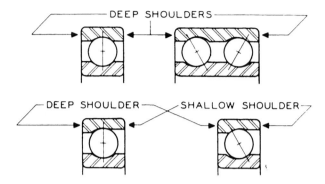

Figure 7-11. Pressure may be applied in either direction with shoulders of equal height.

exert pressure against the tight ring, and the loose ring must be used, it is imperative that the same square and steady pull method be used.

Pressure may be applied in either direction on bearings with shoulders of equal height (Figure 7-11). On counterbored bearings with one deep and one low shoulder, pressure should be applied against the deep shoulder. If pressure is applied against the low shoulder, disassembly of the bearing or serious damage may result. When the pairs of bearings on each end of the shaft are mounted in a back-to-back (DB) relationship, the counterbored outer ring is always exposed. In such cases, it will be necessary to apply the pressure against the low shoulder (counterbored ring) to effect bearing removal from the shaft even though the risk of damage to the inboard bearing is great.

Most machine tool spindles employ Type R or angular-contact bearings (7000 Series) that do not have seals or shields. However, it is possible that a Conrad type bearing equipped with seals or shields may be used in some applications. When using pullers for bearing removal, care must be exercised to avoid damage to the seal or shield (Figure 7-12). If dented and then remounted, an early bearing failure during operation could result.

Bearing removal damage can be caused by the selection of the wrong puller type as easily as it can with improper use of the correct puller. No matter which puller is used, *remember:* if the bearing is not pulled off squarely under steady pressure, it must be scrapped!

Identification and Handling of Removed Bearings

As it is possible that bearings may be suitable for remounting after servicing, it is necessary to replace them in exactly the same position on the shaft. Therefore, each bearing must be specifically tagged to indicate its

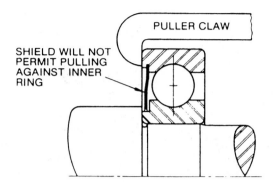

PULLER CLAW

SHIELD WILL NOT
PERMIT PULLING
AGAINST INNER
RING

Figure 7-12. Where a shield or seal does not permit inner ring pressure, use bearing puller with extreme care to avoid denting shield or seal.

proper location. Duplex bearings should be tied together in their proper relationship, DB, DF, or DT and the tag should also indicate the relationship. If a spacer is used between duplex bearings, the tag should indicate its position and relationship to the bearings.

On jobs where the bearing is being removed because performance has not been fully successful, it is often desirable to find out why. Be sure to preserve the bearing until it is practical to examine it. The bearing frequently contains direct evidence as to the cause of failure. It should not be permitted to rust badly and the parts should be abused as little as possible during disassembly.

If the bearing is being removed for reasons other than bearing failure, be certain that it is thoroughly cleaned and oiled immediately after removal. Otherwise there is a good chance that it will get dirty and rusty, which would prevent its reuse.

Bearing Pullers

There are numerous types of bearing pullers on the market, any of which would be satisfactory to use depending upon the dismounting situation encountered. A conventional claw type is used where there is sufficient space behind the bearing puller claws to apply pressure to the bearing. In the illustration (Figure 7-13), the claws are pressing against the bearing preloading spring pack which in turn will force the duplex pair of bearings and spacers from the spindle.

Another type of puller (Figure 7-14) uses a split-collar puller plate (Figure 7-15), the flange of which presses against the inner ring of the bearing. The puller bolts must be carefully adjusted so that the pulling

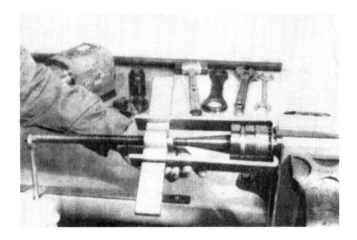

Figure 7-13. Claws pressing against the bearing spring pack will force the duplex pair of bearings and spacers from the spindle.

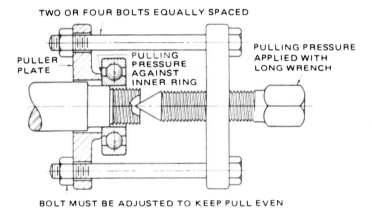

Figure 7-14. Another type of puller. Pulling pressure is applied to inner ring.

pressure is equal all around the ring. The collar must be made in two pieces so that it can be slipped behind the bearing. The collar hole should be large enough so that the two pieces may be bolted together without gripping the shaft.

Most bearing companies do not manufacture bearing pullers, but many bearing distributors stock a variety of the various pullers described above.

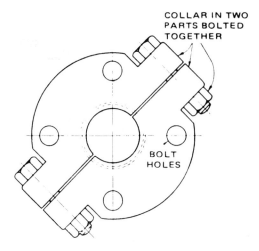

COLLAR IN TWO
PARTS BOLTED
TOGETHER

BOLT
HOLES

Figure 7-15. Split collar puller plate.

Bearing Removal Through Application of Heat

The application of heat via special devices provides a rather straight-forward way of removing inner bearing rings without damaging shafts. The device shown in Figure 7-16 is initially heated by an induction heater (see Figures 7-59 through 7-61, later in this chapter).

To remove the inner ring from a bearing assembly [Figure 7-17(1)], the outer race and rolling elements must first be removed [Figure 7-17(2)]. The device is then heated to approximately 450°C (813°F) and slipped over the exposed ring [Figure 7-17(3)]. By simultaneously twisting and pulling [Figure 7-17(4)], the operator clamps the heated pull-off device onto the ring. Within approximately 10 seconds, the ring will have expanded to the point of looseness [Figure 7-17(5)] and can be removed.

Cleaning and Inspection of Spindle Parts

Insufficient attention is paid to small dust particles which constantly blow around in the open air, But should a particle get in one's eye, it becomes highly irritating. In like manner, when dirt or grit works into a ball bearing, it can become detrimental and often is the cause of bearing failure.

It is so easy for foreign matter to get into the bearing that more than ordinary care must be exercised to keep the bearing clean. Dirt can be introduced into a bearing simply by exposing it to air in an unwrapped

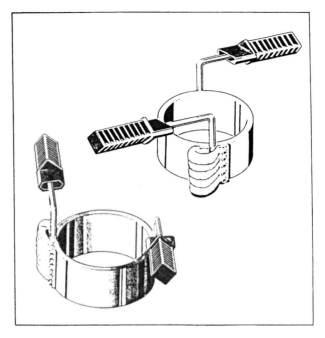

Figure 7-16. Electrically heated "demotherm" device for removal of bearing inner rings from shafts (courtesy Prüftechnik A. G., Ismaning, Germany).

state. Within a short period of time, the bearing can collect enough contaminants to seriously affect its operation. Special care must be taken when the bearing is mounted on a shaft, a time when it is most susceptible to contamination. This cleanliness requirement also extends to the handling of spindle parts, as everything must be clean when replaced in the assembly.

Cleaning the Bearing

During the process of removal from a shaft, the bearing is likely to have become contaminated. The following procedure should be used to clean the bearing for inspection purposes as well as to prepare it for possible remounting on the shaft:

1. Dip the bearing in a clean solvent and rotate it slowly under very light pressure as the solvent runs through the bearing (Figure 7-18). Continue washing until all traces of grease and dirt have been removed. *Do not force the bearing during rotation.*

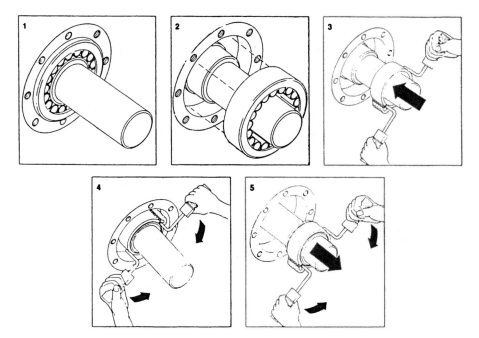

Figure 7-17. An induction-heated pull-off device will effectively remove bearing inner rings from shafts (courtesy Prüftechnik A.G., Germany).

2. Blow the bearing dry with clean, dry air while holding both inner and outer rings to keep the air pressure from spinning them. This avoids possible scratching of balls and raceways if grit still remains in the bearing. A slow controlled hand rotation under light pressure is advisable.
3. After blowing dry, rotate the bearing again slowly and gently to see if dirt can still be detected. Rewash the bearing as many times as necessary to remove all the dirt.
4. When clean, coat the bearing with oil immediately. Special attention should be given to covering the raceways and balls to ensure prevention of corrosion to the highly finished surfaces. Rotate the bearing gently to coat all rolling surfaces with oil.

After cleaning, the bearing should be wrapped with lint-free material such as plastic film to protect it from exposure to all contaminants. Unless this is done, it may be necessary to repeat the cleaning procedure immediately prior to remounting. As other spindle parts are cleaned, they also should be covered to exclude contamination which could ultimately work into the bearing.

Figure 7-18. A metal basket strainer is useful when dipping bearings in clean solvent. Rotate bearing slowly with very light pressure in solvent.

Cleaning the Shaft

The shaft must be cleaned thoroughly with special attention being paid to the bearing seats and fillets. If contaminants or dirt remain, proper seating of the shaft and/or against the shaft shoulder could be impossible. Don't overlook the cleaning of keyways, splines, and grooves.

Cleaning the Housing

Care should be taken to remove all foreign matter from the housing (Figure 7-19). Suitable solvents should be used to remove hardened lubricants. All corrosion should be removed. After cleaning, inspect in a suitable light the bearing seats and corners for possible chips, dirt, and damage, preferably using low power magnification for better results.

The most successful method to maintain absolute cleanliness inside a clean housing is to paint the nonfunctional surfaces with a heat-resisting,

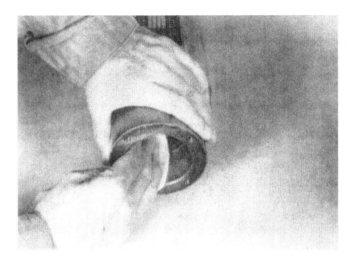

Figure 7-19. Clear bearing seats of housing thoroughly to remove all foreign matter. Then inspect bearing seats and corners for possible damage.

quick-drying engine enamel. Do not paint the bearing seats of the housing. This would reduce the housing bore limits, making it difficult, if not impossible, to mount the bearings properly. Painting seals the housing and prevents loose particles such as core sand from contaminating the bearing lubricants and eventually the bearings. It also provides a smooth surface which helps to prevent dirt from clinging to the surfaces. The housing exterior also may be painted to cover areas where old paint is worn or chipped; but do not paint any of the locating mating surfaces. This type of work should be done in a place outside of the spindle assembly area.

Keep Spindle Parts Coated with Oil

As most of the spindle parts are usually of ferrous material, they are subject to corrosion. When exposed to certain atmospheric conditions, even nonferrous parts may become corroded and unusable in the spindle. Therefore, it is important to make certain that parts are not so affected from cleaning time until they are again sealed and protected in the spindle assembly.

The best protection is to keep parts coated with a light-weight oil, covered, and loosely sealed with a plastic film or foil. Such a covering will exclude contaminants such as dust and dirt. When it is necessary to handle parts for inspection, repair, transportation, or any other purpose,

Figure 7-20. After cleaning, inspect spindle parts by visual means and under magnification. It is important that locating surfaces are free of nicks, burrs, corrosion, etc.

precautions must be taken to ensure they are recoated with oil as some may have rubbed off during handling of the part.

Inspect All Spindle Parts

After the spindle parts have been cleaned thoroughly, the various parts should be inspected visually for nicks, burrs, corrosion, and other signs of damage (Figure 7-20). This is especially important for locating surfaces such as bearing seats, shaft shoulders, faces, and corners of spacer rings if any are used in the spindle, etc.

Sometimes damage may be spotted by scuff marks or bright spots on the bearing, shaft, or in the housing. This scoring may be caused by heavy press fits or build-up of foreign matter drawn onto the mating surfaces. Bright spots may also indicate early stages of "fidgeting" or scrubbing of mating surfaces. The shaft should also be checked for out-of-round and excessive waviness on both two-point and multiple point gauging or checking on centers.

Shaft and Housing Preparation

Bearing Seats on Shaft

The shaft seat for the inner ring of a ball bearing is quite narrow and subject to unit pressures as high as 4,000 lbs per square in. Because of this pressure, particular attention must be paid to the shaft fit to avoid rapid deterioration of the bearing seats due to creepage under heavy load and/or "fretting."

The required fit of the inner ring on the shaft will vary with the application and service. It is dependent on various factors such as rotation of the shaft with respect to the direction of the radial load, use of lock nuts, light or heavy loads, fast or slow speeds, etc. In general, the inner ring must be tight enough not to turn or creep significantly under load (Figure 7-21).

When the bearing has too tight a fit on the shaft, the inner race expands and reduces or eliminates the residual internal clearance between the balls and raceways. Usually bearings, as supplied for the average application, have sufficient radial clearance to compensate for this effect. However, when extremes of shaft fit are inadvertently combined with insufficient radial clearance, extreme overload is caused and may result in heating and premature bearing failure. Tight fits in angular-contact type bearings used for machine tools may cause changes in preload and contact angle, both of which have an effect upon the operating efficiency of the machine. Finally, rings may be split by too heavy a fit.

Excessive looseness under load is also very objectionable because it allows a fidgeting, creeping, or slipping of the inner ring on the rotating shaft (Figure 7-22). This action causes the surface metal of the shaft and bearing to fret, scrub, or wear off which progressively increases the looseness. It has been noticed that, in service, this working tends to scrub

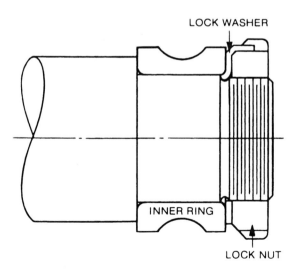

Figure 7-21. To help prevent a heavily loaded bearing from turning on a shaft, a lock nut should be used. The lock nut must be pulled up tight to be effective.

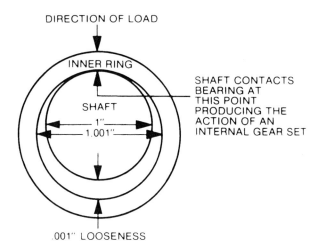

Figure 7-22. Excessive looseness under load allows fidgeting, creeping, or slipping of the inner ring on the rotating shaft.

off fine metal particles which oxidize quickly, producing blue-black and brown oxides on the shaft and/or the bore of the bearing. The bearing should be tight enough on the shaft to prevent this action.

If any of these conditions are noticed on a shaft that has been in service, it may be necessary to repair it to correct size and condition. If the shaft is machined for the bearing seat, it is important not to leave machining ridges, even minute ones. The load very soon flattens down the tops of these ridges and leaves a fit that is loose and will rapidly become looser. For best results, bearing seats should be ground to limits recommended for the bearing size and application.

Shaft Shoulders

Correct shoulders are important because abutment against the shoulder squares the bearing. The bearing is actually squared up when it is pushed home against the shaft shoulder and no further adjustment is necessary.

If a heavy thrust load against the shaft shoulders has occurred during operation, it is possible that the load may have caused the shoulder to burr and push over. Therefore, check the shoulder to make sure that it is still in good condition and square with the bearing seat. If it is not, the condition must be corrected before the spindle assembly operations are begun.

Poor machining practices may result in shaft shoulders that do not permit proper bearing seating.

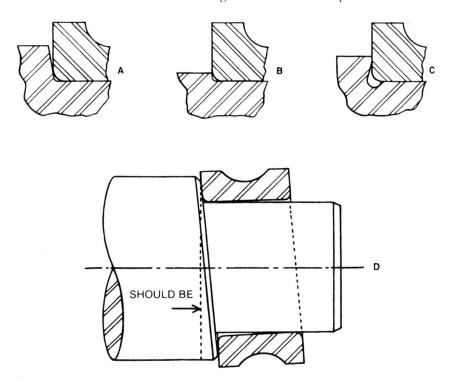

Figure 7-23. Poor seating of the bearing against the corner of the inner ring will result if the shoulder is tapered (A). In (B) the shaft shoulder is so low that it contacts the bearing corner. The condition shown in (C) illustrates that contact between the shoulder and the bearing face is not sufficient. An exaggerated distortion of the inner ring when forced against off-square shoulder is shown in (D).

The shoulder in Figure 7-23A is tapered. This results in poor seating of the bearing against the corner of the inner ring.

The shaft shoulder in Figure 7-23B is so low that the shoulder actually contacts the bearing corner rather than the locating face of the bearing.

With the condition shown in Figure 7-23C, contact between the shoulder and the bearing face is not sufficient. Under heavy thrust loads, the shoulder might break down.

Figure 7-23D is exaggerated to illustrate distortion of the inner ring when forced against off-square shoulder. An off-square bearing shortens bearing life.

Some of these conditions can be corrected when repairs are made on the inner ring seat of the shaft. Such work should be done away from the clean assembly area to avoid possible contamination of the bearing and spindle parts by metal chips or particles from the machining or grinding operations.

The shaft shoulder should not be too high as this would obstruct easy removal of the bearing from the shaft. As described previously, a pulling tool must be placed behind the inner ring and a surface must be left for the tool. Preferably, the inner ring should project somewhat beyond the shaft shoulder to permit pulling the bearing off against this surface. This may not be possible in the case of shielded or sealed bearings where the bearing face is small.

Shaft Fillets and Undercuts

During shaft repair work, it is important to pay attention to the fillet. When it is ground, the fillet frequently becomes larger as the wheel wears, causing an oversize fillet. This in turn locates the bearing on the corner radius instead of the shaft shoulder. In other cases, the corner fillet is not properly blended with the bearing seat or shaft shoulder. This too may produce incorrect axial location of the bearing.

The bearing corner radius originally may be a true 90° segment in the turning, but when the bores, OD's, and faces are ground off, it becomes a portion of a circle less than 90° while the shaft fillet may be a true radius (Figure 7-24A).

Shaft fillet radius specifications are shown in bearing dimension tables with the heading "Radius in Inches" or "Corner Radius." This dimension is not the actual corner radius of the bearing but is the maximum shaft fillet radius which the bearing will clear when mounted. The radius should not exceed this dimension.

The actual bearing corner is controlled so that the above mentioned maximum shaft fillet will always yield a slight clearance. Figure 7-24B illustrates the conventional fillet construction at the shaft shoulder.

Where the shaft has adequate strength, an undercut or relief may be preferred to a fillet. Various types are shown in Figure 7-24 C, D, and E. Where both shaft shoulder and bearing seat are ground, the angled type of undercut is preferred.

Break Corners to Prevent Burrs

When the shaft shoulder or bearing seat is repaired by regrinding, it is desirable to break the corner on the shaft. This will help prevent burrs and nicks which may interfere with the proper seating of the inner ring face

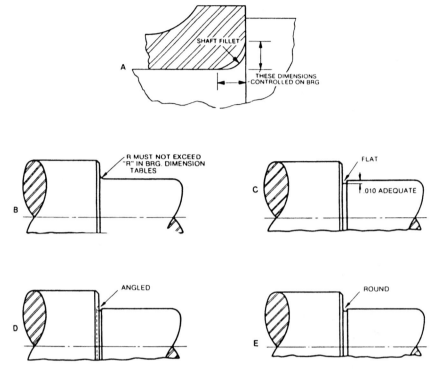

Figure 7-24. When the bores, OD's, and faces are ground off, the bearing corner radius becomes less than 90° as shown in (A). The conventional fillet construction at the shaft shoulder is shown in (B). Various types of relief are shown in (C), (D), and (E).

against the shaft shoulder (Figure 7-25). If left sharp, shoulder corners are easily nicked, producing raised portions which, in turn, may create an off-square condition in bearing location.

The usual procedure to break a corner is to use a file or an abrasive stone. This should be done while the shaft is still in grind position on the machine after regrinding the bearing seat and shoulders. The corner at the end of the bearing seat also should be broken, thus providing a lead to facilitate starting the bearing on the shaft.

If nicks or burrs are found during an inspection and no other work is necessary on the shaft, they can be removed by careful use of a file or stone (Figure 7-26). This work should be done elsewhere than in the clean assembly area. Any abrasive material should be removed from the part before returning it to the assembly area.

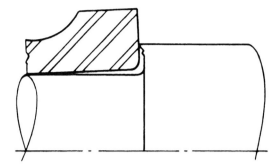

Figure 7-25. Burrs and nicks may interfere with proper seating of the inner ring face against the shaft shoulder.

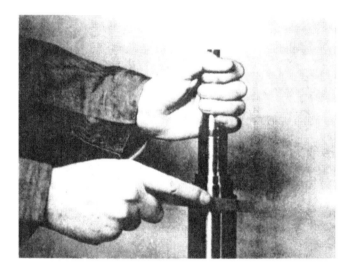

Figure 7-26. A file or stone may be used to remove nicks and burrs.

Check Spindle Housing Surfaces

In many cases, housings will require as much preparatory attention as the shaft and other parts of the spindle. Check the surfaces which mate with the machine mount. Frequently burrs and nicks will be evident and they must be removed before remounting the bearings. Failure to do so may cause a distortion in the bearing, resulting in poor operation and reduced life. These precautions apply to both bearing seats and shoulders.

Shaft and Housing Shoulder Diameters

Recommended shaft and housing shoulders (Figure 7-27) for various sizes of bearings are shown in Table 7-4.

Checking Shaft and Housing Measurements

After all repair work on the shaft has been completed, shafts should be given a final check to make sure the repairs are accurate and within the recommended tolerances. This work may be done with suitable gauging equipment such as an air gage, ten-thousandths dial indicator, electronic comparator, an accurate micrometer, and other instruments as necessary. Accuracies of readings depend on the quality of equipment used, its precision, amplification; and the ability and care exercised by the operator.

It is usually advisable to use a good set of centers which will hold the shaft and permit accurate rotation. The center points should be examined to make sure they are not scored and should be kept lubricated at all times to prevent possible corrosion. Center holes of the shaft must also be of sufficient size, clean and smooth, and free from nicks. Be sure to remove particles of foreign matter that could change the centering of the shaft on the points.

V-blocks will also be helpful to hold the shaft while making various checks. It is important that the V-blocks are clean on the area where the shaft contacts the blocks. Foreign matter and nicks will change the position of the shaft in the blocks and affect any measurements taken.

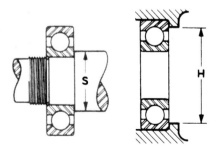

Figure 7-27. Shaft and housing shoulders.

Table 7-4
Shaft and Housing Shoulder Diameters

Shaft and Housing Shoulder Diameters

Shoulder Diameters for Millimeter and Inch-Size MRC brand Bearings

The tables on these pages show Minimum Shaft Shoulder Diameters and Maximum Housing Shoulder Diameters as established by the TRW Bearings Division Product Engineering Department. However, the user may wish to modify these diameters to meet specific requirements such as assembly or disassembly problems or extremely heavy thrust loads.

Basic Brg. No.	S Inch	S MM	H Inch	H MM
34	.22	5.51	.56	14.2
35	.26	6.61	.67	17.0
36	.30	7.61	.67	17.0
37	.34	8.61	.79	20.0
38	.38	9.71	.79	20.0
39	.45	11.4	.92	23.4
100-K	.47	11.9	.95	24.1
101-K	.55	14.0	1.02	25.9
102-K	.67	17.0	1.18	30.0
103-K	.75	19.1	1.30	33.0
104-K	.89	22.6	1.46	37.1
105-K	1.08	27.4	1.65	41.9
106-K	1.34	34.0	1.93	49.0
107-K	1.53	38.9	2.21	56.1
108-K	1.73	43.9	2.44	62.0
109-K	1.94	49.3	2.72	69.0
110-K	2.13	54.1	2.91	73.9
111-K	2.33	59.2	3.27	83.1
112-K	2.53	64.3	3.47	88.1
113-K	2.72	69.1	3.66	92.9
114-K	2.91	73.9	4.06	103.0
115-K	3.11	79.0	4.25	108.0
116-K	3.31	84.1	4.65	118.0
117-K	3.50	88.9	4.84	123.0
118-K	3.84	97.5	5.16	131.0
119-K	4.05	102.0	5.35	136.0
120-K	4.23	107.0	5.55	141.0
121-K	4.53	115.0	5.91	150.0
122-K	4.72	120.0	6.30	160.0
124-K	5.12	130.0	6.69	170.0
126-K	5.51	140.0	7.48	190.0
128-K	5.91	150.0	7.87	200.0
130-K	6.38	162.0	8.39	213.0
132-K	6.77	172.0	8.98	228.0
134-K	7.17	182.0	9.76	248.0
136-K	7.56	192.0	10.55	266.0

Basic Brg. No.	S Inch	S MM	H Inch	H MM
138-K	7.95	202.0	10.95	278.0
140-K	8.35	212.0	11.73	298.0
144-K	9.21	234.0	12.84	326.0
148-K	10.00	254.0	13.62	346.0
152-K	10.95	278.0	15.04	382.0
156-K	11.73	298.0	15.83	402.0
160-K	12.52	318.0	17.40	442.0
164-K	13.31	338.0	18.19	462.0
200	.50	12.7	.98	24.9
201	.58	14.7	1.06	26.9
202	.69	17.5	1.18	30.0
203	.77	19.6	1.34	34.0
204	.94	23.9	1.61	40.9
205	1.14	29.0	1.81	46.0
206	1.34	34.0	2.21	56.1
207	1.53	38.9	2.56	65.0
208	1.73	43.9	2.87	72.9
209	1.94	49.3	3.07	78.0
210	2.13	54.1	3.27	83.1
211	2.41	61.2	3.68	93.5
212	2.67	67.8	3.98	101.0
213	2.86	72.6	4.37	111.0
214	3.06	77.7	4.57	116.0
215	3.25	82.6	4.76	121.0
216	3.55	90.2	5.12	130.0
217	3.75	95.3	5.51	140.0
218	3.94	100.0	5.91	150.0
219	4.21	107.0	6.22	158.0
220	4.41	112.0	6.61	168.0
221	4.61	117.0	7.01	178.0
222	4.80	122.0	7.40	188.0
224	5.20	132.0	7.99	203.0
226	5.67	144.0	8.50	216.0
228	6.06	154.0	9.29	236.0
230	6.46	164.0	10.08	256.0

Basic Brg. No.	S Inch	S MM	H Inch	H MM
232	6.85	174.0	10.87	276.0
234	7.40	188.0	11.50	292.0
236	7.80	198.0	11.89	302.0
238	8.19	208.0	12.68	322.0
240	8.58	218.0	13.47	342.0
242	9.13	232.0	14.13	359.0
244	9.37	238.0	15.04	382.0
246	10.00	254.0	15.63	397.0
248	10.16	258.0	16.61	422.0
250	10.92	277.0	17.07	434.0
252	11.10	282.0	18.03	458.0
256	11.89	302.0	18.82	478.0
260	12.68	322.0	20.39	518.0
264	13.47	342.0	21.97	558.0
300	.50	12.7	1.18	30.0
301	.63	16.0	1.22	31.0
302	.75	19.1	1.42	36.1
303	.83	21.1	1.61	40.9
304	.94	23.9	1.77	45.0
305	1.14	29.0	2.17	55.1
306	1.34	34.0	2.56	65.0
307	1.69	42.9	2.80	71.1
308	1.93	49.0	3.19	81.0
309	2.13	54.1	3.58	90.9
310	2.36	59.9	3.94	100.0
311	2.56	65.0	4.33	110.0
312	2.84	72.1	4.65	118.0
313	3.03	77.0	5.04	128.0
314	3.23	82.0	5.43	138.0
315	3.43	87.1	5.83	148.0
316	3.62	91.9	6.22	158.0
317	3.90	99.1	6.54	166.0
318	4.09	104.0	6.93	176.0
319	4.29	109.0	7.32	186.0
320	4.49	114.0	7.91	201.0

Table 7-4
Shaft and Housing Shoulder Diameters—cont'd

Shaft and Housing Shoulder Diameters

Basic Brg. No.	S		H	
	Inch	MM	Inch	MM
321	4.69	119.0	8.31	211.0
322	4.88	124.0	8.90	226.0
324	5.28	134.0	9.69	246.0
326	5.83	148.0	10.32	262.0
328	6.22	158.0	11.10	282.0
330	6.61	168.0	11.89	302.0
332	7 01	178 0	12 68	322.0
334	7 40	188.0	13 47	342 0
336	7 80	198.0	14 25	362.0
338	8 35	212 0	14 88	378 0
340	8 74	222 0	15 67	398 0
342	9 43	240 0	16 21	412.0
344	9.53	242.0	17.24	438.0
348	10.32	262.0	18.82	478.0
352	11.34	288.0	20.16	512.0
356	12.13	308.0	21.73	552.0
403	.95	24.1	2.17	55.1
404	1.06	26.9	2.56	65.0
405	1.34	34.0	2.80	71.1
406	1.54	39.1	3.19	81.0
407	1.73	43.9	3.58	91.0
408	1.97	50.0	3.94	100.0
409	2.17	55.1	4.33	110.0
410	2.44	61.9	4.65	118.0
411	2.64	67.0	5.04	128.0
412	2.84	72.1	5.43	138.0
413	3.03	76.9	5.83	148.0
414	3.31	84.0	6.54	166.0
415	3.50	88.9	6.93	176.0
416	3.70	93.9	7.32	186.0
417	4.06	103.0	7.56	192.0
418	4.25	108.0	8.15	207.0
419	4.30	109.0	9.10	231.0
420	4.75	120.0	9.64	245.0
421	5.10	129.0	10.52	267 0
422	5.42	137.0	11.57	294.0
1900	.44	11.2	.82	20.8
1901	.52	13.2	.90	22.9
1902	.64	16.3	1.06	26.9
1903	.71	18.0	1.14	29.0
1904	.85	21.6	1.39	35.3
1905	1.05	26.7	1.57	39.9
1906	1.25	31.8	1.79	45.5
1907	1.46	37.1	2.08	52.8
1908	1.68	42.7	2.34	59.4
1909	1.88	47.8	2.5[illegible]	65.3
1910	2.06	52.3	2.74	69.6
1911	2.29	58.2	3.02	76.7
1912	2.49	63.2	3.22	81.8
1913	2.68	68.1	3.42	86.9
1914	2.91	73.9	3.78	96.0

Basic Brg. No.	S		H	
	Inch	MM	Inch	MM
1915	3.09	78.5	3.99	101.0
1916	3.30	83.8	4.18	106.0
1917	3.52	89.4	4.56	116.0
1918	3.71	94.2	4.75	121.0
1919	3.93	99.8	4.93	125.0
1920	4.13	105.0	5.33	135.0
1921	4.33	110.0	5.51	140.0
1922	4.52	115.0	5.70	145.0
1924	4.93	125.0	6.28	160.0
1926	5.36	136.0	6.85	174.0
1928	5.76	146.0	7.24	184.0
1930	6.21	158.0	7.97	202.0
1932	6.62	168.0	8.33	212.0
1934	7.01	178.0	8.74	222.0
1936	7.44	189.0	9.50	241.0
1938	7.83	199.0	9.89	251.0
1940	8.27	210.0	10.61	269.0
R-2 Series				
R-2	.179	...	.325	...
R-2-A	.179	...	.466	...
R-3	.244	...	.466	...
R-4	.310	...	.565	...
R-4-A-4	.322	...	.678	...
R-6	.451	...	.799	...
R-8	.625	...	1.025	...
R-10	.750	...	1.250	...
R-12	.875	...	1.500	...
R-14	1.000	...	1.750	...
R-16	1.125	...	1.875	...
R-18	1.250	...	2.000	...
R-20	1 375	...	2.125	...
R-22	1.500	...	2.375	...
R-24	1.625	...	2 500	...
XLS Series				
XLS-1 3/8	1 9/16	...	2 3/8	...
XLS-1 1/2	1 11/16	...	2 1/2	...
XLS-1 5/8	1 13/16	...	2 11/16	...
XLS-1 3/4	1 15/16	...	2 13/16	...
XLS-1 7/8	2 1/16	...	3	...
XLS-2	2 3/16	...	3 1/8	...
XLS-2 1/8	2 3/8	...	3 3/16	...
XLS-2 1/4	2 1/2	...	3 5/16	...
XLS-2 3/8	2 5/8	...	3 1/2	...
XLS-2 1/2	2 3/4	...	3 5/8	...
XLS-2 5/8	2 7/8	...	3 7/8	...
XLS-2 3/4	3	...	3 7/8	...
XLS-2 7/8	3 3/16	...	4 3/16	...
XLS-3	3 5/16	...	4 3/16	...
XLS-3 1/8	3 7/16	...	4 7/16	...

Basic Brg. No.	S		H	
	Inch	MM	Inch	MM
XLS-3 1/4	3 9/16	...	4 7/16	...
XLS-3 3/8	3 11/16	...	4 11/16	...
XLS-3 1/2	3 13/16	...	4 11/16	...
XLS-3 5/8	3 15/16	...	4 13/16	...
XLS-3 3/4	4 1/16	...	4 13/16	...
XLS-3 7/8	4 3/16	...	5 3/16	...
XLS-4	4 5/16	...	5 3/16	...
XLS-4 1/8	4 7/16	...	5 11/16	...
XLS-4 1/4	4 9/16	...	5 11/16	...
XLS-4 3/8	4 11/16	...	5 15/16	...
XLS-4 1/2	4 13/16	...	5 15/16	...
XLS-4 5/8	4 15/16	...	6 3/16	...
XLS-4 3/4	5 1/16	...	6 3/16	...
XLS-4 7/8	5 3/16	...	6 11/16	...
XLS-5	5 5/16	...	6 11/16	...
XLS-5 1/8	5 1/2	...	6 7/8	...
XLS-5 1/4	5 5/8	...	6 7/8	...
XLS-5 3/8	5 3/4	...	7 1/8	...
XLS-5 1/2	5 7/8	...	7 1/8	...
XLS-5 5/8	6	...	7 3/8	...
XLS-5 3/4	6 1/8	...	7 3/8	...
XLS-5 7/8	6 1/4	...	7 5/8	...
XLS-6	6 3/8	...	7 5/8	...
XLS-6 1/4	6 5/8	...	8 1/8	...
XLS-6 1/2	6 7/8	...	8 3/8	...
XLS-6 3/4	7 1/4	...	8 1/2	...
XLS-7	7 1/2	...	9	...
XLS-7 1/4	7 3/4	...	9 1/4	.
XLS-7 1/2	8	...	9 1/2	...
XLS-7 3/4	8 1/4	...	10	
XLS-8	8 1/2	...	10 1/4	..
XLS-8 1/4	8 3/4	...	10 1/2	..
XLS-8 1/2	9	...	11	...
XLS-8 3/4	9 1/4	...	11 1/4	...
XLS-9	9 1/2	...	11 1/2	...
XLS-9 1/2	10 1/8	...	12 1/8	...
XLS-10	10 5/8	...	12 5/8	...
XLS-10 1/2	11 1/8	...	13 1/8	...
XLS-11	11 5/8	...	13 1/8	...
XLS-11 1/2	12 1/8	...	14 1/8	...
XLS-12	12 5/8	...	15 1/8	...

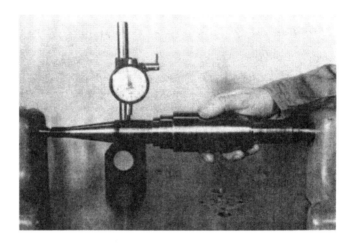

Figure 7-28. A hand gauge may be used to check the bearing seat for out-of-round.

Check Bearing Seat for Out-of-Round

A simple check may be made with a hand gage on the bearing seat (Figure 7-28). This will provide a reading at two points on the shaft 180° apart. However, it does not indicate how those points are related to other points on the shaft.

For a more accurate reading on out-of-round (radial runout) of a bearing seat, mount the shaft between centers and place a suitable indicator in a position perpendicular to the axis of the shaft and contacing the bearing seat. On rotating the shaft slowly by hand, a check is obtained on all points of the shaft which the indicator contacts (Figure 7-29).

Another method of measuring out-of-round is the three-point method using a set of V-blocks and a dial type indicator (Figure 7-30). The shaft should lay in the V-blocks and be rotated slowly while the indicator is centrally located between the points of shaft contact with the V-blocks and perpendicular to these lines of contact. This method will reveal out-of-round which would not have been found by the two-point method of gauging. Therefore, if the equipment is available, it is desirable to check bearing seats using centers or V-blocks as well as two-point gauging.

In all of these checks, the gauge should be placed in different locations on the bearing seat. This will give assurance that the seat is within the recommended tolerances in all areas. While the spindle is mounted on centers, the high point of eccentricity of the bearing seat should be located. Using a dial type indicator, find the point and mark it with a crayon so

Figure 7-29. By rotating the shaft by hand, a check is obtained on all points the indicator contacts.

Figure 7-30. The three-point method.

that it can be easily located when the bearing is to be remounted. The high point of eccentricity is covered in more detail later.

Check Shoulders for Off-Square (Figure 7-31)

The shaft shoulder runout should be checked with an indicator contacting the bearing locating surface on the shaft shoulder while the shaft

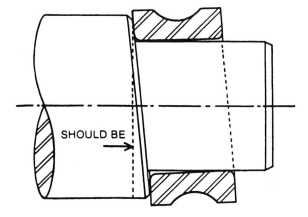

Figure 7-31. Checking shoulders for off-square.

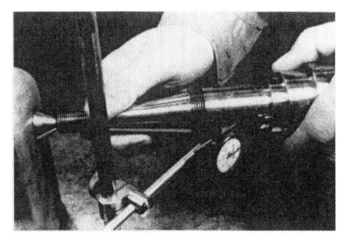

Figure 7-32. If runout is outside established tolerances, the inner ring of the bearing will be misaligned.

is still supported on centers (or V-blocks) with the center of the shaft against a stop. Tolerances have been established for this. If the runout is outside these tolerances, the inner ring of the bearing will be misaligned causing vibrations when the spindle is in operation (Figure 7-32).

Check Housing Bore Dimensions

The housing bore dimensions and shoulder should be checked to make sure that they are within the recommended tolerance for size, out-of-

Figure 7-33. Indicator-type gauge commonly used to check housing bore dimensions.

round, taper, and off-square. The gauge commonly used for this purpose is an indicator type (Figure 7-33).

Recheck Dimensions if Necessary

It is important to be absolutely sure that all dimensions are correct before any assembly is begun. If there is any question, a recheck should be made. If variations are noted, the shaft should be repaired to obtain the correct measurements and then rechecked for accuracy and compliance with the recommended tolerances.

Duplex Bearings

Many methods are used to mount bearings because of various machine tool spindle designs. The simplest spindles incorporate two bearings, one at each end of the shaft. Others are more complicated using additional bearings mounted in specific combinations to provide greater thrust capacity and shaft rigidity. As duplex bearings are usually used in these instances, mounting arrangements for duplex bearings should be understood before actual spindle assembly is begun.

Duplex bearings are produced by specially grinding the faces of single-row bearings with a controlled relationship between the axial location of the inner and outer ring faces. A cross section of a duplex bearing set in back-to-back relationship is shown in Figure 7-34. Note that it consists of

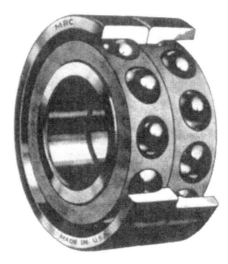

Figure 7-34. Duplex bearings set in back-to-back relationship.

two identical bearings placed side by side. The two units of the pair are clamped tightly together on the shaft with adjacent backs (or faces if DF type) of the inner and outer in actual contact.

Certain definite characteristics and advantages are derived from this mounting which make duplex bearings particularly applicable to several kinds of difficult service and loading conditions. They are recommended for carrying pure radial or thrust loads, or combined radial and thrust loads. Through their use, it is possible to minimize axial and radial deflections thereby, for example, increasing the accuracy of machine tool spindles.

Before explaining the basic mounting methods, it is necessary to understand the difference between the face and back of a bearing as well as to know what a contact angle is.

Referring to the bearing drawing in Figure 7-35, note that the counter-bored low shoulder side of the bearing is called the "face" side. The deep shoulder side (also called high side), stamped with the bearing number and other data, is designated as the "back" side.

As defined by AFBMA standards, a contact angle is the nominal angle between the line of action of the ball load and a plane perpendicular to the bearing axis (Figure 7-36). Essentially, this means that when a load is applied to a bearing, it forces the balls to contact the inner and outer raceway at other than a right angle (such as in a Type S bearing).

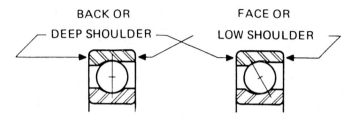

Figure 7-35. The counterbored low shoulder side of the bearing is called the "face" side. The deep shoulder or "high" side, stamped with bearing number, is designated as the "back" side.

Figure 7-36. A contact angle is the nominal angle between the line of action of the ball load and a plane perpendicular to the bearing axis.

Basic Mounting Methods

Duplex bearings can be mounted in three different ways to suit different loading conditions. The three positions bear the symbols, "DB," "DF," and "DT."

DB—Back-to-Back bearings are placed so that the stamped backs (high shoulders) of the outer rings are together. In this position, the contact angle lines diverge inwardly (Figure 7-37).

DF—Face-to-Face bearings are placed so that the unstamped face (low shoulders) of the outer rings are together. Contact angle lines of the bearing will then converge inwardly, toward the bearing axis (Figure 7-38).

DT—Tandem bearings are placed so that the stamped back of one bearing is in contact with the unstamped face of the other bearing. In this case, the contact angle lines of the bearings are parallel (Figure 7-39).

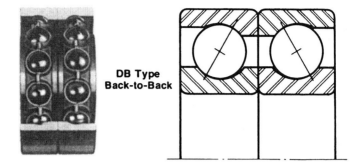

Figure 7-37. Back-to-back bearings are placed so the high shoulder of the outer rings are together.

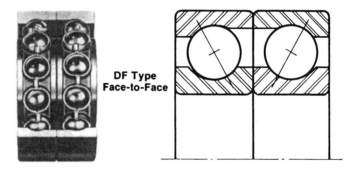

Figure 7-38. Face-to-face bearings are placed so the low shoulder of the outer rings are together.

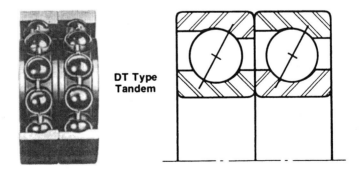

Figure 7-39. Tandem bearings are placed so that the stamped back of one bearing is in contact with the unstamped face of the other bearing.

Sometimes when duplex bearings are used in a number of arrangements, it is desirable to eliminate the need to stock duplex bearings ground specifically for DB, DF, or DT applications. Two types of single bearings, also called a $\frac{1}{2}$ pair, can be used in such circumstances.

DS—Bearings are ground with special control of faces to provide either specific preloading or end play. Preloads can be light, medium, heavy, or special while end plays are always special. Normally MRC brand "DS" replacement duplex bearings will be supplied ground with predetermined light preload for universal mounting. It is important not to mix DS bearings with other types. Be sure that bearings which are used together have identical markings on their individual boxes.

DU—Bearings are free-running and have no end play. The inner and outer rings are ground flush and may be matched in DB, DF, or DT mountings, thus permitting complete interchangeability with like bearings. As with DS bearings, they should be paired only with bearings which have identical box markings.

Packaging

All MRC brand DB, DF, and DT duplex bearings are banded in pairs in the manner in which they are to be mounted on the shaft. The duplex set is then packaged and the box stamped with the appropriate symbol.

Universally ground DS and DU bearings may be packaged separately or two to a box (Figure 7-40). Bearing number, tolerance grade, cage type and preload are shown on each box. It is this information that must be used when pairing MRC brand DS and DU bearings for mounting.

Spacers Separating Duplex Bearings

Equal length spacers mounted between the two inner and outer rings of a duplex pair of bearings are intended to provide greater rigidity to the assembly and incidentally may increase the rigidity of the shaft.

The relative rigidity of DB and DF mountings compared to the DB pair with a spacer is indicated by the heavy black bar (moment arm) between the extended lines of the bearing contact angles (Figure 7-41). Within reasonable limits, the longer the moment arm, the greater resistance to misalignment. In the DF arrangement, space between the converging contact angles is short and shaft rigidity is relatively low. However, this mounting permits a greater degree of shaft misalignment. As the angles are spread

Figure 7-40. Universally ground DS and DU bearings may be packaged separately or two to a box.

by the DB mounting to cover a greater space on the shaft, rigidity is correspondingly increased. With spacers between the DB pair, greatest resistance of misalignment is obtained. Figures 7-42 and 7-43 show typical examples of mounting bearings for high speed operation using a pair of DB type bearings and a pair of DT type bearings with spacers.

Faces of these spacers must be square with the bores and OD's of their respective locating surfaces. During removal of bearings from the shaft, spacers and bearings must be identified as to radial position so that they may be remounted in exactly the same relationship as removed. Changing position of any of the elements is likely to change the balance of the assembled spindle.

If it is necessary to replace the spacers for any reason, the new ones must be exactly the same length as the original pair. Faces must be parallel and square with the bore of OD depending on whether it is the inner or outer spacer. In addition, the bore and outside diameter of the inner and outer spacers, respectively, should have dimensions and tolerances nearly the same as the bearings they separate. The spacers will then be properly centered on the shaft and in the housing, preventing an unbalance of the assembled spindle.

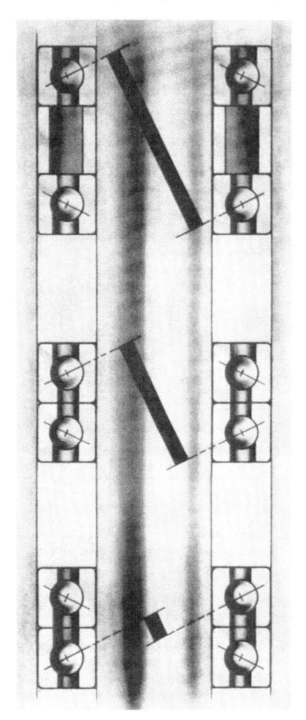

Figure 7-41. The heavy black bar indicates the relative rigidity of DB and DF mountings compared to the DB pair with a spacer.

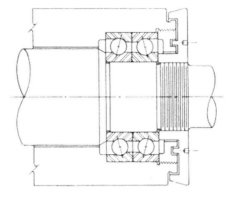

Figure 7-42. Duplex pair mounted in back-to-back (DB) arrangement without spacers.

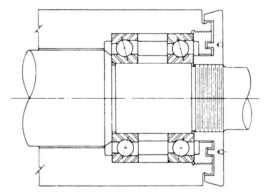

Figure 7-43. Duplex tandem pair separated by equal length spacers between inner and outer rings.

Hints on Mounting Duplex Bearings

If duplex bearings are mounted in any combination other than the one for which they were originally ground, the following conditions may occur:

1. Preload bearings may lose their preload or be greatly overloaded, respectively causing poor performance and premature failure.
2. If mounted DB instead of DF, the bearing will not take care of possible misalignment.
3. If mounted DF instead of DB, the bearing will not give the proper rigidity to the shaft.
4. If mounted DF with the outer rings floating instead of DB, the bearing may be loose and have no preload. In addition, the balls

might run over the low shoulder, causing extreme localized loading and premature failure.

5. If mounted DT in the wrong direction, the bearings may support excessive thrust load against the counterbore or the low shoulder of the outer ring.

Do Not Use Two Single-Row Bearings as Duplex Bearings Unless Properly Ground or Shimmed

Two ordinary angular-contact or radial single-row bearings generally cannot be combined to make a duplex pair. Duplex bearings are usually produced with extreme accuracy, and the twin units are made as identical as possible. Pairing of unmatched single-row bearings will result in any one of a variety of conditions: excessive or inadequate preload, too much end play, internal looseness, etc. In any case, spindle operation will be affected and early bearing failure could occur. However, it is permissible to separate a duplex bearing into two halves (each a single-row bearing) and use them as separate bearing supports.

Fit on Shaft

Duplex bearings generally have a looser fit on the shaft than other standard types of bearings. "Push" fits (finger pressure fits), are generally employed. This helps prevent a change of internal characteristics and facilitates removal and remounting of the bearings. Where heavier fits are employed, special provisions must be made internally in the bearing.

Faces of Outer Rings Square with Housing Bore

It is very important that, in the back-to-back (DB) position, the faces of the outer rings be perfectly square with the housing bore. It is possible that the units of a duplex bearing can be tilted even though preloaded, thus introducing serious inaccuracies into the assembly. The primary cause of tilting is an inadequate and/or off-square shoulder contacting the low shoulder face. This forces the outer ring to assume an incorrect position in respect to the inner ring resulting in excessive bearing misalignment. Localized overloading is caused and generally results in early failure. Foreign matter between the bearings or between the shaft and housing shoulders and faces, inaccurate threads and off-square face of nut with respect to threads also are contributing causes of misalignment that may result in premature failure.

Dismounting and Remounting of Duplex Bearings

Duplex bearings must be kept in pairs as removed from the spindle. Tag the bearings to indicate which end of the shaft and in which position they were so that they can be replaced in the same position when reassembling.

When new bearings are to replace old ones, they should be the exact equivalent of those removed from the shaft and must be mounted in the same relationship. Even if only one old bearing in a set requires replacement, it is recommended that all bearings be replaced at the same time. This will avoid the dangers involved when trying to match two bearings, one of which has unknown characteristics and unknown life expectancy.

Preloading of Duplex Bearings

Bearings are made with varying degrees of internal looseness. This allows for expansion of the inner ring and increases the capacity of a bearing when it is subjected to a thrust load. An excessive amount of interference between the inner ring bore and the shaft seat or a higher-than-anticipated temperature differential between the inner and outer rings of the bearing will reduce internal clearance in the bearing below the optimum value, creating a detrimental effect on bearing life and performance.

In a bearing subjected to a load with a very small or no thrust component, an excessive amount of internal clearance may result in poor shaft stability and high heat generation due to ball skidding in the unloaded segment of the bearing. A thrust load applied to the bearing by preloading, or applying a spring load, will alleviate both of these undesirable conditions.

When a load is applied to a bearing, deflection occurs in the contact area between the balls and races due to the elastic properties of steel. The relationship between load and deflection is not a straight line function. For example, if the load applied to a bearing is multiplied by a factor of two, the deflection in the contact areas between the balls and races is multiplied by a factor considerably less than two. As the load on a bearing is increased, the lack of linear relationship between the load and the deflection becomes very pronounced. A point is reached where rather large increases in load result in very insignificent increases in deflection.

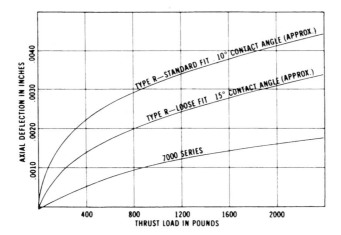

Figure 7-44. Axial deflections.

In Figure 7-44, the three curves show the difference in the axial deflections among Type R bearings with standard (AFBMA Class 0) and loose (AFBMA Class 3) internal fits and the 7000 Series angular-contact bearings (29° contact angle). Using the Type R standard fit curve as a basis of comparison, the deflection for Type R loose fit bearings with 100 pounds of thrust load is approximately 50 percent less while the 7000 Series is about 85 percent less. As the thrust load is increased, the abrupt rise in deflection shown for lower thrust loads is nearly eliminated. The leveling out of curves continues and, at 2,400 pounds, the deflection rate is reduced 25 percent and 58 percent, respectively. This ratio between the curves remains fairly constant for loads above this point.

This comparison shows that bearings which have a low degree of contact angle (Type R standard fit) usually have the highest rate of axial deflection. The greatest increase takes place under low thrust load. Bearings with a high angle of contact, such as the 7000 Series, tend to retain a more even rate of deflection throughout the entire range of thrust loads.

Figure 7-45 illustrates the difference in radial deflection among the same types of bearings in Figure 7-44. The amount of radial deflection for all three types is more closely grouped with small differences between each type for the amount of radial load applied. In contrast to Figure 7-44, radial deflection increases in relation to the degree of contact angle in the bearing type. The Type R standard fit bearing which has an initial contact angle of approximately 10° has a lower rate of radial deflection than the 7000 Series bearing with an initial contact angle of 29°.

Figures 7-46, 7-47, and 7-48 illustrate the effects of light, medium, and heavy preloads on bearings of each type. The top curve on each chart is

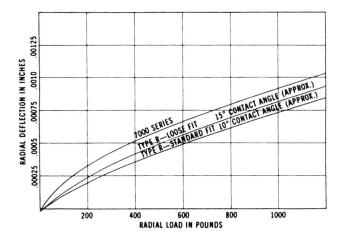

Figure 7-45. Radial deflections.

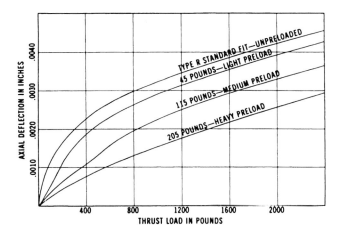

Figure 7-46. Axial deflections for type R standard fit bearings.

for duplexed unpreloaded bearings and is the same curve used on Chart 1, for a single bearing of the same type. The light, medium and heavy preload curves show reduction of axial deflection that can be obtained for bearings of each type. It is interesting to note that, in all cases, the axial deflection for preloaded types is reduced throughout the entire curve. At the low end of the applied loads, the increase is considerably less than in unpreloaded bearing types and the deflection rate levels off throughout the entire curve.

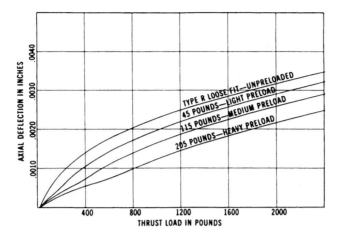

Figure 7-47. Axial deflections for type R loose fit bearings.

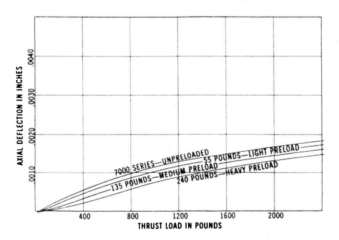

Figure 7-48. Axial deflections for 7000 series bearings.

The deflection curves presented on these charts represent calculations determined from 207-R and 7207 bearings. These specific axial and radial deflection conditions occur only in these bearings. However, in general, these curves do indicate what may occur in other bearings in the same series. Both axial and radial deflection characteristics will change in general proportion as the bearing size in the same series is increased.

It is the relationship between load and axial and radial deflections that frequently makes it desirable to preload bearings. Preload refers to an initial predetermined internal thrust load incorporated into bearings for

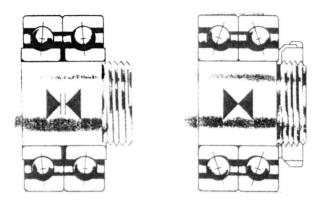

Figure 7-49. Duplex pair with preload mounted back-to-back.

the purpose of obtaining greater axial and radial rigidity. By careful selection of bearing type and amount of preload, axial and radial deflection rates best suited to a specific application can be obtained.

When a duplex pair with preload is mounted back-to-back (Figure 7-49) there is a gap between the two inner rings. As the two bearings are clamped together, the two inner rings come in contact to eliminate the gap. This changes the ball position to contact both inner and outer raceways under load establishing the basic contact angle of the bearings. The centerline on the balls shows this change. Bearings duplexed back-to-back greatly increase the effective shaft rigidity especially to misalignment. When equal (and square) pairs of spacers are used with these bearings, effective rigidity is increased still further.

In the face-to-face arrangement (Figure 7-50), the preload offset is between the outer rings. After clamping, the balls come into contact with both inner and outer races at the basic contact angle. Effective radial rigidity of the shaft is equal to that of the back-to-back arrangement; but less rigidity is given to conditions of misalignment.

Preloading is accomplished by controlling very precisely the relationship between the inner and outer ring faces. A special grinding procedure creates an offset between the faces of the inner and outer rings of the bearing equal to the axial deflection of the bearing under the specified preload.

When two bearings processed in an identical manner are clamped together, the offset is eliminated, forcing the inner and outer rings (depending on where the offset occurs) to apply a thrust load on the balls and raceways even before rotation is started. This results in deflection in the contact areas between the balls and races that corresponds to the amount of preload that has been built into the bearings. Balls are forced to contact

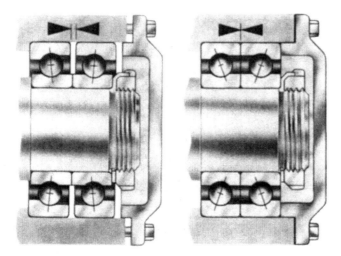

Figure 7-50. Face-to-face arrangement.

the raceways immediately upon clamping the bearings together, thus eliminating the internal looseness. An additional load applied to the set of bearings will result in deflections of considerably smaller value than would be the case if the bearings were not preloaded.

Preload Offset

The relationship of the inner and outer ring faces of DB and DF pairs of bearings duplexed for preload is shown in Figures 7-49 and 7-50. Note the gap between the inner rings of the DB pair and the outer rings of the DF pair. This is referred to as preload offset. In the illustrations the offset has been greatly exaggerated to show the action that takes place. In most cases the offset is so small that it cannot be detected without the proper gauging equipment.

DTDB and DTDF Sets

Tandem duplex bearings may be preloaded under certain conditions. These are applications where a tandem set of two or more bearings is assembled either DB or DF with a single bearing (Figure 7-51) or another tandem set of two or more bearings. Preload in these sets is the clamping force, applied across outboard sides of the set, necessary to bring all mating surfaces in contact.

Figure 7-51. A tandem set of two or more bearings is assembled DB or DF with single bearing.

Preload for individual tandem bearings in a set must be equal to the preload of the set divided by the number of bearings on each side.

Example: Three DT bearings are matched DB with two DT bearings. The set has a 600 lb preload.

Each of these DT bearings (one side) must have preload of 600/3 or 200 lbs.

Each of the two DT bearings (other side of set) must have 600/2 or 300 lbs.

Importance of the Correct Amount of Preload

Since the deflection rate of a bearing decreases with increasing load as shown in Figures 7-44 and 7-45, it is possible, through preloading, to eliminate most of the potential deflection of a bearing under load. It is important to provide the correct amount of preload in each set of duplex bearings to impart the proper rigidity to the shaft. However, rigidity is not increased proportionately to the amount of preload. Excessive preload not only causes the bearings to run hotter at a higher speed but also reduces the operating speed range. As machine tools must perform many types of work under varying conditions, the proper preload must be provided for each bearing to meet these conditions while retaining operating temperatures and speed ranges to which the bearings are subjected.

Duplex bearings are generally manufactured so that the proper amount of preload is obtained when the inner and outer rings are simply clamped together. If the duplex bearing has the correct preload, the machine will function satisfactorily with the proper shaft rigidity and with no excessive operating temperature. Any change in the initial preload is generally undesirable and should be made only if absolutely necessary. This is especially true for machine tool spindle bearings that are made to extremely fine tolerances. Any attempt to change the initial preload in these bearings is more likely to aggravate the faulty condition than correct it.

Factors Affecting Preload

There are various conditions which may adversely affect the initial preload in duplex bearings:

- Inaccurate machining of parts can produce a different preload than originally intended, either increasing or decreasing it depending upon the nature of the inaccuracy
- Use of spacers that are not equal in length or do not have the faces square with the reference diameter (OD or ID) can produce an improper preload
- Foreign matter deposited on surfaces or lodged between abutting parts as well as nicks caused by abuse in handling may produce cocking of the bearing and misalignment. Either condition can result in a variation of the preload or binding in the bearing.

The following precautions should be taken to avoid distortion when the parts are clamped together.

- Make a careful check of the shaft housing shoulder faces and the end cover surfaces abutting the bearing to see that they are square with the axis of rotation
- Make sure that the end surfaces of each spacer are parallel with each other and square with the spacer bore
- Carefully inspect the lock nut faces for squareness
- Inspect all contacting and locating surfaces to make sure they are clean and free from surface damage

Preload Classifications

MRC brand Type R and 7000 Series angular-contact ball bearings are available with any of three classes of preloads—light, medium, or heavy. The magnitude of the preload depends upon the speed of the spindle and required operating temperatures and rigidity requirements.

Preloaded Replacement Bearings

Normally replacement duplex bearings will be supplied universally ground with predetermined light preload. These are designated as "DS" bearings. If preload recommendations are desired when ordering bearings, all data possible, such as the equipment in which the spindle is used, spindle speeds, loads, and lubrication, should be supplied.

Preloaded Bearings with Different Contact Angles

Less than 5 percent of all pump bearings reach their calculated life. Compared to the *average* calculated thrust bearing life of 15 to 20 years, actual application life for pump bearings in the hydrocarbon processing industry (HPI) is only 38 months or less based on 2004 data.

Preloaded bearings with different contact angles can significantly increase the service life of bearings in many pump applications. The key to their superior performance lies in the system's directionally dissimilar yet interactive spring rates. One such bearing system, MRC's "PumPac," consists of a matched set of 40° and 15° angular contact ball bearings with computer-optimized internal design. It is designed to interact as a system, with each component performing a specific function.

By using this special set of bearings, ball skidding and shuttling are virtually eliminated. The result: lower operating temperatures, stable oil viscosity, consistent film thickness, and longer service life.

Figure 7-52 depicts a shaft equipped with MRC's "PumPac." The two bearings are mounted back-to-back, with the apex of the etched "V" pointing in the direction of predominant thrust.

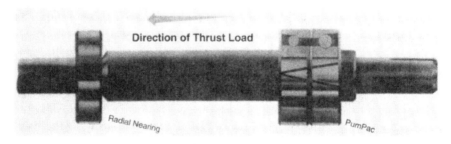

Figure 7-52. Preloaded thrust bearing set with different contact angles counteracts skidding of rolling elements (courtesy MRC Bearings, Jamestown, New York).

Assembly of Bearings on Shaft

Bearing Salvage vs. Replacement Considerations

The final decision now must be made whether to reuse the bearings removed from the spindle or to replace them with new bearings. The choice probably will be self-evident, especially after the visual inspection mentioned in item #5 on the checklist in Table 7-5.

If the bearing has defects that will affect its operation, it must be replaced with a new bearing of the same size and tolerance grade.

Experience will be a guide in determining if the bearing is to be replaced. The apparent condition of a bearing will not be always a deciding factor. Bearings can still be used if they are not badly pitted or brinelled on nonoperating surfaces. This also applies to bearings that do not show excessive wear or signs of overheating. There are some instances where the boundary dimensions may have been affected by operation. Where possible, they should be checked to determine if they are within the desired tolerances.

Often a simple check on a bearing's internal contact surfaces can be made by spinning the bearing by hand. This may be done after the bearing has been thoroughly cleaned to eliminate possible harmful grit inside it. If the bearing has some imperfect contact surface, this can be felt when

---

Table 7-5
Spindle Servicing Checklist

At this point, all cleaning and repair work on the shaft and spindle parts should have been completed. A review of all steps taken in the servicing of a spindle are listed here for checking purposes.

1. Remove shaft and bearings from the housing.
2. Dismount bearing from shaft using arbor press or bearing puller.
3. Tag bearings and spacers (if any) for identification and proper location when remounting on the shaft.
4. Clean bearings and spindle parts.
5. Make visual inspection of all spindle parts for nicks, burrs, corrosion, other signs of damage.
6. Prepare shaft for remounting of bearings. Make any repairs necessary on bearing seat, shaft shoulders, fillets, etc.
7. Prepare housings by making any required repairs on machine mounting surfaces, Paint non-functional surfaces as necessary.
8. Check shaft and housing measurements for bearing seat out-of-round, off-square shoulders, housing bore, etc.

spinning the outer ring slowly while holding the inner ring (Figure 7-53). This test should be made under both lubricated and dry conditions. However, when dry, extreme care must be taken when spinning the bearing as the rolling surfaces of the balls and raceways are even more sensitive to possible scratching by grit.

Another point to consider is anticipated bearing life. If a bearing has been in service for a long time and, according to the records, is nearing the end of its natural life, it should be replaced with a new bearing. If a longer life can be expected, then an evaluation must be made comparing the cost of a replacement bearing against the remaining life of the old bearing and its later replacement. Also, the evaluation should take into account the possibility of *new* bearings in certain services having a statistically provable *higher* failure rate than bearings that have been in successful short time service.

If a replacement bearing is to be used, it should be understood that dimensional interchangeability does not necessarily guarantee functional interchangeability. In certain applications, there are other characteristics

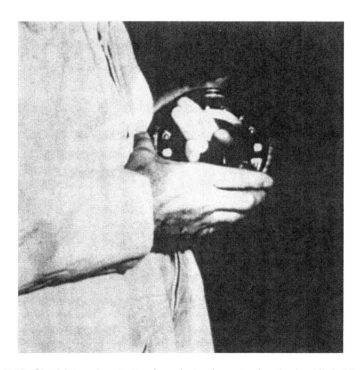

Figure 7-53. Check internal contact surfaces by turning outer ring slowly while holding inner ring.

such as internal fit, type and material of cage, lubricant, etc., that are of vital importance. If you have questions about the selection of the correct ball bearing replacement, it is always wise to consult the product engineering department of capable major bearing manufactuers.

Cautions to Observe During Assembly of Bearings into Units

Whether using the original bearing or replacing it with a new one, care must be taken to avoid contamination when mounting the bearing. A critical period in the life of a bearing starts when it leaves the stockroom for the assembly bench where it is removed from its box and protective covering. This critical period continues until the bearing passes its first full-load test after assembly. Here are a few rules that should be observed during this crucial period.

1. Do not permit a bearing to lie around uncovered on work benches (Figure 7-54).
2. Do not remove a bearing from its box and protective covering until ready for installation.
3. When handling bearings, keep hands and tools clean.
4. Do not wash out factory-applied lubricant unless the bearing has become exposed to contamination.
5. If additional lubrication must be applied, be sure it is absolutely clean. In addition, the instrument used for application must be clean, and chip and splinter proof.

Figure 7-54. Keep unboxed bearings covered until ready for mounting.

Figure 7-55. Cover subassemblies, especially those with mounted bearings, with plastic material while waiting to assemble into housing.

   6. If subassemblies are left for any length of time, they should be lightly covered with clean, lintless material (Figure 7-55).

   Some other precautions to be exercised during assembly were discussed under "Cleanliness and Working Conditions" earlier. If there is any chance that the bearing may have become contaminated, don't take any chances—wash the bearing again following the procedure outlined in that section.
   In summary, many precautions have been taken by the bearing manufacturer to make sure that the bearings are delivered in a clean condition. In a few seconds, carelessness can destroy the protective measures of the manufacturer . . . shorten the life of the bearing . . . jeopardize the reputation of the organization for which you work. It pays to do everything possible to prevent abrasive action caused by dirt in a bearing. But assembly precautions do not stop here. The user must resist his inclination to "clean" a bearing by removing the preservative coating applied by the bearing manufacturer. Prelubrication is not usually necessary and extreme vulnerability can he introduced by precoating certain rolling element bearings with extreme light viscosity or inferior quality oils.

High Point of Eccentricity

   When remounting bearings with tolerance grades of ABEC-5, ABEC-7, or ABEC-9, it is essential to orient them on the shaft with reference to the "high point of eccentricity." Super-precision bearings are usually marked to indicate this detail.

The high point of eccentricity of the outer ring is the highest reading obtained when measuring its radial runout. It is found by placing the bearing on a stationary arbor and applying an indicator directly over the ball path on the outside diameter of the outer ring. When the outer ring is rotated, the difference between the highest and lowest reading is the amount of radial runout of the outer ring. The high point of eccentricity of the inner ring is determined in the same manner except that the inner ring is rotated. To indicate the high point of eccentricity, a dot is burnished on both inner and outer rings (Figure 7-56) on Type R and 7000 Series angular-contact bearings of ABEC-5 or higher tolerance grades.

The burnished dots are applied to the rings so that the bearings can be mounted to reduce or cancel the effects of shaft seat runouts. When mounted, the dots should be 180° from the high point of eccentricity of the bearing seat on each end of the shaft. The high point of eccentricity of the shaft also should be determined and marked when the shaft is on centers (or V-blocks). This method of mounting will help keep radial runout of the spindle assembly to a minimum. This is important especially in high speed applications. Matching the burnish marks in duplexed bearings will reduce internal fight between bearings.

Thrust Here

The words "Thrust Here" are stamped on the back of the outer ring of all MRC brand Type R and angular-contact bearings. This serves as a guide when mounting the bearing so that the shaft thrust carries through the bearing (Figure 7-57). Note that in the "right" method of mounting,

Figure 7-56. Burnished dots show high point of eccentricity.

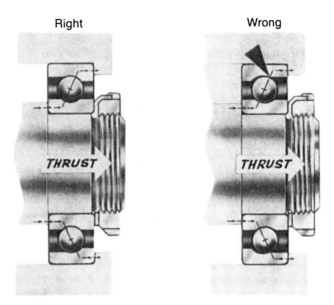

Figure 7-57. "THRUST HERE" on outer ring shows the side of the ring to which shaft thrust is to be imposed. Improper mounting may force balls to ride the edge of the low shoulder.

the thrust is along the shaft, through the inner ring along the angle of contact of the balls, through the heavy shoulder of the outer ring (stamped "Thrust Here") to the shoulder of the housing.

If the bearing position were reversed as shown in the "wrong" method, the shaft thrust would follow the angle of contact through the low shoulder side of the outer ring. As the outer ring will not carry loads of any magnitude, it is likely that the thrust would then force the balls to ride the edge of the low shoulder. This could cause early failure due to concentrated loads at the race-shoulder intersection and possibly even cause cracking of the balls.

Mount Bearings with Push Fit

Precision bearings used in the machine tool industry normally are mounted on the shaft with a push fit, that is, pressing the bearing in place (Figure 7-58) with hand pressure. In some cases, the original bearings may be used again. No difficulty should be encountered while mounting them. However, if a new bearing is to be used, the proper tolerance grade bearing must be selected so that a push fit results. An application of light oil on the shaft will increase ease of mounting.

Figure 7-58. Shaft is held in a vise when mounting bearing with a push fit. Cover vise jaws with wood or soft metal.

Mounting with Heat

If a bearing is to be mounted with a tighter than "push" fit, a convenient and acceptable method of mounting is to expand the rings by moderate heating. To make sure the bearing is not overheated, a thermostatically controlled heat source should be used. An inexpensive type of household oven will usually serve the purpose satisfactorily. The oven also protects the bearing from contamination while being heated.

Before heating, the bearing must be removed from its plastic packing bag or other wrapping as the temperature reached may melt the material. If gloves are used when handling the bearing, they should be made of a lint-free material such as nylon or neoprene. Set the thermostat to a temperature between 175°F and 200°F (80° to 94°C). In most cases, this will be adequate to expand the bearing without overheating it. Heat the bearing a sufficient amount of time to allow for ring expansion. Upon removal, slip the bearing on the shaft immediately with full regard for direction of thrust as well as orientation of the high points of eccentricity of both bearing and shaft. Certain bearings not employing cages, seals, or lubricants susceptible to damage at the higher temperature may be heated to 275°F (136°C).

Induction heaters (Figures 7-59 through 7-61) offer several different heating programs to suit user requirements in a variety of situations. The heaters are suitable for continuous use, and their compact temperature probes ensure exceptionally exact temperature control. Most importantly,

Figure 7-59. Typical induction heater for mounting rolling element bearings (courtesy Prüftechnik A. G., Ismaning, Germany).

Figure 7-60. Large induction heater used for bearing assembly on machinery shafts (courtesy Prüftechnik A. G., Ismaning, Germany).

Figure 7-61. Compact bearing induction heater (courtesy Prüftechnik A. G., 8045 Ismaning, Germany).

properly engineered, state-of-the-art induction heaters completely demagnetize the bearing automatically at the end of the heating cycle, because, otherwise, the magnetized bearing would literally act as a trash collector for ferritic particles, leading to an untimely demise of the bearing.

Large induction heaters include features such as swivel-arm crossbar design and a pedal-operated crossbar lift, which allows one-man operation even when mounting extremely large workpieces, while the heavy, welded-steel carriage provides on-site portability. Further features might include auto demagnetization, automatic temperature probe recognition, dynamic heating power regulation, and suitability for continuous operation.

Typical technical data of large units are as follows:

Power consumption:	14 kVA max.
Heating capacity:	approx. 400 kg (880 lb)
Heating duration:	10 sec.–1 hr
Precision, Time:	+1 sec.
Temperature:	+2°C (3.6°F)

More compact and still robust, smaller units may offer such standard features as microprocessor-controlled heating by time or temperature, auto demagnetization, automatic temperature probe recognition, and dynamic heating power regulation. Technical data of the more compact units are typically as follows:

Power consumption:	3.5 kVA max.
Heating capacity:	approx. 15 kg (33 lb)
Heating duration:	10 sec.–1 hr
Precision:	better than 3°C (5.4°F)
Time:	+1 sec.
Temperature:	+2°C (3.6°F)

Other Mounting Methods

A variation of the dry heat method to expand the bearing involves the use of infrared lamps inside a foil-lined enclosure. The lamps should be focused on the inner ring. Care must be taken to keep the temperature below 200°F (94°C), except as noted previously (Figure 7-62).

It also is possible to use dry ice to cool the shaft which will then contract sufficiently to permit mounting of the bearing with the proper finger pressure. In this method, special precautions should be used to prevent resulting condensation from producing corrosion on the bearing components.

An arbor press and a hollow tube (Figure 7-63) are frequently employed to mount bearings on shafts in those cases where press fits are involved.

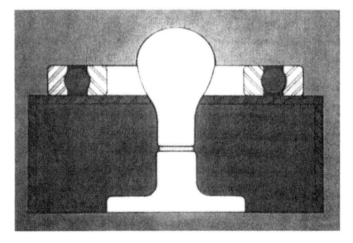

Figure 7-62. A variation of the infrared lamp method suited for large size bearings. Take care not to overheat the bearing.

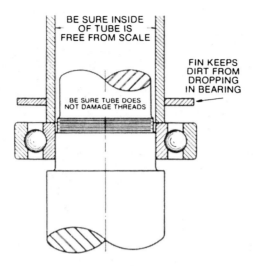

Figure 7-63. Arbor press and hollow tube method used to mount bearings on shafts when press fits are involved.

When doing so, the press and the tube should be completely clean to avoid possible contamination of the bearing. The tube must contact the inner ring when pressing the bearing on the shaft (Figure 7-64). This will avoid possible brinelling of the bearing which might occur if pressure were applied on the outer ring. The press ram, tube, and bearing axis should be in good alignment. If abnormal pressures are required, the alignment probably is not good enough. In this case, alternate application of pressure and relief may help ease the pressure required.

Exercise Caution When Starting Bearing On Shaft

To start the bearing, the shaft should have a lead and the bearing face should be square with the shaft (Figure 7-64). Pressure should be in line with the shaft and must be uniform against the face when pressing the bearing into place. If excessive binding occurs, it generally indicates an off-square condition, a burr, high spots, a tapered shaft, dirt or chips wedged under the bearing. When sticking or binding occurs, the bearing should not be forced on the shaft as the hard inner ring is apt to cut the softer metal of the shaft and raise a ridge or burr (Figure 7-65). Remove the bearing from the shaft to determine and correct the problem.

Figure 7-64. Bearing may be pressed on shaft using a tube and arbor press.

Checking Bearings and Shaft After Installation

After the bearings have been assembled on the shaft, a number of points should be checked to make certain the bearings have been correctly installed. These include visual checks and the use of various gauges to determine the accuracy of the mounting. It also may be necessary to balance the shaft assembly before insertion into the housing.

Check for Internal Clearance

The outer rings should rotate freely without binding except in unusual cases where a tight fit has been specified or where preloaded duplexed

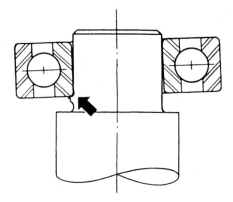

Figure 7-65. Result of starting bearing off-square.

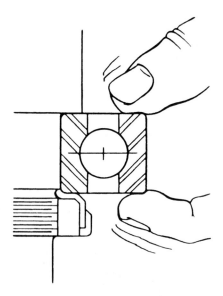

Figure 7-66. Check a mounted bearing for internal clearance by rocking outer ring back and forth.

bearing sets are used. Residual internal clearance can be felt by holding the outer ring between the thumb and forefinger (Figure 7-66) and rocking it back and forth. If the bearing is free, the ring will have a slight axial freedom of movement or "rock." This applies to all single-row bearings except a single bearing of the angular-contact type which is very loose.

Make Visual Check of Bearing

Be sure that the bearing is flush against the shaft shoulder all the way around. The best simple check for this is to hold the shaft in front of a light source such as a window or an electric light. If no light shows between the inner ring face and the shaft shoulder, the bearing may be considered in proper position. This check should be made all the way around the shaft. If light is visible at any point, carefully remove the bearing and recheck the shoulder and fillet for burrs, out-of-round or too large a radius on the fillet.

Check the height of the shoulder against the height of the inner ring. In general, it should be a minimum of about half the width of the inner ring face. If heavy thrust pressures are involved, the shoulder should be higher.

Check for Bearing Squareness on the Shaft

Check the face of the outer ring for squareness of the inner ring with the shaft using a suitable indicator (Figure 7-67). The assembled front and rear bearings or sets of bearings should be placed in V-blocks with an indicator point contacting the face of the outer ring. When the shaft and inner rings are turning, any off-square condition is transmitted through the balls to the outer ring. This will cause the outer ring to rock or tilt and shows up as a variation on the indicator reading. Readings in excess of the bearing tolerances indicate an effective misalignment, resulting from an off-square condition.

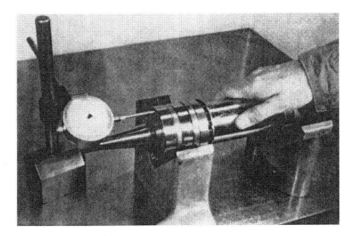

Figure 7-67. Checking for squareness of bearing on shaft.

Foreign matter between the bearings and shaft shoulder, fillet interference, a nick on the shaft shoulder, raised metal from the bearing seat, a nick on the spacer rings, and many other causes will produce this off-square condition. Under these circumstances, remove the bearing from the shaft to determine the actual cause and make the necessary repairs.

Balancing the Shaft Assembly

After the bearings and other units such as pulleys, etc., are properly seated on the shaft, the assembly should be balanced, preferably dynamically, to obtain a smooth-running spindle (Figure 7-68). All parts that rotate with the assembled spindle should be included in the balancing operation with whatever is applied to retain the bearings on the shaft.

Common Causes of Unbalance in Shaft Assemblies

Unbalance is commonly introduced through an eccentricity in some portion of the shaft assembly that has not been properly finished. Eccentricities may be present in the components affixed to the shaft. Ground bearing seats may not be concentric with turned portions of the assembly. Strains may develop in a shaft that has been heat-treated, causing warp in the shaft that may create misalignment of the bearings. In other cases, a thread may not be true to the shaft center. If the bore of the bearing spacer is too loose when used, it may not be properly centered with respect to the shaft axis. Causes of unbalance could include other factors in addition to these items.

Correction of Unbalanced Shaft Assemblies

There are several methods to correct unbalance in either hardened or soft shafts. In general, balancing consists of removing material from the heavy side of the shaft or adding material to the light side of the shaft. Any balancing operations should be conducted in an area removed from the clean assembly area.

To balance a hardened shaft, sufficient material usually is removed to create proper balance by grinding on the heavy side of the shaft nearest the end which needs to be balanced. The grinding is usually done on a portion of the shaft where the largest diameter occurs.

Several methods may be used to bring a soft shaft into balance. It is possible to apply a correct amount of weight. It may be preferable to drill

Figure 7-68. Dynamics of balancing spindle assembly. This equipment is used by TRW's Spindle Maintenance Department, but other types are available which will balance assemblies as accurately as necessary.

a hole of the correct size and depth in the shaft. Metal generally is removed from the shaft in the area of the largest diameter and at the greatest possible distance from the center of the shaft toward the end which must be balanced. Extreme care must be taken to prevent the removed metal from being introduced into the bearings mounted on the shaft.

Protect Bearings and Shaft Assembly from Contamination

After a ball bearing has been mounted on a shaft, often there is a time interval, possibly overnight, before final installation in the housing can be

Figure 7-69. Protect bearings and shaft from contamination until assembled and completely sealed in housing. Use plastic film to cover spindle.

started. In such cases, it is advisable to wrap the bearings and shaft in plastic film to protect them from contamination (Figure 7-69). If the bearings are left exposed on the shaft, or even when installed in the housing, dust and/or other contaminants may enter the bearing. If installed in the housing, always be sure to cover the open end with film until all assembly work is completed and the housing is completely closed.

Assembly of Shaft and Bearings into Housing

After the bearings have been assembled on the shaft, checked and balanced, the entire assembly is ready for insertion into the housing (Figure 7-70) in accordance with the methods outlined in the manufacturer's manual. During this operation, care must be taken to start the bearings into the housing seats squarely to avoid damage to the bearings or housing. Any force exerted on the shaft passes through the bearings and, if excessive, can cause bearing damage.

The outer ring generally should have a slightly loose fit in the housing. This is necessary to permit the bearing axial movement to assume its normal operating position regardless of temperatures which occur during operation. If the bearing is too tight, axial movement is prevented and violent overloads could result in nonfixed radial positioning of the shaft.

Figure 7-70. Start bearings into housing seat squarely to avoid damage to either bearings or housing. A "push" fit should be used.

This causes noisy operation, excessive fretting and pounding out of the housing seat and, occasionally, excessive spinning in the housing.

Testing of Finished Spindle

When the spindle has been completely assembled, a final check of the eccentricity should be made as follows. Place an indicator point against the center of the shaft extension on the work end of the spindle and rotate the spindle by hand (Figure 7-71). Where possible, the opposite end of the spindle also should be checked. Sometimes it is possible to detect roughness or vibration in the spindle when turning the shaft by hand. Roughness may be felt as a hitch or click. Do not attempt to run the spindle if these conditions are major. Dismantle it and find the cause of the roughness. The cause may be the application of excessively dirty lubricant or dirt that has worked its way into the bearing during assembly. Under conditions of slow rotation, necessary cage looseness also may create some binding which disappears when running to speed and under load.

Vibration is usually detected when the spindle reaches its normal operating speed. Causes of vibration include excessive runout of the pulley, a loose cap on the spindle assembly, bearings damaged in assembly, or possibly loose bearing spacers. In any case, when vibration is detected, spindle run-in should be discontinued to investigate the cause.

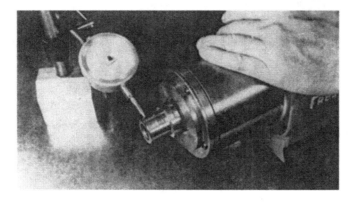

Figure 7-71. Check eccentricity of assembled bearing with indicator gage on shaft extension.

During run-in, a close check should be kept on the temperature attained, especially in the first part of the run. This is particularly true for spindles which are grease-packed. If the heat becomes excessive, over 140° to 150°F (60° to 66°C), it is usually advisable to stop the spindle and permit it to cool off. This type of excessive heat is commonly caused by insufficient channeling of the lubricant in the spindle. Stopping the spindle will allow the temperature to equalize, reducing the risk of radial or axial preloading. When restarted, the spindle usually will run at temperatures within the recommended range unless excessive quantities of grease are in the housing or if overloads are present. If the spindle continues to heat excessively, checks should be made to determine the cause.

Maintain Service Records on All Spindles

The maintenance department in any company should keep records regarding the history of each shaft or spindle serviced (Figure 7-72). All particulars should be recorded from the day the spindle was placed in operation until retirement. Such a record may be as complete and detailed as possible. It may simply record dates when the spindle or shaft was checked to correct certain conditions. In either case, it is recommended that the record state clearly the corrective actions taken to place the spindle in proper operating condition.

Shop records also will enable the maintenance department to keep a close periodic check on spindles thoughout the plant. It is possible to establish a regular inspection procedure which will help assure continued

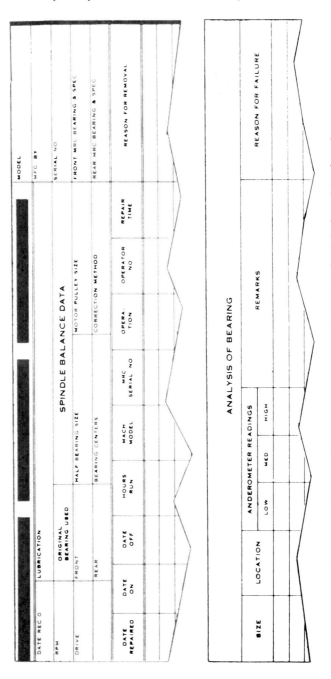

Figure 7-72. Form provides space to record complete history of spindle from date of purchase.

spindle operation. This will reduce machine downtime, always a factor in maintaining a low cost operation.

Shaft and Housing Shoulder Diameters

Table 7-4 shows minimum shaft shoulder diameters for general purpose installations.

Bearing Maintenance Checklist

Finally, we direct your attention to Table 7-6, which summarizes all necessary bearing maintenance steps in checklist form.

Table 7-6
Bearing Maintenance Checklist

Don't	Recommended Action
Don't open the bearing box until you are ready to install the bearing.	Leave the bearing in the manufacturer's box and protective wrapping until it is necessary to remove it for installation. The package is designed to protect the bearing from contamination and keep it in factory fresh condition.
Don't work in an area having a hostile environment.	Ideally areas designated for bearing installation should have controlled humidity, controlled temperature and clean air to help prevent bearing exposure to particulate or corrosive contamination.
Don't work on bench tops that are not in good condition.	The working surface should be material that is not likely to produce contamination by abrading or chipping. Splinter-free wood, phenolic composition, hard rubber or smooth metal are satisfactory surfaces for this kind of work.
Don't work on a dirty bench top.	Make certain that the working surface is free from contamination that could find its way into the bearing or corrode the bearing surfaces.
Don't force the bearing off the shaft by uneven pressure or with hammer blows if you plan to reuse the bearing.	Use a bearing puller or an arbor press with adapters which will exert a balanced load on the inner ring face. This avoids a static load on the balls and possible shaft damage.
Don't place the shaft assembly in a dirty area when it has been removed from the housing.	After the shaft assembly has been removed from the unit it should be protected from contamination and corrosion by placing it in a clean covered container, or covered with a lint and dirt-free material such as plastic film. Rotating a contaminated bearing will damage the ball and race surfaces.
Don't strike or apply pressure to the ball cage while the bearing is being dismounted.	Select the proper removal tools and adapters to insure that a load is not applied to the balls through the cage.

(Table continued on next page)

Table 7-6
Bearing Maintenance Checklist—cont'd

Don't	Recommended Action
Don't use improper tools to remove the bearing from the shaft.	Study the assembly and select the removal method, tools and fixtures that are least likely to cause damage to the bearing or the shaft.
Don't apply pressure to the outer ring of the bearing when removing it from the shaft if it can be avoided.	Remove bearings from the shaft with pressure on the inner ring face to prevent loading the balls. In those applications where the only surface available for pulling the bearing is the outer ring face, care should be taken to pull evenly and smoothly. The bearing should then be examined for possible damage.
Don't damage seals or shields of the bearing while it is being removed.	Place bearing puller jaws on adapter or inner ring as required by the method of mounting. This will avoid damage to the seals or shields caused by possible contact with the puller.
Don't push a bearing from the housing with pressure on the inner ring in applications where the bearing is a tight fit in the housing unless it cannot be avoided.	Use a section of rod or tubing slightly smaller than the bore of the housing. This will apply pressure to the outer ring face and will not brinell the bearing by actually loading the ball assembly.
Don't overlook scuff marks, burrs, nicks or other indications of damage on the bearing locating surfaces during inspection of the shaft and housing.	Carefully inspect the bearing locating surfaces and correct any damaged areas that could prevent intimate contact between the bearing and the locating surfaces.
Don't make shaft or housing repairs which may cause contamination or produce wear products in the clean assembly area.	Remove any parts from the clean assembly area when performing procedures that produce wear products, such as reworking with abrasives. Make certain the parts are clean before reintroducing them into the clean assembly area.
Don't allow too large a fillet radius between the shaft seat and the shaft shoulder or the housing seat and the housing shoulder.	Make sure the radius in each location is no larger than recommended in the bearing manufacturer's catalog. If the radius is large enough to act as a bearing locating surface it is very likely to result in bearing misalignment or ring distortion.
Don't regrind shaft shoulders without breaking corners.	Break corners while the shaft is still mounted on centers. Removal of burrs, etc. on the corners will permit proper seating of the inner ring against the shaft shoulder. Also it makes the shaft less susceptible to damage.
Don't nick or scratch housing bearing locating surfaces.	The housing bearing locating surfaces should be examined for indications of damage. Any damage found that could prevent proper location of the bearing outer ring in the housing must be corrected.
Don't wash the original lubricant from a new bearing.	New bearings are either prelubricated with grease or protected with a preservative that is compatible with most bearing lubricants. It will continue to protect the bearing until the unit is put into service.

Table 7-6
Bearing Maintenance Checklist—cont'd

Don't	Recommended Action
Don't allow a bearing to spin rapidly when drying it with compressed air.	When drying a bearing with compressed air, hold both the inner and outer rings to prevent them from rotating. Spinning an unloaded bearing in this manner can result in ball and race surface damage caused by the balls skidding against the race surfaces.
Don't neglect to clean the shaft and housing and flush out oil lubricating systems.	Remove both free and caked-on deposits from all parts of the shaft and housing assembly. This will provide clean parts which will not contaminate bearings with dirt, etc. when the spindle is reassembled and run.
Don't paint bearing seats in housing.	It is a good practice to seal the internal surfaces of the housing using a sealer that is made for this purpose. It will reduce the presence of core sand and other particles which could contaminate the bearing lubricant. The bearing locating surfaces should not be painted.
Don't mount bearings until the high point of eccentricity of the shaft has been determined.	Align bearing high points of eccentricity and mount them 180° from the high point of eccentricity on the shaft. This practice reduces the radial runout to a minimum.
Don't mount counterbored type bearings in a manner that will allow the thrust load to be imposed on the low shoulder side of the outer ring race.	A thrust load applied to a counterbored type bearing must be applied in such a manner that it will be carried on the high shoulder side of the outer ring race. The face of the outer ring on the thrust-carrying side will be marked "THRUST HERE."
Don't use two or more bearings as a pair or set of bearings unless they have been properly duplex ground for use in this manner.	MRC brand bearings that have been duplex ground to be used as a component in a pair or set of bearings are identified by the letter "D" as a suffix to the bearing number. The suffix will be DU for bearings that have been duplexed to have no end play and no preload when mounted in a back-to-back or face-to-face arrangement. The suffix DS identifies the bearings duplex ground for either preload or negative preload (end play) when matched either back-to-back or face-to-face. Bearings mounted in a tandem arrangement must also have the same suffix.
Don't mix duplex bearings and spacers which have been removed from the shaft if old bearings are to be reused.	Replace bearings and spacers in exactly the same position as they were before removing them from the shaft. Any change in position of the components is likely to result in a change in preload in the bearing pairs or sets.
Don't use spacers which are not equal in length.	The preload ground into a pair or set of bearings will be altered in applications where spacers are used between bearings if the spacer between the inner ring faces is not exactly the same length as the spacer between the outer ring faces. It is also essential that the faces of the spacers must be ground parallel and remain free from damage.

(Table continued on next page)

Table 7-6
Bearing Maintenance Checklist—cont'd

Don't	Recommended Action
Don't attempt to adjust preload which is built into the bearings.	Use only bearings that have been duplexed for the proper amount of preload for the application. Attempts to adjust preload by using spacers that are not the same length are likely to result in unsatisfactory bearing operation or possibly premature bearing failure.
Don't use worn V-blocks or centers for measuring the geometry of shaft bearing seats.	Use a good set of centers and clean center holes for accuracy. Worn centers or V-blocks can result in misleading results when measuring the concentricity of the shaft seats to the shaft center.
Don't force a bearing on the shaft seat in applications where an interference fit between the bearing shaft seat is not specified.	In applications where an interference fit between the bearing and the shaft is not called for, either strong finger pressure or a very light press is all that should be necessary to mount the bearing. If more pressure is required the shaft bearing seat diameter should be measured to insure that it is not oversized. Too tight a fit on the shaft will remove an excessive amount of internal clearance, or may alter the preload in a pair of bearings.
Don't overheat the bearings when trying to expand the inner rings to facilitate mounting them on to the shaft seat.	Use a thermostatically controlled heat source to heat bearings. Excessive temperatures can result in softening of the parts, loss of dimensional stability and damage to non-metallic separators.
Don't hold the shaft in a metal vise without protecting the shaft from the vise jaws.	Cover the vise jaws with wood or soft metal. This will avoid marring or nicking of the shaft.
Don't press bearings onto the shaft seat in an off-square manner.	Be very careful to start the bearing onto the shaft seat squarely and to use tooling or fixtures that are accurate enough to maintain squareness between the bearing and the shaft center. Off-square mounting is likely to damage the shaft seat or the bearing. Shaft seats that are damaged in this manner would not allow the bearing inner ring to have intimate contact with the ring locating surfaces.
Don't balance spindle with parts missing that will be rotating with the spindle in the final assembly.	Balance shafts with the entire bearing and pulley assembly as well as assembled accessories. This will eliminate any possible unbalance when the unit is completely assembled.
Don't leave bearings unprotected for any significant length of time.	After the bearings have been removed from the manufacturer's package they should never be left unprotected for a significant length of time. If the bearings are not mounted they can be repackaged or protected by wrapping them in a suitable material. If they have been mounted, the entire assembly should be wrapped or covered with a suitable material such as plastic film.
Don't overlubricate the bearings.	Too much grease or oil will result in excessive heat generation and can lead to unsatisfactory operation of the bearings or shorten the bearing life.
Don't use a lubricant that hasn't been protected from contamination.	Contaminated lubricant is a major cause of premature bearing removal. Lubricants should always be stored in a clean, covered container.

Chapter 8

Repair and Maintenance of Rotating Equipment Components

Pump Repair and Maintenance*

Sealing performance, bearing, and seal life will depend to a great extent upon the operating condition of the equipment in which these components are used. Careful inspection of the equipment will do much to minimize component failure and maintenance expenses.

Following is a list of the major trouble spots.

1. **Seal housing.** The seal housing bore and depth dimensions must match those shown on the seal's assembly drawing within ±0.005 in. (±0.13 mm). Shaft or sleeve dimensions must be within ±0.001 in. (±0.03 mm). See Figure 8-1 for complete seal housing requirements.
2. **Axial shaft movement.** Axial shaft movement (end play) must not exceed 0.010 in. (0.25 mm) Total Indicator Reading (T.I.R.) To measure axial movement, install a dial indicator with the stem bearing against the shaft shoulder as shown in Figure 8-2. Tap the shaft—first on one end then the other—with a soft hammer or mallet, reading the results.

Excessive axial shaft movement can cause the following problems:

- Pitting, fretting, or excessive wear at the point of contact between the seal's shaft packing and the shaft (or sleeve) itself. It is sometimes helpful to replace any PTFE shaft packings or secondary sealing elements with those made of the more resilient elastomer materials to reduce fretting damage.

---

*Courtesy of Flowserve Corporation, Kalamazoo, Michigan 49001.

447

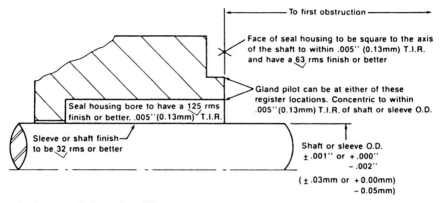

To first obstruction

Face of seal housing to be square to the axis of the shaft to within .005'' (0.13mm) T.I.R. and have a 63 rms finish or better

Gland pilot can be at either of these register locations. Concentric to within .005''(0.13mm) T.I.R. of shaft or sleeve O.D.

Seal housing bore to have a 125 rms finish or better. .005''(0.13mm) T.I.R.

Sleeve or shaft finish to be 32 rms or better

Shaft or sleeve O.D. ±.001'' or +.000'' −.002''

(±.03mm or +0.00mm) −0.05mm)

- Bearings must be in good condition.
- Maximum lateral or axial movement of shaft (end play) − .010'' (0.25mm) T.I.R.
- Maximum shaft runout at face of seal housing − .003'' (0.07mm) T.I.R.
- Maximum dynamic shaft deflection at face of seal housing − .002'' (0.05mm) T.I.R.

Figure 8-1. Seal housing requirements.

Radial Bearing

Thrust Bearing

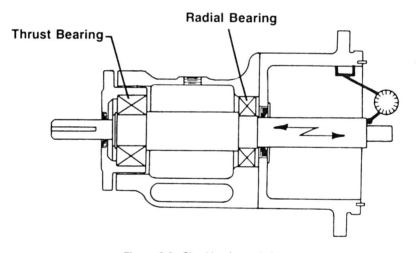

Figure 8-2. Checking for end-play.

- Spring overloading or underloading and premature seal failure.
- Shock-loaded bearings, which will fail prematurely.
- Chipping of seal faces. Carbon and silicon carbide faces are especially vulnerable to axial shaft movement.

3. **Radial shaft deflection.** Radial shaft deflection at the face of the seal housing must not exceed 0.002 in. (0.05 mm) T.I.R. To measure radial movement, install a dial indicator as close to the seal housing

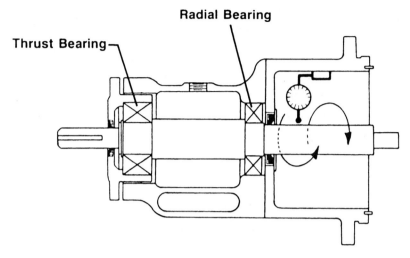

Figure 8-3. Checking for whip or deflection.

face as possible (see Figure 8-3). Lift the shaft or exert light pressure at the impeller end. If movement is excessive, examine for damaged radial bearings and bearing fits—especially the bearing cap bore.

Excessive radial shaft movement can cause the following problems:

- Fretting of the shaft or sleeve
- Excessive leakage at the seal faces
- Excessive pump vibration, which can reduce seal life and performance

4. **Shaft sleeve run-out.** Shaft run-out (bent shaft) must not exceed 0.003 in. (0.07 mm) at the face of the seal housing. Clamp a dial indicator to the pump housing as shown in Figure 8-4, and measure shaft run-out at two or more points on the outside dimension of the shaft. Also measure the shaft run-out at the coupling end of the shaft. If run-out is excessive, repair or replace the shaft.

    Excessive run-out can shorten the life of both the radial and the thrust bearings. A damaged bearing, in turn, will cause pump vibration and reduce the life and performance of the seal.

5. **Seal chamber face run-out.** A seal chamber face which is not perpendicular to the shaft axis can cause a serious malfunction of the mechanical seal. Because the stationary gland plate is bolted to the

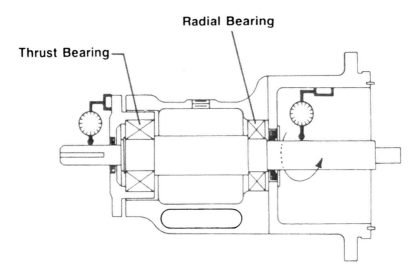

Figure 8-4. Checking for run-out.

face of the seal housing, any misalignment will cause the gland to cock, which causes the stationary element to cock and the entire seal to wobble. This condition is a major cause of fretting wear where the mechanical seal shaft packing contacts the shaft or sleeve. A seal that wobbles can also cause wear or fatigue of metal bellows or drive pins, which can cause premature seal failure.

To measure seal chamber face squareness, leave the housing bolted in place and clamp the dial indicator to the shaft as shown in Figure 8-5, with the stem against the face of the housing. The total indicator run-out should not exceed 0.005 in. (0.13 mm) T.I.R.

6. **Seal chamber register concentricity.** An eccentric chamber bore or gland register can interfere with the piloting and centering of the seal components and alter the hydraulic loading of the seal faces, resulting in reduction of seal life and performance.

To measure chamber-bore concentricity, leave the housing bolted in place and insert the dial indicator stem well into the bore of the housing. To measure the gland register concentricity, the indicator stem should bear on the register O.D. The bore or register should be concentric to the shaft within 0.005 in. (0.13 mm) of the T.I.R.

If the bore or register are eccentric to the shaft, check the slop or looseness in the pump-bracket fits at location "A" as shown in Figure 8-6. Corrosion, whether atmospheric or due to leakage at the gaskets, can damage these fits and make concentricity of shaft and housing

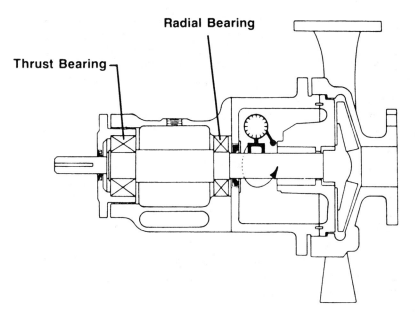

Figure 8-5. Checking for seal chamber face run-out.

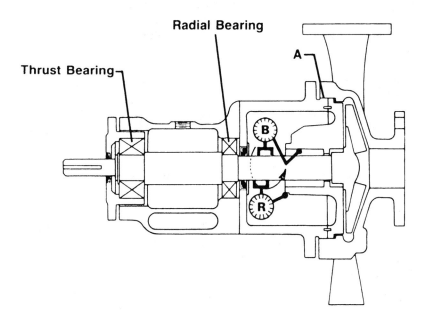

Figure 8-6. Checking for seal chamber bore concentricity.

bore impossible. A remedy can sometimes be obtained by welding the corroded area and re-machining it to proper dimensions, or by replacing the damaged parts. If this is not practicable, it may help to center the entire housing and dowel it in place.

7. **Driver alignment and pipe strain.** Regularly scheduled inspections are absolutely essential to maintain proper coupling and driver alignment. Follow the recommendation of the coupling manufacturer to check coupling alignment. Because temperature can affect coupling alignment due to thermal growth of pump parts, be sure to check pump coupling alignment at the operating temperature.

Pipe strain can cause permanent damage to pumps, bearings, and seals. Many plants customarily blind the suction and discharge flanges of their inactive pumps. These blinds should be removed before aligning the pump driver. After the blinds have been removed, and while the flanges on the suction and discharge are being connected to the piping, read the dial indicator at the O.D. of the coupling half as the flanges are being secured. Any fluctuation indicates that pipe strain is present.

Seal Checkpoints

Modern process plants use only factory-reconditioned, or brand-new cartridge and/or cassette-type mechanical seals. Therefore, only a few points need to be prechecked on both new and factory-reconditioned assemblies:

- Make sure that all parts are clean, especially the mating faces of the assembly
- Check the seal rotary unit and make sure it is free to rotate
- Check the setscrews in the rotary unit collar to make sure they are free in the threads. *Note:* Setscrews should be replaced after each use.
- Check the thickness of all accessible gaskets against the dimensions specified in the assembly drawing. An improper gasket thickness may be a safety hazard
- Check the fit of the gland ring to the equipment. Make sure there is no interference, binding on the studs or bolts, or other obstructions. Be sure any gland ring pilot has a reasonable guiding fit for proper seal alignment.
- Make sure all rotary unit parts of the seal fit over the shaft. Particular care should be given to elastomeric secondaries.

Installation of the Seal

Many seal failures can be traced to installation errors. Careful installation is a major factor in the life of a seal.

1. Read the instruction booklet and review the drawing that accompanies each cartridge or cassette-type assembly.
2. Remove all burrs and sharp edges from the shaft or shaft sleeve, including sharp edges of keyways and threads. Replace worn shaft or sleeves.
3. Make sure the seal housing bore and face are clean and free of burrs.
4. Prior to seal assembly, lubricate the shaft or sleeve lightly with silicone lubricant. Do not use oil or silicone at the seal faces. Keep them untouched and do not disassemble the unit.
5. Flexibly mounted inserts should be lightly oiled and pressed in the gland by hand pressure only. Where an insert has an O-ring mounting on the back shoulder, it is usually better to nest this O-ring into the gland cavity and then push the insert into the nested O-ring.
6. Strictly follow the manufacturer's installation instructions. They vary for different types of seals.

In the rare instances when an old-style, non-cartridge seal is used, and when the seal drawing is not available, the proper seal setting dimension for inside seals can be determined as follows for seals configured as shown in Figure 8-7.

A - Stuffing Box Face

B - Spring Gap

C - Seal Setting from
    Reference Mark

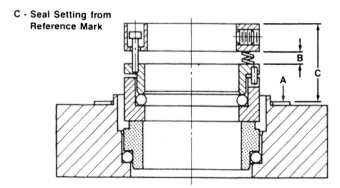

Figure 8-7. Calculation of seal collar setting.

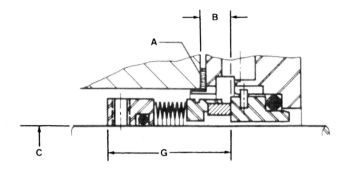

Figure 8-8. Calculation of seal collar setting.

1. Establish a reference mark on the shaft or sleeve flush with the face of the stuffing box (point "A" in Figure 8-8).
2. With the insert in place, stack up the gland and the rotary unit on a clean bench.
3. Compress the seal rotary unit until the spring gap, dimension "B" in Figure 8-8, equals the dimension stamped on the collar for pusher-type seals. This can be done by inserting an Allen wrench or piece of tool stock of the proper dimension between the collar and the compression unit before compressing the rotary unit.
4. Measure the distance "C" between the gasket face and the end of the collar. This is the collarsetting dimension.

If the seal is configured as shown in Figure 8-8, proceed as follows:

1. Measure the distance from the reference mark "A" to the face of the stationary insert in the gland plate. Be sure to include any gland gasket for proper measurement. This is shown as dimension "B" in Figure 8-8.
2. Refer to the manufacturer's instructions for the correct installed length of the rotary unit (Figure 8-8, dimension "G").
3. Subtract dimension "B" from dimension "G". This is the collar setting dimension.

Outside seals (rarely acceptable in modern process plants) should be set with spring gap ("A" in Figure 8-9) equal to the dimension stamped on the seal collar.

Cartridge seals, Figure 8-10, are set at the factory and are installed as complete assemblies. No setting measurements are needed. These assemblies contain centering tabs or spacers that must be removed after the seal assembly is bolted in position and the sleeve collar is locked in place.

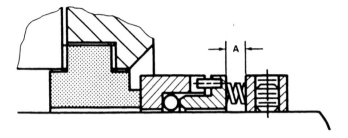

Figure 8-9. Setting for outside seal.

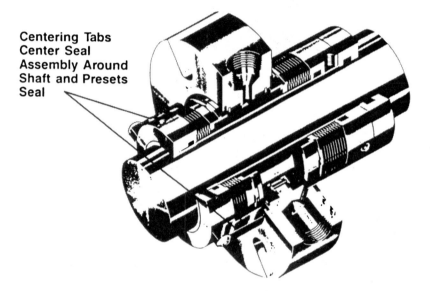

Figure 8-10. Preset cartridge seal.

Retain the tabs to be used when resetting the seal for impeller adjustments or when removing the seal for repairs.

Optical Flat

Since this text no longer advocates in-house repairs of mechanical seals, optical flats are only considered useful tools for "postmortem" failure identification and troubleshooting (see Volume 2 of this series). An optical flat is a transparent quartz or pyrex disc having at least one surface flat within 0.00001 in. to 0.00005 in. (0.025 to 0.125 microns) (see Figure

Figure 8-11. Optical flat.

8-11). The least expensive optical flats, those in the 0.000005 in. (0.025 microns) accuracy range, are suitable for measuring seal face flatness.

Select an optical flat with a diameter at least equal to the diameter of the part being measured. Optical flats in sizes up to 8 in. (200 mm) in diameter are available from most seal manufacturers. Care must be taken not to slide or lay the flat side of an optical flat on any rough surface, because it is easily scratched.

Monochromatic Light. White light from the sun is actually a combination of several colors, each of which represents a different wavelength of electromagnetic energy. If sunlight were used to measure seal face flatness, each color would generate its own pattern of bands on the optical flat. A far more practical light source for this purpose is one that provides light of only one color—a monochromatic light source.

Helium gas in a tube, when excited by an electrical charge, emits fewer colors than sunlight. One of these colors—yellow orange—is so prominent it overrides all the others. The yellow-orange wavelength is measurable and constant at 23.13 millionths of an inch (0.58 microns). A helium lamp, therefore, is described as emitting a monochromatic light of that wavelength.

Flatness Readings. After a seal part has been polished, it is placed under a monochromatic light, and an optical flat is positioned over its surface. Both the surface of the optical flat and the surface of the part must be absolutely dry and free from any particles of dirt, dust, or lint.

A pattern of light and dark lines appears where the reflection of the surface of the part and the reflection of the surface of the optical flat meet. The dark lines, called "light interference bands," will be visible every half-wavelength. Therefore, one interference band equals one-half wavelength or 11.6 micro-inches (0.29 microns). These bands are used to measure the degree of flatness of the surface. The presence either of several straight interference bands, or of a single circular interference band, indicates a surface that is flat within 11.6 micro-inches (0.29 microns).

When several curved interference bands appear, the degree of flatness is measured by plotting an imaginary straight-line tangent with one of the curved interference bands. If this straight line intersects only one interference band, the lapped surface is said to be flat within one light band. When two or more curved interference bands are intersected by this line, the degree of flatness is determined by multiplying the number of intersected bands by 11.6. Thus, if three interference bands are intersected by a straight tangent line, the surface is out-of-flat by three light bands or 34.8 micro-inches (0.87 microns).

Although a maximum tolerance of three light bands is considered adequate for proper seal face performance, seal faces from experienced manufacturers consistently exhibit a flatness between one and two light bands. Special concave or convex geometries are available if special conditions warrant.

A seal face with a diameter greater than 10 in. (250 mm) and a thin cross-section is difficult to measure precisely for flatness.

See Figures 8-12 and 8-13 for sample light-band readings. Note that when the band pattern is inconsistent or seems to be missing some bands, as in Figure 8-13(e), (f), and (g), the tangent line "AB" is drawn to connect the two points at which two imaginary radii—90° apart and perpendicular to the axis of the part—intersect with the outer circumference of the seal face.

Once flatness has been achieved, a seal face must be kept clean while being returned to service.

Installation of Stuffing Box Packing*

Maintenance of a stuffing box consists primarily of packing replacement. This may sound simple, but certain rules must be followed if the best results are to be obtained from the packing.

The following procedure should be preceded by a careful study of the pump manufacturer's instruction manual to determine the correct type and

---

*Source: Packing Pump Packings, H. A. Schneller, Allis Chalmers Mfg. Co.

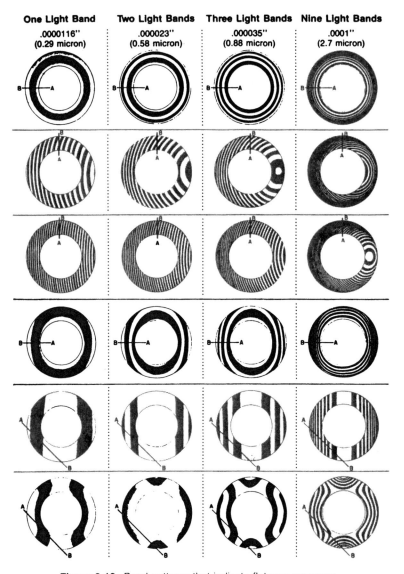

Figure 8-12. Band patterns that indicate flatness accuracy.

size, number of packing rings, location of lantern ring, and possible special features of construction, operation, or maintenance.

1. Remove gland and packing. If the box contains a lantern ring, make certain that all the packing inboard of it is also removed. Flexible packing hooks are available.

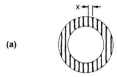

(a)

FLAT—The most prominent seal face band pattern produced at Durametallic is flat to within one light band. Distance "X" is dependent on the amount of air between the optical flat and the face, and has nothing to do with flatness.

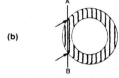

(b)

Bands bend at outer edges and indicate wash out of the periphery due to the polishing process. This is a normal pattern and is related to flatness. Line AB intersects **one** black band. The areas contacted by AB show the face out **1** light band.

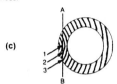

(c)

Bands bend on one side and show an out of flat condition of 3 light bands. Since line AB intersects 3 bands as illustrated by arrows, this pattern is out of flat beyond the acceptable quality level imposed on *Dura Seal* seal faces.

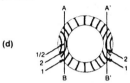

(d)

2 bands are intersected by Line AB and fall between 2 at the center of the ring indicating 2½ light band curvature. Line A'B' intersects 2 bands that curve in opposite direction to those intersected by Line AB. This indicates an egg shaped curvature of 2½ light bands.

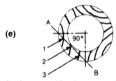

(e)

Bands show a saddle shape out of flat condition of 3 light bands.

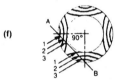

(f)

Bands again show a saddle shape out of flat condition. However, in this illustration, we have 6 bands intersected or 6 light bands out of flatness.

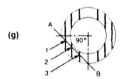

(g)

Band pattern shows a cylindrical shaped part with a 3 light band reading error.

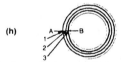

(h)

Band symmetrical pattern indicates a concave or convex flat. In this case, we count the total number of bands on the part. Line AB intersects 3 bands. Light pressure on the part will cause the symmetrical lines to move. If the lines move toward the center the surface is concave. If the lines move toward the edge the surface is convex.

Figure 8-13. Lightband readings.

2. Inspect the shaft or shaft sleeve for score marks or rough spots. A badly worn sleeve or shaft must be replaced; minor wear must be dressed smooth and concentric. Inspect lantern ring to make sure that holes and channels are not plugged up.
3. Clean bore of box thoroughly and be sure sealing-fluid passages are open.
4. Depending on operating conditions, it is recommended that at least the outside diameter of the replacement packing rings be

lightly oiled or greased. Start by installing one end of the first ring in the box and bring the other end around the shaft until it is completely inserted. *Note:* Preformed or die-molded rings will ensure an exact fit to the shaft or sleeve and stuffing box core. A uniform packing density is an added benefit since molded rings are partially compressed. If the packing comes in continuous coil form, make sure that the ends are cut square, on a correctly sized mandrel.

5. Push the packing ring to the bottom of the box with the aid of a split bushing. Leave the bushing in place, and replace the gland. Pull up on the gland and seat the ring firmly and squarely. Ideally, the split bushing should be a slip fit over the shaft and in the bore. This precaution will prevent the formation of a lip on the packing being seated.

6. Repeat this method for each ring, staggering the joints 90°. It is especially important to seat the first few bottom rings firmly, otherwise the rings immediately under the gland will do most of the sealing.

7. The location of the lantern ring, if used, should be predetermined. This can be done via the pump manufacturer's instruction manual or by counting the number of packing rings on the inboard and outboard side of it while removing them. Proper location at the point of fluid-seal is necessary.

8. Replace the gland and tighten the gland bolts. Make sure that the gland enters the stuffing box squarely. A cocked gland causes uneven compression and can damage the shaft or sleeve when the pump is placed in operation.

9. Keep the packing under pressure for a short period of time, say 30 seconds, so that it can cold flow and adjust itself.

10. Loosen the gland; allow packing to fully expand with no fluid pressure in the pump, then tighten the gland and bring it up evenly to the packing, but this time only finger tight. The packing should be loose enough to enable you to turn the pump shaft by hand, assuming that the pump is not too large to turn by hand.

11. Precheck the lines to the lantern ring or quench gland for flow and proper pressure.

12. Start the pump. The leakage may be excessive, but do not take up on the gland bolts for the 20–30 minute run-in period. If too tight, shut pump down and repeat steps 8–11 or keep pump running and loosen gland bolts a couple of flats.

13. If leakage is still more than normal after the run-in period, tighten the gland bolts evenly, one flat or a sixth of a turn at a time. This should be done at 20–30 minute intervals until leakage is reduced

to normal. This may take several hours, but will pay for itself many times over in maximum packing and sleeve life.

Steady Leakage Flow Is a Must

Leakage must be sufficient to carry away the packing friction heat. Without sufficient leakage—a steady flow, as opposed to a drip—the packing will burn and the shaft or sleeve will be scored. It follows, then, that a planned maintenance program must be developed around the amount of leakage that is considered normal. Pumps in continuous service may require a daily, or even an hourly inspection. Inspection need be no more than a visual check to determine any deviation from the normal leakage, and make slight corrections as required.

Correct procedure and scheduled maintenance will make the repacking of a pump a predictable job rather than an emergency repair.

Consider Upgrading to Mechanical Seals

It should be pointed out that conversion from packing to well-designed mechanical seal systems is often feasible and has proven to be a cost-effective reliability improvement step. We strongly advocate such upgrading wherever feasible; it usually *is* feasible!

Welded Repairs to Pump Shafts and Other Rotating Equipment Components*

This section will highlight some of the technical aspects of welded repairs to rotating equipment, with particular emphasis on pump shafts. We believe this to be a reasonable alternative to new part installation.

Welded repairs to rotating equipment can be an extremely useful maintenance method; however, it should be emphasized that welding is *not* a panacea for all problems. In fact, many people refuse to recommend or even consider welded repairs on any rotating components. This reluctance is probably due to unsuccessful experiences in controlling distortion. However, other methods of reconditioning, such as sleeving, plating, and spraying also have their limitations.

---

*Courtesy of Mr. Thomas Doody, Aramco, Dhahran, Saudi Arabia. From a paper presented at the National Petroleum Refiners Association, Plant Maintenance Conference, February 8–11, 1983.

The primary problems with these alternative repair methods are the limitations on allowable coating thickness and lack of bonding with the substrate. For example, sleeving has both minimum and maximum thickness limitations and is also limited to the stepped end areas of a shaft.

The advantages of welded repairs are:

1. Full fusion of the overlay with the base material is achieved.
2. Shaft strength is maintained.
3. The corrosion or wear resistance can be improved over the original material.
4. The overlay thickness is essentially unlimited (minimum or maximum).

The only real disadvantage of welded repairs is the possibility of distortion. For many engineering applications, this is of negligible concern. However, for most rotating equipment components, this is the major concern and the reason many people recommend against attempting welded repairs. If the distortion problem can be eliminated, then welded repairs can be used with confidence as a standard maintenance procedure. The techniques discussed here are designed primarily to minimize distortion.

How to Decide if Welded Repairs Are Feasible

These questions must be answered in determining if welded repairs are suitable for a particular situation:

1. *What is the damage and what type of repair is required?*

As previously mentioned, the advantages of welded repairs must be balanced against the increased risk of shaft warpage. However, there are situations where welding is the only possible method of saving a shaft and returning it to service. A good example is damage to a coupling area, where all other repair techniques are unsuitable because of the mechanical strength and bonding requirements.

In general, the damaged area should not have reduced the shaft diameter by more than 15 percent and should not exceed 10 percent of the shaft length. The repair should be limited generally to one area or two well-separated areas. These figures are only guidelines; there is no hard information on which these values are based. Situations exceeding these restrictions may be repairable but a more careful consideration of shaft strength, possible distortion, and economics needs to be made on an individual basis.

2. *What is the shaft material?*

The shaft material must have good weldability. Some common shaft materials that can be included in this category are:

304 Stainless Steel
316 Stainless Steel
Monel 400
Monel K-500
Nitronic 50 (XM-19)
Ferralium
A 638 Gr 660
Inconel 625

Typical compositions are listed in Table 8-1.

Table 8-1
Nominal Composition of Materials Used for Rotating Equipment Weld Repairs

	C	Mn	Fe	Cr	Ni	Mo	Cu	Other
304	.08	2.0	bal.	18.0–20.0	8.0–10.5	—	—	—
316	.08	2.0	bal.	16.0–18.0	10.0–14.0	2.0–3.0	—	—
MONEL* 400	0.12	2.0	2.5	—	63.0–70.0	—	bal.	
MONEL* K-500	0.13	1.5	2.0	—	63.0–70.0	—	bal.	Al 2.8 Ti 0.6
NITRONIC* 50 (XM-19)	0.06	4.0–6.0	bal.	20.5–23.5	11.5–13.5	1.5–3.0	—	Cb 0.1–0.3 V 0.1–0.3 N 0.2–0.4
FERRALIUM*	0.04	0.8	bal.	25.5	5.2	3.5	1.7	N 0.17
A 638 Gr 660	.08	2.0	bal.	13.5–16.0	24.0–27.0	1.0–1.5	—	Al .35 B .001–.01 Ti 1.9–2.35 V .1–.5
INCONEL* 625	.10	0.5	5.0	20.0–23.0	bal.	8.0–10.0	—	Cb 3.15–4.15 Ti 0.40 Al 0.40

*Trademarks: Monel and Inconel–International Nickel Co.
Nitronic–ARMCO Inc.
Ferralium–Langley Alloys Ltd.
Stellite–Cabot Corp.

All of these materials are very ductile, do not exhibit hardened heat-affected zones, and will usually maintain adequate corrosion resistance in the "as welded" conditions.

There does appear to be a common problem with dimensional stability of the Ferralium and the 300 series stainless steel. These materials appear to creep, even under static conditions. The other materials listed do not seem to be quite as sensitive to this problem. It is obvious that welded repairs will exacerbate this situation, but it is not possible to address this problem at this time.

Shaft materials that should *not* normally be considered for welding are:

4140
4340
410 Stainless Steel

Although many companies have probably done emergency or spot repairs on shafts of these materials, it should be emphasized that welded repairs are not generally recommended.

3. *What is the overlay material?*

Given the previously mentioned shaft materials that would be considered for welded repairs, Inconel 625 (AWS A5.13 ERNiCrMo-3) would normally be selected as the filler metal. Although other materials could be used, it is easiest to standardize on one filler metal that gives a deposit of exceptional corrosion resistance. The deposit may or may not match the tensile strength of the original shaft material, but since the deposit is extremely ductile and has complete fusion at the interface, it has only a marginal effect on total shaft strength. It should be remembered that alternative repair methods require undercutting the shaft diameter, thus permanently and perhaps significantly reducing the total shaft strength.

For other components, such as wear rings, Stellite 6 (AWS A5.13 RCoCr-A) can be used to give a hard, wear resistant, anti-galling deposit. The Stellite overlay requires a more careful application because the deposit is crack sensitive. As a result, the depth of overlay is restricted and repairs to the overlay may be difficult.

Other overlay materials might be required for special cases but these would need to be considered on an individual basis.

Repair Techniques

1. *Shaft Preparation* (Figure 8-14)

The damaged area should be undercut by machining the shaft, but the amount of material removed should be limited as much as possible. For a

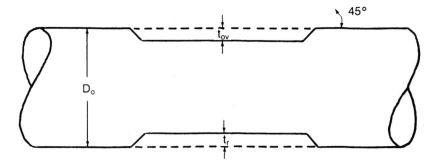

t_{ov} = $1/32$–$1/16$-inch (0.8–1.6 mm) for corrosion or wear resistant overlays
t_r = 7.5% of shaft diameter (D_o) for repair of mechanical damage

Figure 8-14. Shaft preparation by machining.

shaft with only surface-type damage—such as pitting corrosion—the depth of cut should be between $1/32$ and $1/16$ in. (0.8 to 1.6 mm). The minimum depth of cut is specified to avoid having the fusion line positioned directly on the final machined surface. The maximum depth is limited in order to reduce distortion. For mechanical damage, the depth should not have reduced the shaft diameter by more than 15 percent, i.e., depth not to exceed 7.5 percent × D. The edges of all machined areas should be tapered at 45° to ensure good sidewall fusion.

The area to be welded must be thoroughly cleaned and degreased.

2. Welding Procedure

The GTAW (TIG)* process should always be used in order to limit the heat input and reduce the possibility of weld defects. The welding current should be reduced to where good fusion and adequate bead thickness are still obtained but without resorting to long dwell times. There is a trade-off between current, travel speed, and filler rod diameter. These variables need to be adjusted to give the lowest heat input in order to control distortion.

In general, a current of 100A or less (using a $3/32$ in. EWTh-2 tip) and a filler rod size of $3/32$ or $1/8$ in. (2.5 or 3.2 mm) diameter should be used.

3. Welding Technique

The shaft should be well-supported on rollers and mounted in a turner. The shaft is to be rotated at all times during welding and for 30 minutes

---

* Gas-Tungsten-Arc-Welding (Tungsten-Inert-Gas).

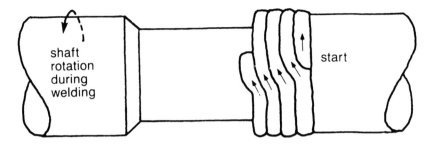

Figure 8-15. Spiral welding sequence for shafts.

after completion. The rotational speed should be set for the welding speed (2–3 in. per minute). This will usually be about 0.1–0.2 revolutions per minute for most large shafts.

The welding is always done in a spiral pattern (Figure 8-15). The undercut depth is limited in order to obtain the required thickness in one thin pass. This helps to minimize distortion by limiting the volume of weld metal and reducing the heat input. The maximum bead width should be limited to $^3/_8$ in. (10 mm). As a minimum, one complete circumferential bead should be completed before stopping or interrupting the welding sequence. In general, welding is started on the edge to be repaired closest to the middle of the shaft and should proceed toward the shaft end.

The maximum interpass temperature is limited to 350°F (175°C). This is of primary importance since the thermal profile of the heat-affected zone is a major determinant of residual stress and distortion. As heat build-up occurs, the width of the heat-affected zone increases, which increases shrinkage.

In one case, the shaft runouts were monitored during a portion of the welding. It was found that shaft end deflections (the weld area was 20 in. from the end) of up to 0.015 in. (0.38 mm) occurred during the actual welding but would return to less than 0.005 in. (0.13 mm) during cooling periods.

Some cold straightening may be required to correct any residual distortion, but this has not usually been a difficult problem.

The finish-machined shaft surface should be completely free of any defects, such as porosity or lack of fusion. Other components, such as hard-facing on wear rings or impellers, are not as critical and an acceptance criterion for rounded indications (porosity) has been adopted.

Case Histories

A number of related experiences are summarized below:

Pump shafts, all overlaid with Inconel 625:

1. *Water injection pump* (Figures 8-16 and 8-17)—Monel K-500 shaft, 5-in. dia, approximately 27 in. length overlaid on coupling end, $^1/_{16}$ in. deep; approximately 21 in. length overlaid on thrust end, $^1/_{32}$–$^1/_{16}$ in, deep. Successful.
2. Numerous other water injection pumps (identical to 1)—small areas on shaft ends: Locknut areas, O-ring seal areas, etc. All successful.
3. *Seawater vertical lift pumps shafts*—Monel K-500 shaft, 5-in. dia. Overlaid at both ends (coupling and bearing area) and center bearing. All successful.
4. *Water injection pump*—A 638 Gr 660 shaft, 5-in. dia. Repair of mechanical damage ($^3/_4$ in. wide, $^3/_{16}$ in. deep). Successful.
5. *Brine injection pump*—XM-19 shaft, 5-in. dia. Numerous areas with corrosion damage, of which 8 were impeller fit areas; up to $^1/_8$ in. deep. Unsuccessful.

Figure 8-16. General view of repaired shaft during machining.

Figure 8-17. Edge of weld repair area in the rough machined condition.

Since any unsuccessful attempt should generate as much useful information as a successful result, it is worthwhile to discuss the lessons learned from this last case:

a. The undercut depth may have been excessive (specified at $^1/_{16}$ to $^1/_8$ in.), which when combined with excessive and unnecessary overfill, caused excessive residual stress and distortion.

b. A large number of separate repairs on the same shaft can create complex distortions that are difficult to correct by straightening. A single repair, even if over a large area, will usually create only a simple bend that can be easily machined and straightened. These particular repairs were closely spaced with critical tolerance areas between them. It was not possible to mechanically straighten the shaft to correct the variety of distortions in these critical areas.

Other Components

1. *Impellers*

Water injection pump impellers (CF8M) are routinely repaired by welding, such as for cavitation damage and bore dimension buildup. A modification has now been instituted to eliminate the impeller

Figure 8-18. Impeller with direct overlay of Stellite to replace wear rings.

wear rings by direct Stellite overlay on the impeller. If, for example, the pump is a 10-stage design and over 50 pumps are in operation, any potential savings for even one part are well amplified. In addition, the elimination of the wear ring also eliminates the problems of stellited wear ring installation and fracture during operational upsets.

The basic procedure involves building up the impeller shoulder with E316L electrodes (SMAW* process) to the specified wear ring diameter, machining 0.060 in. undersize on the diameter (0.030-in. cut), Stellite 6 overlay (GTAW process), and final machining to size (Figures 8-18 and 8-19).

Using this procedure, matched spare sets of impellers and case wear rings are produced, which are exchanged as a complete set for the existing components during a pump rebuild.

---

* Submerged Metal Arc Welding.

Figure 8-19. Impeller with direct Stellite overlay in final machined condition.

As mentioned previously, an important part of the procedure is to limit the heat input, particularly during the buildup of the shoulder using SMAW electrodes. If this is not controlled, distortion of the impeller shrouds can occur. In order to prevent this, $^1/_8$-in, diameter electrodes, a stringer bead technique, and a maximum interpass temperature of 350°F are specified.

2. *Water Injection Pump Case*

Due to a combination of the water chemistry and the pump design, the carbon steel pump cases were experiencing interstage leakage due to erosion/corrosion under the case wear rings and along the case split line faces. The repair procedure developed consists of undercutting ($^1/_8$ in. deep) the centerline bore and the inner periphery of the split line face. These areas are overlaid with Inconel 182 (AWS A5.11 ENiCrFe-3). After rough machining, the cases are stress relieved and then machined to final dimensions (Figures 8-20 and 8-21). The erosion/corrosion problem has been effectively eliminated while providing a significant savings compared to the cost of a stainless or alloy replacement case.

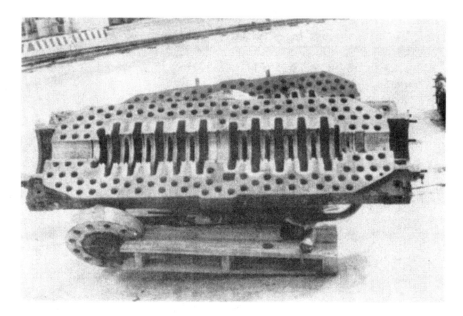

Figure 8-20. Pump case with overlay along centerline bore and edge of split line face.

Figure 8-21. Close-up of pump case overlay in the partially machined condition.

3. *Seal Flanges*

The Monel seal flanges (glands) on a water injection pump were experiencing pitting corrosion on the sealing faces. A localized overlay using Inconel 625 (Figure 8-22) has eliminated the problem.

4. *Impeller Wear Rings*

Prior to the decision to hardface directly on the impeller, attempts were made to fabricate replacement wear rings. The first attempts used core billets as raw stock, however, it appears easier to use solid bar stock. The OD is overlaid before drilling the center bore.

Unsolved Problems

1. Split bushings have not yet been successfully overlaid. This is due to the nonuniform stresses that are created. The distortion resulting from these unbalanced stresses can be enormous. These stresses also change significantly during machining; thus, it is extremely difficult to obtain the proper dimensions.
2. Materials such as 4140, 4340, and 410 SS have not been included in this discussion, although some 4140 shafts have been welded for emergency repairs. For these materials, the primary concern is the possibility of cracking in the hard heat-affected zone formed during welding. Cracking can occur either during (or slightly after) welding due to delayed hydrogen cracking or during service. If a temper bead

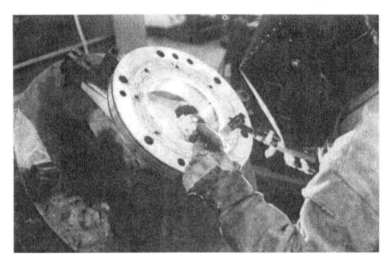

Figure 8-22. Overlaying of seal flange faces.

technique can be effectively developed or if a vertical localized post-weld heat treatment could be accomplished without shaft distortion, then welded repairs to these materials might also become feasible.

Outlook and Conclusions

1. The possibility of using a low temperature stress relief of 600° to 800°F (315° to 425°C) for several hours has been considered for the impeller and wear ring repairs; however, this has not yet been tried on a controlled basis in order to judge its effectiveness.
2. The use of heat absorbing compounds may be tried in order to minimize heat buildup for more critical components, such as shaft repairs.

We conclude:

- Experience has shown that welded repairs to shafts and other rotating equipment components can be successfully accomplished.
- Welding techniques and procedures must be selected in order to minimize distortion. This includes the use of low heat inputs and special sequences.
- Filler metal selection can provide improved properties, such as corrosion and wear resistance, over the original base metal.

High Speed Shaft Repair

In the foregoing we saw several successful pump shaft repair techniques described. Quite often the restoration of low speed shafts with less damage than we saw previously does not represent any problems. Flame spraying by conventional oxyacetylene methods most often will lead to satisfactory results. The market abounds in a variety of flame spray equipment, and most in-house process plant maintenance shops have their preferred makes and techniques. We would now like to deal with the question of how to repair damaged journals, seal areas, and general geometry of high speed turbomachinery shafts. We will mainly focus on centrifugal compressor and turbine rotor shafts in excess of 3,600 rpm.

Four repair methods can generally be identified: Two, that result in the restoration of the original diameter, i.e.,

1. Flame spraying—hard surfacing.
2. Chemical plating.

The other two methods result in a loss of original diameter. They are:

1. Polishing.
2. Turning down the diameter.

Chemical Plating. Later, in Chapter 10, we will discuss the technique of industrial hard chrome plating of power engine cylinders. Worn bearing journals, shrink fit areas of impellers and turbine wheels, thrust collar areas and keyed coupling hub tapers have been successfully restored using industrial hard chrome. We do not see much benefit in describing hard chrome specifications. We recommend, however, that our readers always consult a reputable industrial hard chrome company.

Since chrome plating is too hard to be machined, grinding is the only suitable finishing process. Again, experience and skill of the repair organization is of the utmost importance: Soft or medium grinding wheels should be applied at the highest possible, but safe speeds. Coolant must be continuous and copious. Only light cuts not exceeding 0.0003 in. (7.5 μm) should be taken, as heavy cuts can cause cracking and heat checks.

As a rule of thumb, final ground size of a chrome plated shaft area should not exceed 0.007 to 0.010 in. Chrome plating for radial thickness in excess of these guidelines may require more than one chrome plating operation coupled with intermediate grinding operations. Knowing this, it would be well to always determine the required time for a shaft chrome plating project before a commitment is made.

Flame Spray Coatings. The available flame spray methods will be described later. For practical reasons the detonation gun, jet gun, plasma arc, and other thermal spray processes may suit high speed machinery. There is, however, reason to believe that other attractive techniques will become available in the future.

We believe that coatings applied by conventional oxyacetylene processes tend to have a weaker bond, lower density, and a poorer finish than other coatings. Further, there are too many things "that can go wrong," a risk to which we would not want to subject high speed machinery components. The authors know of an incident where a critical shaft had been allowed to be stored several hours before oxyacetylene metallizing. Dust and atmospheric humidity subsequently caused a problem with the coating well after the machine was up and running. In conclusion, we think that the occasional unavailability of D-gun or plasma coating facilities and the high cost of these methods far outweigh the risk that is inherent in applying oxyacetylene flame sprays.

Shaft Repair by Diameter Reduction. In polishing up the shaft journal, minor nicks and scratches can be dressed up by light stoning or strapping. It goes without saying that depth of scratches, affected journal area, roundness and taper—or shaft geometry—are factors that should be considered when making the repair decision. Generally, scratch depths of 0.001 in. or less are acceptable for use. A good method is to lightly run the edge of a coin over the affected area in order to obtain a feel for scratch severity. Deeper scratches, from 0.001 in. to approximately 0.005 in. must be strapped or stoned. Usually scars deeper than 0.005 in. should call for a clean-up by machining of the shaft.

Strapping. This is done with a long narrow strip of #200 grit emery cloth. The strap is first soaked in kerosene and abraded against a steel surface to remove sharper edges of the abrasive material. It is then wrapped around the journal at least two times and pulled back and forth in order to achieve a circumferential polishing motion. This can best be accomplished by two persons—one on each end of the strap. The amount of material removed from the journal diameter must not exceed 0.002 in.

Stoning. This consists of firm cutting strokes with a fine grit flat oil stone following the journal contour. The stone is rinsed frequently in diesel oil or cleaning solvent to prevent clogging. To avoid creating flat spots on the journal, stoning should be limited to removing any raised material surrounding the surface imperfection.

If the journal diameter is 0.002 in. or more outside of the tolerance, then journal, packing ring, and seal surfaces can be refinished to a good surface by turning down and grinding to the original finish. This introduces the need for special or nonstandard bearings or shaft seals. Stocking and future spare parts availability become a problem. Machining of shaft diameters for nonstandard final dimensions can therefore only be an emergency measure.

Generally, the diameters involved should be reduced by the minimum amount required to clean up and restore the shaft surface. For this the shaft must be carefully set up between centers and indicated to avoid eccentricity. "Standard" undersize dimensions are in 0.010 in. increments.

The maximum reduction is naturally influenced by a number of factors. It would mainly depend on the original manufacturer's design assumptions. Nelson[1] quotes the U.S. Navy cautioning against reducing journal diameters by more than $^1/_4$ in., or beyond that diameter which will increase torsional shear stress 25 percent above the original design, whichever occurs first. Table 8-2 shows this guideline.

Finally, the assembled rotor should be placed in "V" blocks and checked for eccentricity. Table 8-3 shows suggested guidelines for this check.

Table 8-2
Limiting High Speed Shaft Journal Reductions[1]

Original Design Diameter	Minimum Diameter to Which Shaft May Be Reduced
Less than 3.6 inches	93 percent of original design diameter
3.6 inches or greater	Original design diameter less $\frac{1}{4}$ inch

Table 8-3
Recommended Eccentricity Limits for High Speed
Turbomachinery Rotors

Surface	Tolerance (in.)
Impeller eye seal	0.002
Balance piston	0.002
Shaft labyrinth	0.002
Impeller spacer	0.002
All other	0.0005

Shaft Straightening*

Successful straightening of bent rotor shafts that are permanently warped has been practiced for the past 40 or more years, the success generally depending on the character of the stresses that caused the shaft to bend.

In general, if the stresses causing the bend are caused from improper forging, rolling, heat treating, thermal stress relieving, and/or machining operations, then the straightening will usually be temporary in character and generally unsuccessful.

If, however, a bent shaft results from stresses set up by a heavy rub in operation, by unequal surface stresses set up by heavy shrink fits on the shaft, by stresses set up by misalignment, or by stresses set up by improper handling, then the straightening will generally have a good chance of permanent success.

---

*From "Repair Techniques for Machinery Rotor and Case Damage," by H. A. Erb, Elliott Co., Greensburg, Pennsylvania. *Hydrocarbon Processing*, January 1975. By permission.

Before attempting to straighten a shaft, try to determine how the bend was produced. If the bend was produced by an inherent stress, relieved during the machining operation, during heat proofing, on the first application of heat during the initial startup, or by vibration during shipment, then straightening should only be attempted as an emergency measure, with the chances of success doubtful.

The first thing to do, therefore, is to carefully indicate the shaft and "map" the bend or bends to determine exactly where they occur and their magnitude. In transmitting this information, care should be taken to identify the readings as "actual" or "indicator" values. With this information, plus a knowledge of the shaft material available, the method for straightening can be selected.

Straightening Carbon Steel Shafts

Repair Techniques for Carbon Steel Shafts

For medium carbon steel shafts (0.30 to 0.50 carbon), three general methods of straightening the shaft are available. Shafts made of high alloy or stainless steel should not be straightened except on special instructions that can only be given for individual cases.

The Peening Method. This consists of peening the concave side of the bend, lightly hitting it at the bend. This method is generally most satisfactory where shafts of small diameters are concerned—say shaft diameters of 4 in. (100 mm) or less. It is also the preferred—in many cases, the only—method of straightening shafts that are bent at the point where the shaft section is abruptly changed at fillets, ends of keyways, etc. By using a round end tool ground to about the same radius as the fillet and a $2^{1}/_{2}$-lb machinist's hammer, shafts that are bent in fillets can be straightened with hardly any marking on the shaft. Peening results in cold working of the metal, elongating the fibers surrounding the spot peened and setting up compression stresses that balance stresses in the opposite side of the shaft, thereby straightening the shaft. The peening method is the preferred method of straightening shafts bent by heavy shrink stresses that sometimes occur when shrinking turbine wheels on the shaft. Peening the shaft with a light ($^{1}/_{2}$ lb) peening hammer near the wheel will often stress-relieve the shrink stresses causing the bend without setting up balance stresses.

The Heating Method. This consists of applying heat to the convex side of the bend. This method is generally the most satisfactory with large-diameter shafts—say $4^{1}/_{2}$ in. (~112.5 mm) or more. It is also the preferred method of straightening shafts where the bend occurs in a constant diameter portion of the shaft—say between wheels. This is generally not applicable for shafts of small diameter or if the bend occurs at a region of rapidly changing shaft section. Because this method partially utilizes the compressive stresses set up by the weight of the rotor, its application is limited and care must be taken to properly support the shaft.

The shaft bend should be mapped and the shaft placed horizontally with the convex side of the bend placed on top. The shaft should be supported so that the convex side of the bend will have the maximum possible compression stress available from the weight of the rotor. For this reason, shafts having bends beyond the journals should be supported in lathe centers. Shafts with bends between the journals can usually be supported in the journals; however, if the end is close to the journal, it is preferable to support the shaft in centers so as to get the maximum possible compression stress at the convex side of the bend. In no event should the shaft be supported horizontally with the high spot on top and the support directly under the bend, since this will put tension stresses at the point to be heated, and heating will generally permanently increase the bend. Shafts can be straightened by not utilizing the compressive stress due to the weight of the rotor, but this method will be described later.

To straighten carbon steel shafts using the heating method, the shaft should be placed as just outlined and indicators placed on each side of the point to be heated. Heat should be quickly applied to a spot about two to three in. (~50–75 mm) in diameter, using a welding tip of an oxyacetylene torch. Heat should be applied evenly and steadily. The indicators should be carefully watched until the bend in the shaft has about tripled its previous value. This may only require perhaps 3 to 30 seconds, so it really is very important to observe the indicators. The shaft should then be evenly cooled and indicated. If the bend has been reduced, repeat the procedure until the shaft has been straightened. If, however, no progress has been made, increase the heat bend as determined by the indicators in steps of about 0.010–0.020 in. (0.25–0.50 mm) or until the heated spot approaches a cherry red. If, using heat, results are not obtained on the third or fourth try, a different method must be tried.

The action of heat applied to straighten shafts is that the fibers surrounding the heated spot are placed in compression by the weight of the rotor, the compression due to expansion of the material diagonally opposite, and the resistance of the other fibers in the shaft. As the metal is heated, its compressive strength decreases so that ultimately the metal in the heated spot is given a permanent compression set. This makes the fibers on this side shorter and by tension they counterbalance tension stresses on the opposite side of the shaft, thereby straightening it.

The Heating and Cooling Method. This method is especially applicable to large shafts that cannot be supported so as to get appreciable compressive stresses at the point of the bend. It consists of applying extreme cold—using dry ice—on the convex side of the bend and then quickly heating the concave side of the bend. This method is best used for straightening shaft ends beyond the journals or for large vertical shafts that are bent anywhere.

Here, the shaft side having the long fibers is artificially contracted by the application of cold. Then this sets up a tensile stress in the fibers on the opposite side which, when heated, lose their strength and are elongated at the point heated. This now sets up compressive stresses in the concave side that balance the compressive stresses in the opposite side. Indicators should also be used for this method of shaft straightening—first bending the shaft in the opposite direction from the initial bend, about twice the amount of the initial bend—by using dry ice on the convex side—and then quickly applying heat with an oxyacetylene torch to a small spot on the concave side.

Shafts of turbines and turbine-generator units have been successfully straightened by various methods. These include several 5,000-kw turbine-generator units, one 6,000-kw unit, and many smaller units. Manufacturers of turbines and other equipment have long used these straightening procedures, which have also been used by the U.S. Navy and others. With sufficient care, a shaft may be straightened to 0.0005 in. or less (0.001 in. or 0.025 mm total indicator reading). This is generally satisfactory.

Casting Salvaging Methods

Repair of Castings. Quite often cast components of process machinery cannot be repaired by welding. We will now deal briefly with these salvaging methods:

1. Controlled-atmosphere furnace brazing.
2. Application of molecular metals.
3. Metal stitching of large castings.

Braze repair of cavitation damaged pump impellers is an adaptation of a braze-repair method originally developed for jet engine components[2].

The first step is rebuilding the eroded areas of the impeller blades with an iron-base alloy powder. The powder is mixed with an air-hardening plastic binder and used to fill the damaged areas. Through-holes are backed up with a temporary support and packed full of the powder/binder mixture. After hardening, the repaired areas are smoothed with a file to restore the original blade contour.

A nickel-base brazing filler metal in paste form is then applied to the surface of the repaired areas and the impeller is heated in a controlled atmosphere furnace. In the furnace, the plastic binder vaporizes and the brazing filler metal melts, infiltrating the alloy powder. This bonds the powder particles to each other and to the cast iron of the blade, forming a strong, permanent repair.

After the initial heating, the impeller is removed from the furnace and cooled. All nonmachined surfaces are then spray coated with a cavitation-resistant nickel-base alloy and the impeller is returned to the furnace for another fusion cycle. After the treatment, the impeller will last up to twice as long as bare cast iron when subjected to cavitation.

Because the heating is done in a controlled-atmosphere furnace, there is no localized heat build-up to cause distortion and no oxidation of exposed surfaces. Unless the machined surfaces are scored or otherwise, physically damaged, repaired impellers can be returned to service without further processing.

An average impeller can be repaired for less than a third of the normal replacement cost.

Molecular metals have been applied successfully to the rebuilding and resurfacing of a variety of process machinery components. Molecular metals[3] consist of a two-compound fluidized metal system that after mixing and application assumes the hardness of the work piece. The two compounds are a metal base and a solidifier. After a prescribed cure time the material can be machined, immersed in chemicals, and mechanically or thermally loaded.

Molecular metals have been used to repair pump impellers, centrifugal compressor diaphragms, and engine and reciprocating compressor water jackets damaged by freeze-up.

Metal stitching is the appropriate method to repair cracks in castings. One reputable repair shop describes the technique[4].

1. The area or areas of a casting suspected of being cracked are cleaned with a commercial solvent. Crack severity is then determined by dye penetrant inspection. Frequently, persons unfamiliar with this procedure will fail to clearly delineate the complete crack system. Further, due to the heterogeneous microstructure of most castings, it is quite difficult to determine the paths the cracks have taken. This means that the tips of the cracks—where stress concentration is the highest—may often remain undiscovered. This also means that cracks stay undiscovered until the casting is returned to service,

resulting in a potential catastrophe. It takes an experienced eye to make sure that the location of the tips is identified.

2. To complete the evaluation of the crack system, notice is taken of the variations in section thickness through which the crack or cracks have propagated. This step is critical because size, number, and strength of the locks and lacings—see Figure 8-23—are primarily determined by section thickness. Where curvatures and/or angularity exist, the criticalness of this step is further increased.

3. Metallurgical samples are taken to determine the chemical composition, physical properties, and actual grade of casting. This enables the repair shop to select the proper repair material. And this, along with the cross-sectional area of the failure, determines how much strength has actually been lost in the casting.

4. After these decisions have been taken, the actual repair work is started.

   a. Repair material is selected. This material will be compatible with the parent material, but greater in strength.

   b. The patterns for the locks are designed onto the casting surface.

   c. These patterns are then "honey combed" using an air chisel. This provides a cavity in the parent metal that will accept the locks. Improper use of these tools produces a cavity which is not prop-

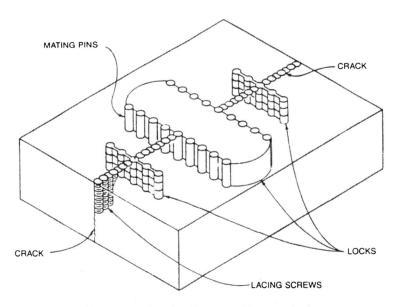

Figure 8-23. Metal locking a machinery casting[4].

erly filled by the lock. The result is a joint that lacks strength and from which new cracks may emanate.

d. Assuming the lock is properly fitted, a pinning procedure is now undertaken. This consists of mating the lock to the parent metal by drilling holes so that one half of the hole circles are in the parent metal and the remaining halves are in the locks. High alloy, high strength, slightly over-sized mating pins are driven into these holes with an air gun. This produces a favorable residual stress pattern: In the immediate area of the lock, tensile stresses exist which change to desired compressive stresses as one moves out into the parent metal. This is to prevent future crack propagation. Additionally, these pins prevent relative movement between the locks and the parent metal.

e. The final repair step aside from dress-up is the insertion of high strength metallic screws into previously drilled and tapped holes along the cracks paths in between the locks. To clarify, it should be noted that the orientation of the locks is such that the longitudinal axis of the locks is perpendicular to the path of the crack. Thus, between locks, the lacing screws are used to "zipper-up" the crack. Care must be exercised to make sure each lock is properly oriented. Care must also be exercised so that, when the lacing screws are driven to their final positions, a harmonious blending with the parent metal is achieved.

The entire repair sequence can be easily visualized by referring to Figure 8-24. An amazing variety of machines have been successfully repaired using metal stitching techniques (Table 8-4).

Contact with Service Shops[5]

The person or persons responsible and accountable for machinery repair and maintenance should establish contact with service shops. This is best done by visiting them and judging their facilities, "track record" and personnel. This could lead to a numerical rating on a scale of one to ten to help with the final decision.

It goes without saying that quotations for new equipment prices should be obtained, so that the practicality of a rebuild or repair order can be ascertained. For instance, as a rule of thumb it would not be advisable to have an electric motor rewound if costs exceeded 70 percent of a new equivalent replacement, or if higher efficiency replacement motors are available. Also, if time is available, the purchase of surplus equipment may

(Text continued on page 487)

Table 8-4
Typical Field and Shop Repair Services Offered by Process Machinery Repair Shops[6,7]

Machinery Repairs—Field & Shop Work / Components	Grouting	Alignment	Straightening	Machining	Milling	Boring	Line-Boring	Grinding	Honing	Metal Locking	Babbitting	Plasma Sprayed Coating	Welding	Fabricating	Chrome Plating	Micro Finishing	Replacement	Rebuilding	Dynamic Balancing	Visual Inspection	Physical Inspection	N.D.T.*	Vibration Analysis	Field Service
Process Pumps																							•	•
Centrifugal Compressors																								
Rotary Blowers																								
Fans																								
Steam Turbines																								
Large Motors (Rotors)																								
Gears & Transmissions																								
Turbo Chargers																								
Bearings						•					•						•			•	•	•		
Shafts			•	•				•												•				
Wheels/Impellers								•				•	•	•	•			•	•	•	•	•		
Rotors												•						•	•					
Diaphragms/Diffusers				•	•	•				•		•	•					•		•		•		
Casings				•	•	•	•			•		•	•	•				•		•		•		

Table 8-4—cont'd
Typical Field and Shop Repair Services Offered by Process Machinery Repair Shops[6,7]

Machinery Repairs— Field & Shop Work / Components	Grouting	Alignment	Straightening	Machining	Milling	Boring	Line-Boring	Grinding	Honing	Metal Locking	Babbitting	Plasma Sprayed Coating	Welding	Fabricating	Chrome Plating	Micro Finishing	Replacement	Rebuilding	Dynamic Balancing	Visual Inspection	Physical Inspection	N.D.T.*	Vibration Analysis	Field Service
Reciprocating Compressors																								
Power Pumps																								
Foundation	•																	•		•				•
Base										•				•				•						
Frame		•	•	•	•	•				•								•		•				
Crankshaft		•		•				•							•	•	•	•	•	•		•		
Main Bearings				•							•		•					•		•		•		
Main Brg. Saddles				•	•	•												•		•		•		
Journals																				•		•		
Valves								•							•	•	•			•	•	•		
Cylinders				•		•		•	•	•					•		•	•		•	•	•		
Cylinder Liners				•		•		•	•						•		•	•		•		•		
Plungers				•				•				•					•			•				
Pistons				•													•			•	•	•		

Piston Rings

Piston Rods

Connecting Rods

Coolant Passages

Diesel & Gas Engines

Foundation

Base

Frame

Crankshaft

Main Bearings

Main Brg. Saddles

Journals

Cylinder Heads

Cylinders

Cylinder Liners–Wet/Dry

Pistons

Piston Rings

Connecting Rods

Inlet Valves

Exhaust Valves

Gas Injection Valves

Valve Train

Cooling Water Systems

Manifolds (Int./Exh.)

Fuel & Ignition Systems

Auxiliaries

* *Non-Destructive Testing:*
—Ultrasonic   *—Radiography*   *—Eddy Current*
—Magnaflux   *—Dye Penetrant*   *—Strain Gauging*

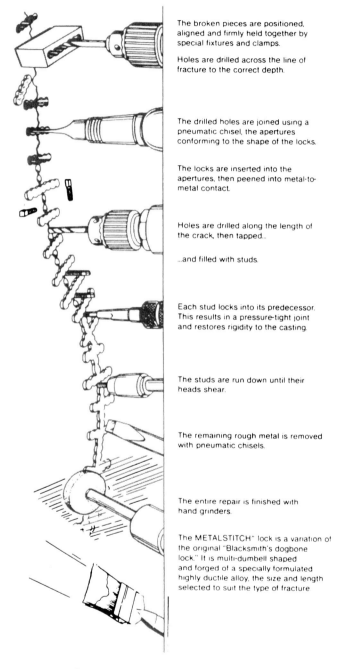

The broken pieces are positioned, aligned and firmly held together by special fixtures and clamps.

Holes are drilled across the line of fracture to the correct depth.

The drilled holes are joined using a pneumatic chisel, the apertures conforming to the shape of the locks.

The locks are inserted into the apertures, then peened into metal-to-metal contact.

Holes are drilled along the length of the crack, then tapped...

...and filled with studs.

Each stud locks into its predecessor. This results in a pressure-tight joint and restores rigidity to the casting.

The studs are run down until their heads shear.

The remaining rough metal is removed with pneumatic chisels.

The entire repair is finished with hand grinders.

The METALSTITCH* lock is a variation of the original "Blacksmith's dogbone lock." It is multi-dumbell shaped and forged of a specially formulated highly ductile alloy, the size and length selected to suit the type of fracture.

Figure 8-24. Metalstitch® process of casting repair (courtesy In-Place Machining Company, Milwaukee, Wisconsin)[8].

(Text continued from page 482)

be considered. It would be advisable to maintain a subscription to at least one used or surplus equipment directory for that purpose.

The machinery maintenance person, during his facility visit, should gather information as to what procedures the shop uses to comply with plant specifications. Obviously, a final sourcing decision should be made only after analyzing all available data and after the visit. The analysis can be made in form of a spreadsheet, using a marking pen to highlight pertinent facts and color-coding prices by relative position. In essence, this rigorous procedure is similar to a formal bid evaluation process and would rank the bidders by shop capacity, experience, reputation, recent performance, order backlog, or even labor union contract expiration date and the like.

OEM vs. Non-OEM Machinery Repairs

Equipment users are inundated with reams of technical information concerning machinery in the purchasing phase. Yet, seldom do operators/users get an opportunity to ask some very basic questions that deserve to be answered to run their business, and even less information is available on repairing. The questions presented in this segment of our text were elicited by the Elliott Company* from a group of users, and answers to these questions are fundamental in helping to keep machinery running. Basic questions of what, why, when, and especially how to repair instead of buying new are considered. It is a simple guide to what the buyer of repair services should ask.

When to Consider Repairing a Worn or Damaged Component or Assembly Instead of Buying New

It is always worthwhile to ask an expert repair company about the repairability of a worn or damaged component, and the advice is usually free. Fortunately, most turbomachinery components can be repaired at lower cost and shorter lead time than buying new. Only in the case of small, inexpensive, mass-produced components is repair not worthwhile.

Usually repair is considered to reduce delivery time and costs while maintaining product integrity (Figure 8-25).

---

*Elliott Company Reprint R240, Antonio Casillo, "Twenty Questions About Repairing Machinery," reprinted by permission of both the author and Turbomachinery International, ©1990, Business Journals, Inc. This material was originally presented at the Fifth Turbomachinery Maintenance Conference in London, U.K., September 1989.

Figure 8-25. Gas expander blades of superalloy are typical examples of new parts deliveries that can exceed 10 months. These blades can be repaired with controlled welding, heat treatment, and coating processes in weeks.

The term *expert repairer* is meant to indicate a dedicated repair facility of an original equipment manufacturer. This type of facility provides rapid action required by an after-sales service organization while at the same time having available the experienced engineering department and know-how of an original equipment manufacturer.

How to Find Out if the Component Is Repairable

A phone call to an expert repairer with a description of the component and of the problem will often result in an answer (Figure 8-26). For bigger problems users can ask the repairer to conduct an inspection of the component at site.

Figure 8-26. Classical repair problem on all types of rotating machinery is the scoring of shaft journals. Welding can be used to repair damage to any depth. Formerly journal repair was limited to allowable chrome plating thicknesses. Thrust collars can similarly be repaired. Quotations for this type of repair can be made quickly.

What Components Can Be Repaired

Practically any part of a rotating machine can be repaired. Any list of parts that are repairable would be lengthy and still be incomplete. But just to give an idea, machinery that can be repaired includes pumps, compressors, steam turbines, gas turbines, mixers, and fans.

Repairs can be effected for breakage, wear, erosion, corrosion, galling, fretting, cracking, bending, and over-temperature, to name but a few types of the many conceivable problems (Figures 8-27 and 8-28). Among the numerous components that have been successfully repaired we find rotors of all types, impellers, blades, disks, shafts, bearings, diaphragms, stators, seals, vanes, buckets, combustion chambers, casings, nozzles, valves, equipment casings, and gears.

Knowing How to Manufacture a Component that Is Totally Destroyed

The simplest way of reconstructing a destroyed component is to use a spare part from stores as a guide. In many instances the component is also contained in a spared machine, which can then be used as a model. As a last resort, a part can be redesigned from the space created by the surrounding parts. Great care must be used with this method. The repairer needs to be an original equipment manufacturer as well as a repair specialist. This means that he is fully familiar with the function of the part and the engineering principles and tools necessary to reconstruct it.

Figure 8-27. Another classical problem is erosion of steam turbine casings in the grooves that retain the diaphragms. These can be restored to original dimensions using a combination of welding and mechanical techniques.

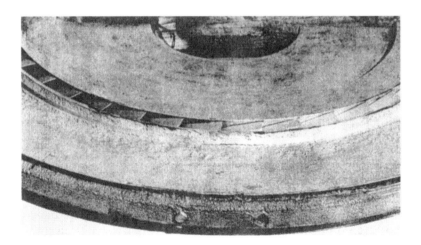

Figure 8-28. Turbine diaphragm for the casing shown in Figure 8-27. Steam erosion, which is very evident on the outer diameter, is repaired by welding. Nozzles can also be repaired, although these are in good condition.

Will a Repairer Manufacture Spare Parts?

The supply of spare parts is a specialized function requiring the know-how of the original equipment manufacturer (OEM). Another OEM can indeed reconstruct spare parts manufactured by others. However, he has to spend engineering hours to create a drawing and to specify the processes and quality controls. This means that it is often more economical to buy parts from the original manufacturer.

An expert repairer who is also an OEM has the capability of designing and manufacturing a spare part required to complete a repair and will do so if needed. One should be wary of any repairer offering only spare parts since these will often not meet the original specifications and quality controls.

Proof of Repairability

An expert repairer keeps careful records of his activities. One of the quality records maintained is a job folder for each repair containing all the information pertaining to that repair. This includes dimensional checks, nondestructive examination, specifications, in-progress quality records, and a detailed list of the work carried out.

Another set of records contains certified repair procedures carried out on certain classes of parts. This, in effect, is a repair manual containing all the necessary procedures, specifications, and quality checks to carry out a repair. This is also a record of repairs actually carried out, proved, and documented. By repeating the exact same process on a similar part, the repairability is 100 percent assured.

In the rare case of an absolutely untested repair, the proposed procedure is first used on a model of the failed component. The parameters of the procedure are recorded. The model is then examined destructively to provide positive proof of the repair. Samples are provided to the owner of the part to make his own tests and reach the same results as those of the repairer (Figures 8-29 and 8-30).

Ascertaining Integrity of the Repair Process

A reputable repairer will take every precaution to determine the material, metallurgical, and physical state of the part prior to commencement of the repair. Only known and controlled processes are applied to the part prior to the repair. There should thus be no doubt as to the expected results of the process.

Figure 8-29. Close-up view of a centrifugal compressor impeller. The damage to the impeller eye, cover, and disk outer diameter was repaired by welding. Even though the procedures were certified, a sample weld was made on similar material and given to the customer for his metallurgy department prior to commencement of work.

The repair process is controlled through work instructions in which every step is detailed. Each step calls out the necessary tools and specifications required to perform that step. Each step is inspected and verified before the next step is performed (Figure 8-31). Final inspection and tests confirm the quality that has been built into the repair process each step of the way. A final report must record the history of the repair as well as the verified results of tests and inspections.

A formal and active quality system is mandatory for the repair facility. This means an all-encompassing system to control all the activities of an organization that ensures that what is shipped is exactly what was ordered. Included in such a system are the following as a minimum:

- Formal organization and control
- Control of documents
- Calibration of instruments
- Training and qualification of personnel
- Product identification and traceability
- Corrective action

Figure 8-30. The impeller shown in Figure 8-29 after weld repair and final machining. The impeller was heat treated prior to machining and overspeed tested at 115 percent of maximum continuous speed in accordance with API specifications.

By the use of certified procedures carried out in a certified facility, it can be assured that no harm will come to an owner's part during the repair. Furthermore, the part will perform exactly as predicted after the repair.

Specifications Applied to the Process

In general, the same specifications that were applied in originally making the part are applied to the repair. If the part is from a compressor or steam turbine of a petroleum refinery, for example, American Petroleum Institute (API) standards are applied. In actual practice once a repair facility is capable of working to these high standards they will be applied to all parts whether they come from a refinery or not.

Specifications are applied to individual processes as needed. For example, welders and weld procedures must be qualified to American Society of Mechanical Engineers (ASME) XIII and IX. Nondestructive examinations are subject to Society for Non-Destructive Testing (SNT) standards. Material specifications are used in the selection and

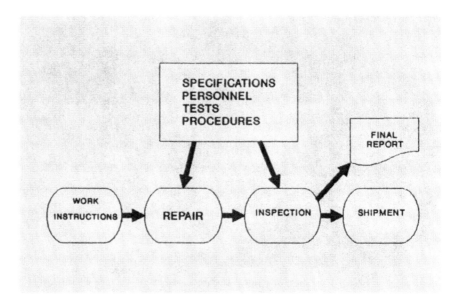

Figure 8-31. ISO 9002 requires implementation of quality assurance system control procedures, qualification of personnel, and calibration of instruments used. The aim is that the customer gets exactly what he ordered, precisely when promised.

acceptance of materials as, for example, American Society for Testing Materials (ASTM), British Standards (BS), or Deutsche Industrie Normen (DIN).

An overall specification must be applied to the repair facility and this is International Standards Organization (ISO) 9002 quality systems model for quality assurance in production and installation. This standard controls every facet of an operation. It includes control of calibration, order processing, documents, materials, personnel and procedures qualifications, and a method of eliminating root causes of problems. This international and demanding standard has as its objective that the customer should get exactly what he bought, precisely on time.

Resolving Different Opinions: Scrap vs. Repair

It could happen that the OEM recommends buying new while an expert repairer proposes a repair. It does not mean that the OEM is remiss or callous but only that he is not familiar with the repair technology. All OEMs are capable of designing and manufacturing new equipment. Only a very few have applied any research effort in developing repair

Figure 8-32. The foreign-object damage on these turbine blades was repaired by welding. The original manufacturers offered only new blades. The repair was made at a fraction of the cost of new blades and in a matter of weeks.

technology. Therefore, they simply are incapable of carrying out a repair, although they can manufacture a new part quite readily (Figure 8-32).

With regard to the repair, the owner can assure himself of the security of the repair by asking to see references, certified procedures, and, if necessary, tests prior to committing to the repair. The savings in cost and delivery make it worthwhile to consider repair.

Knowledge Base of Repairers

Machines come in a wide variety of shapes, makes, models, and materials. Nonetheless, they are all subject to the laws of nature as interpreted through the science and art of engineering. Any complex problem or machine can be broken down into its component elements and the laws of engineering applied to it. An expert repairer who is also an OEM is familiar with basic engineering principles and can apply them to any type of machine and problem. For example, rotor dynamic design for a pump, a compressor, and a turbine follow the same principles. Another example is seal clearance for an unfamiliar machine, which can be determined by knowledge of the fluid and its pressure, and the physical dimensions of the seal. Thus by knowing how to design a turbine or a compressor, an expert repairer can repair any type or make.

Another means of repairing an unfamiliar machine is by technology transfer. The repair procedures can be transferred from one type of machine to an entirely different type. For example, a steam turbine and a mixer may seem two entirely dissimilar machines. However, the shafts perform the same basic function and can be made of the same type of material. Thus a journal or coupling repair developed for a steam turbine can be used with perfect assurance on a mixer shaft.

Therefore, by having the experience of designing and manufacturing machinery, by the use of basic engineering, and by technology transfer, an expert repairer accumulates a wide reference list of repairs on all types of machines. This experience is greatly augmented if the repairer has a number of facilities around the world and these facilities freely exchange their information.

Cost and Delivery Issues

With few exceptions, the cost of repairing is a fraction of buying new. This cost is known prior to commencement of the work and indeed is quoted as a firm price to the owner.

One of the big advantages of repair is that it can be done in a fraction of the time required to make a new component. This means that machines can be put back in operation within days or weeks instead of months. Even if the component is a spare, the owner has the security of having the repaired part close at hand sooner.

By the foregoing discussion it is apparent that repairing is not a hit-or-miss proposition but a controlled science. By defining the work scope, the processes, and the specifications, a repairer can absolutely determine and guarantee the cost. Similarly, the time required to repair is calculated, and when combined with facility capacity and load, a delivery date is determined. Normal working hours are from 8 in the morning to 6 the following morning. In an emergency, around-the-clock working can be instituted.

Repair Guarantees and Insurance Issues

All repairs are guaranteed for material and workmanship as if they were new and this guarantee can be obtained in writing. This naturally follows from the discussion above, that repairing is scientific; consequently, the performance of the repaired component is generally predictable and guaranteeable. Only in rare instances, damage may be just beyond the reach of guaranteed repairability. However, for operational reasons the owner may need to have the equipment running quickly. Under these circumstances a repair may be effected on a best effort basis.

Insurance companies that provide machinery breakdown coverage are intensely interested in repairs. Their interest is based on the ability to reduce the cost of a breakdown and to avoid the introduction of potential risks through the repair process. Consequently, they have become increasingly involved with owners, repairers, and researchers in qualifying repair methods. The principal focus has been in the weld repair of highly loaded rotors.

Insurance companies are promoting the repair of components, provided (and it's a very important proviso) that the risks of repair can be assessed and controlled. This control can be exercised through the specifications, qualified procedures, and facilities discussed above.

Initiating the Repair Sequence

The repair process can be initiated by simply telephoning an expert repairer, describing the problem, and asking for an opinion. If the repairer needs more information, he will conduct a visual inspection either at the owner's site, if the part is large, or at the repair facility (Figure 8-33). This often results in a formal proposal to repair. Should a more detailed examination be required to assess the extent of the damage, it would be conducted at the repairer's facility where the necessary equipment is available. An owner can ask that the repairer formally quote the price and work scope of this examination. In any case, the cost is very small and usually worthwhile.

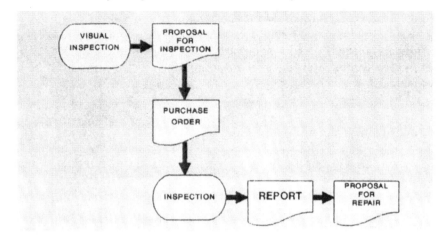

Figure 8-33. The process of finding out the cost of repairing a component is easy. Most of the time a visual inspection can result in a price and delivery. The expert repairer will give an opinion free of charge.

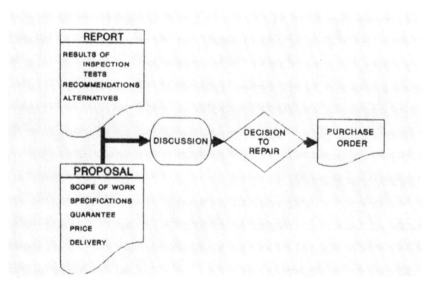

Figure 8-34. Detailed inspections result in a quotation and inspection report. These form a basis for discussion between the owner and the expert repairer. Only when the owner is fully satisfied about the security of the repair does the actual repair work commence.

Following the tests, the repairer will present a report of the findings and a proposal for repairs (Figure 8-34). At this point a discussion can be held between repairer and owner so that the owner can be clearly informed of the proposed methods and select alternatives, if proposed. If the owner decides to proceed, the repairer creates work instructions detailing procedures and specifications to be used in doing the repair and the quality checks required. As the work is done, careful records are maintained. Shipment of the repaired component is followed by a report containing the quality records of the work.

Installation vs. Reinstallation

A part whose repair is guaranteed is indistinguishable from a new part. Where it is different than a new part is when additional work is carried out to make it better than a new part. Therefore, whether new or repaired, a part can be reinstalled. The owner should consider that the equipment has failed, and the failure mode must be examined to find and address the root cause of failure. It is possible to set up a monitoring system to make sure that the root cause has been eliminated. This monitoring should check the replaced part as well as any parts that are functionally related. This approach is valid regardless of whether the part is repaired or replaced with a new one.

Shipments to Other Countries

Components being temporarily exported are not subject to import duties, and shipping documents should be marked accordingly. Shipping documents should indicate that the component is being temporarily exported for repair and will be returned to the country of origin. The procedure is simpler than for exporting new machinery.

Transporting Damaged or Repaired Components

The repairer can provide expert advice not only in export/import documentation but in arranging quick methods of transportation. Low-cost air freight, good highways, and roll-on, roll-off vessels have made possible extremely short transit times.

References

1. Nelson, W. E., and Wright, R. M., Amoco Oil Co., "Reengineering of Rotating Equipment through Maintenance," presented at the ASME Petroleum Division Conference and Workshop, Dallas, Texas, September 1981.
2. Wall Colmonoy Corporation, Detroit, Michigan. Reprint in *Welding Journal*, April 1971.
3. Technical Bulletin by Belzona Molecular Ltd., Harrogate, Yorkshire, U.K., publication No. 15673.
4. Harris, E. W., "Procedure for Repairing Large Castings or Forgings," Casting Repair Service, Inc., Center, Texas 75935, 1983, Pages 1–3.
5. Boak, J. D., "Selecting a Motor Repair Shop," *Plant Services*, June 1983, Pages 64–65.
6. Advertising Bulletin, Peacock Brothers Limited, Toronto, Ontario, Canada.
7. Byron Jackson Marketing/Sales Department, Los Angeles, California, "The Advisor," Volume 1, No. 4.
8. Technical Bulletin by In-Place Machining Company, Milwaukee, Wisconsin, 1990.

Chapter 9

Centrifugal Compressor Rotor Repair

Turbomachinery rotor repair is a complex business. It involves knowledge of design concepts, good machinery practices, and above all, patience and keeping track of details.

Repair work on the rotating elements of compressors and turbines has traditionally been the field of the original equipment manufacturer, not the equipment owner. However, a number of independently owned or original equipment manufacturer (OEM)-owned dedicated repair facilities are available to users worldwide. These should be considered for compressor rotor repair and reliability enhancements since their competence is often as good as, or even better than, the OEM's main factory.

Figures 9-1 through 9-3 illustrate work in progress at one such non-OEM location.

However, a reliability-focused user will seek involvement in the compressor repair process. Equipped with basic knowledge and comparison standards (checklists and procedures, etc.), the user is in a good position to ask relevant questions of those responsible for turbomachinery repair.

Compressor Rotor Repairs*

There are two basic types of compressor rotors: the drum type and the built-up type. There are two variations of each style.

---

* Material contributed by W. E. ("Ed") Nelson (†) and gratefully acknowledged by the authors.

Figure 9-1. Compressor repair and run-out verification at a major independent repair facility. (Source: Hickham Industries, La Porte, Texas.)

Built-Up Rotor

This style is used for virtually all centrifugal compressors and some axial designs.

1. *Heavy-Shrink Style*—The impellers are usually shrunk to the shaft with an interference fit of $^3/_4$ to $1^1/_2$ mils/in. of shaft diameter. In other words, a six in. shaft may have an interference of four to nine mils total before the impellers are placed on the shaft. This requires heating of the wheels to 400°–600°F for assembly. A small axial gap of about four to six mils is necessary between elements, consisting of sleeves and impellers, to allow for expansion during a rub and the flexing of the rotor during rotor mode shifts at critical speeds.

   Sleeves can be used at the interstage seals to protect the shaft from heavy rubs. The sleeve material can be selected with excellent rubbing and heat dissipation characteristics. Sleeves can also be replaced readily.

2. *"Stacked" Rotors*—This design is similar to the shrunk rotor, except that light shrink or press fits are used on the impellers. A large nut

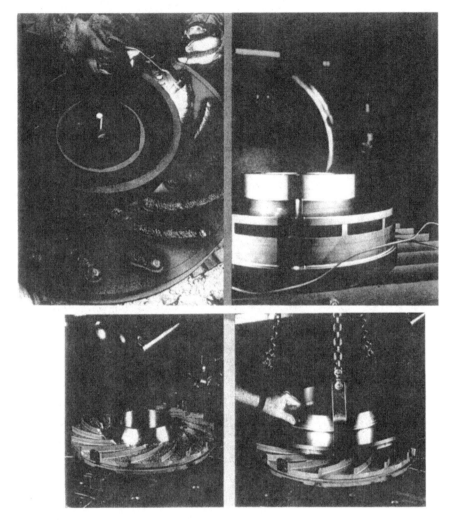

Figure 9-2. Impeller construction in progress at a major independent repair facility. Clockwise from upper left: welding; brazing; machining; assembly. (Source: Hickham Industries, La Porte, Texas.)

on one or both ends hold the impellers and sleeves together axially. This design permits very low tip speeds so it is usually limited to blowers that operate below their first critical speed.

Drum Type Rotor

This type is used only for axial flow-type rotors.

Figure 9-3. Turbomachine being reassembled at a major independent repair facility. (Source: Hickham Industries, La Porte, Texas.)

Impeller Manufacture

There are several manufacturing techniques for impellers.

1. The oldest form of impeller, produced by riveting blades of simple curvature between a disc and a shroud of contour-machined, forged steel, is still in use on multistage compressors of moderate tip

speeds—600–800 ft/sec. With rivets milled integrally out of blade stock, and with alloy steel discs of 120,000 psi tensile stress or higher, tip speeds up to 1,000 ft/sec have been obtained operationally.

2. Fabricated shrouded impellers, assembled with the aid of welding. This permits greater aerodynamic refinement in blade curvature design and, with newly developed alloy steels and welding techniques, operational tip speeds up to 1,100 ft/sec.

3. Cast aluminum or cast steel impellers, both shrouded and open, are used where high production rates justify the pattern cost.

4. Aluminum alloy impellers produced by plaster mold or similar precision casting techniques are used on turbochargers up to 1,300 ft/sec tip speed.

5. For extreme tip speeds, 1,200–1,600 ft/sec, radially bladed, semi-shrouded impellers are contour-milled out of aluminum, magnesium, or steel alloy forgings, sometimes combined with separately machined or cast inducers at the inlet end. Also, EDM (electric discharging machining) is used to fabricate impellers from a single forging.

A controlling design factor is the rotor velocity expressed in the form of tip speed or peripheral speed of the impeller. The head produced by a compressor of given geometry will be a square function of the tip speed alone, quite independent of the size of the machine. Tip speed rather than shaft speed in rpm controls the mechanical stresses of a given rotor configuration.

The tip speed of a conventional impeller is usually 800 to 900 ft/sec. This means that an impeller will be able to develop approximately 9,500 ft of head (the resulting pressure depends on the gas being compressed). Multistage compressors are needed if duties exceed this value. Heavy gases such as propane, propylene, or Freon (mol wt 60 or above) require a reduction in tip speeds due to lower sonic velocities of these gases when compared with air. For these gases, the relative Mach number at the inlet side of the impeller generally is limited to 0.8.

Overspeed Test of Impellers

Overspeed tests are carried out at 115 percent of maximum operating speed (132 percent of operating stress) on most milled or fabricated steel impellers. Cast impellers are sometimes spun at 120–140 percent of design speed (that is, 144–200 percent of design stress levels). Overspeeding is done in a heavily shielded "spin pit" that can safely contain, if necessary, a bursting impeller. The pit is evacuated for each run to minimize the windage loss of the spinning impeller. Bursting speeds may range from 1,200 to 2,000 ft/sec (tip speed) on fabricated and cast impellers, and may be even higher on forged and milled wheels.

Impellers of a ductile material will deform long before reaching burst speed. If the operating tip speeds must be as high as 1,200–1,600 ft/sec, a 15 percent overspeed test will cause local yielding of the most highly stressed region near the hub bore. In a material of sufficient ductility—over about 8 percent—this local yield will redistribute the stresses and improve dimensional stability in later operation; the impeller will then have to be finish-machined after the spin test.

Impeller Materials

AISI 4140 is the most prevalent steel alloy used for impellers and shafts in petrochemical gas compression services. For hydrogen sulfide laden gases, corrosion engineers recommend that the heat treatment be held to 22 Rockwell "C," which results in a yield stress of 90,000 psi.

Impellers in relatively clean service such as in ammonia plants often exceed this value and 30–35 Rockwell "C" hardness is allowed here.

Working stresses of most impellers are about 50,000 psi.

Impeller Attachment

A half-sectional profile of an impeller disk is a right angle triangle shape having a peripheral angle of about 16°. The hub design is significant in that it constitutes the impeller suction eye configuration, transmits the driving torque from the shaft, provides the necessary disk and rigidity, and fixes the impeller balance position. Most U.S. manufacturers use keys staggered 90° to complement the balance of multi-stage machines. The impeller hubs are reamed to a shrink interference fit of 0.75 mil/in. to 1.5 mil/in. A sharp-cornered keyway with a snug fit can develop stress densities two to three times the nominal stress. A loose fit would compound this stress concentration. The use of two close fitted feather keys has merit. The thickness of a feather key is one-fourth of the width (peripheral) dimension. All keyways should have well-rounded fillets.

The principal source of rotor internal friction is the interference shrink fit of rotor elements on the shaft. The friction effect on built-up rotors could be minimized by making the shrink fit as short and heavy as possible. This led to the universal and most important practice of undercutting the bore of all elements mounted on the shaft, as shown in Figure 9-4.

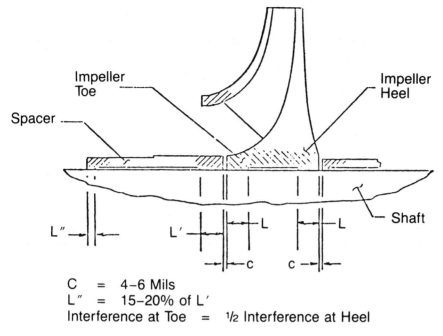

C = 4–6 Mils
L″ = 15–20% of L′
Interference at Toe = ½ Interference at Heel

Figure 9-4. Typical arrangement of impeller and spacer mounted on the shaft.

The impellers are locked against axial movement by various methods: split rings with locking bands, threaded locknuts, or a combination of both designs. If threaded nuts or sleeves are used, a lock ring is needed. Special care is needed in setting the lock ring:

1. Most lock rings are AISI 410 stainless material.
2. Hardness should be dead soft, under Rockwell C-20 (255 Brinell).
3. The locking notches or grooves in the sleeve, impeller, or balancing drum should be deburred.
4. The locking ring should be installed, the sleeve tightened, and the location of the locking notches or grooves marked on the ring.
5. The ring should be removed and nicked or saw cut about one-third of the way through at the marks.
6. The ring and sleeve should be reinstalled with the cuts in the rings aligned with the locking notches or grooves.
7. A soft round-nosed tool should be used to bend the tabs (alternately) into the notches.
8. Do not overbend the tabs.
9. Locking rings should not be reused. *Important!*

Compressor Impeller Design Problems

The high speed rotation of the impeller of a centrifugal compressor imparts the vital dynamic velocity to the flow within the gas path. The buffeting effects of the gas flow can cause fatigue failures in the conventional fabricated shrouded impeller due to vibration-induced alternating stresses. These may be of the following types:

1. Resonant vibration in a principal mode.
2. Forced-undamped vibration, associated with aerodynamic buffeting or high acoustic energy levels.

The vibratory mode most frequently encountered is of the plate type and involves either the shroud or disc. Fatigue failure generally originates at the impeller OD, adjacent to a vane. The fatigue crack propagates inward along the nodal line, and finally a section of the shroud or disk tears out.

To eliminate failures of the plate type, impellers operating at high gas density levels are frequently scalloped between vanes at the OD (Figure 9-5). The consequent reduction in disc friction may cause a small increase in stage efficiency.

Several rotors have been salvaged by scalloping the wheels after a partial failure has occurred. Scalloping describes a machining procedure which removes material from the impeller periphery between adjacent vanes. The maximum diameter from vane tip to vane tip 180° apart remains unchanged by scalloping. One eight-stage, 6,520 hp, 10,225 rpm compressor in low molecular weight service, had 54 scallops done to each wheel during an emergency shutdown. To accomplish this, the rotor was unstacked. Each wheel was set up in the milling machine and scalloped. Then each wheel was individually balanced on a mandrel. The rotor was restacked and the machine returned to service in slightly over a week.

Impeller Balancing Procedure

The quality of any dynamic balancing operation depends upon the following:

1. The control of radial run-out.
2. The elimination of internal couples along the length of the rotor.

Figure 9-5. Compressor impeller "scalloping" via removal of material from both disk and cover.

Individual balancing of impellers is vital. The following procedures are a *must*:

1. Prepare half-keys as required for the balancing of individual impellers on a mandrel. These keys must precisely fill the open keyways at the impeller bores.
2. An impeller precision balancing mandrel must be made; the actual geometry should match the minimum pedestal spacing and the roller configuration of the balancing machine.

Guidelines for designing the mandrel are:

1. The mandrel should preferably be made of low alloy steel, i.e., AISI 4140 or AISI 4340, which has been suitably stress-relieved.
2. The journal surfaces should preferably be hardened and ground, with a finish not poorer than 16 rms.
3. All diameters must be concentric within 0.0001 in. TIR.
4. The diameter of the section where the impeller is to be mounted, should be established on the basis of heating the impeller hub to a temperature of approximately 300°F for installation and removal.
5. Keyways are not incorporated in the mandrel.
6. The impeller balancing mandrel should be checked to assure that it is in dynamic balance. Make corrections on the faces, if required.
7. The impeller balancing mandrel should be free of burrs and gouges.
8. Mount each individual impeller, together with its half-key, on the balancing mandrel. A light coating of molybdenum disulfide lubricant should be first applied at the fits. Such mounting requires careful and uniform heating of the impeller hub, using a rosebud-tip torch, to a temperature of approximately 300°F; a temperature-indicating Tempil® stick should be used to monitor the heating operation. Install the mandrel, with the mounted impeller, in the balancing machine. Cool the impeller by directing a flow of shop air against the hub while slowly rotating the assembly by hand or with the balancing machine drive. When the impeller and mandrel have cooled to approximately room temperature, proceed to identify the required dynamic corrections with the balancing machine operating at the highest permissible speed.
9. Make the required dynamic corrections to the impeller by removing material over an extended area, with a relatively fine grade grinding disc, at or near the impeller tip on both the cover and disc surfaces.

Other Compressor Balancing

1. Inspect all impeller spacers to assure that they are of uniform radial thickness in any plane perpendicular to their axis; minor axial taper variations are of no concern. The spacers may be mounted on a machining mandrel for concentricity checking.
2. The balance piston is considered to be the equivalent of an impeller.

Marking of Impellers

Stamping of numbers on impellers should never be allowed. Use a high speed pencil grinder, also called rotating pencil, in low stress areas. Burn marks from the rotating pencil are not as harmful as stamping marks. However, they do create stress raisers, which have occasionally caused fatigue failures. Therefore, even these marks should be used only where absolutely necessary, and in carefully selected locations.

Critical Areas

Sharp edges on impeller bore: These edges dig into the shaft, causing high stress concentrations and possible starting points for shaft cracks. Polished radii should be provided.

Sharp corners in keyways: Keyway radii should be about $^1/_4$ of the keyway depth. Two keyways are preferable because single keyways will cause nonuniform heating of the shaft and also shaft warpage because there is no radial shrink stress at the keyway region, while full radial shrink stress acts on the shaft opposite to the keyway. This will cause a shaft bow which changes with shrink stress, i.e., with temperature and speed. The keyway in the impeller should be as shallow as possible, since this key is not meant to transmit torque, but to act only as a positioning and safety feature.

Two shallow keys instead of one deep key are expensive, because keyways must be positioned very accurately, and depth must be equal. But it eliminates a host of problems (cranking effect, contact-pressure effect, component balance problems, thermal bow), and when comparing cost of problems and balancing difficulties against the higher manufacturing cost of double keys, these will probably be more economical, even for initial cost, but certainly when field problems and production loss are included.

A makeshift solution can be applied in the field, by relieving the shaft surface opposite the single keyway to get equal contact pattern. The relief need not be deep (say 20 mils or so). This will at least improve the contact pressure and thermal distortion.

Another possibility presently gaining favor is to eliminate the keys entirely, depending 100 percent on the shrink fit engagement. The parts are mounted with a very high shrink (2.5 mils/in. or more). Special methods must be used if disassembly becomes necessary, or the parts may be severely damaged. The possibility of stress-corrosion comes to mind with these high shrink stresses (~70,000 psi) but actually the stress in a keyway corner is much higher, usually exceeding yield strength, even with a moderate shrink fit (1 mil/in. or less).

Rotor Bows in Compressors and Steam Turbines

Rotors sometimes will operate very satisfactorily for years, then upon restarting after a shutdown, excessive vibration occurs at the first lateral critical speed. This vibration problem may originate from the following:

1. The unit is tripped at rated speed and the rotative speed abruptly drops.
2. The effective interference fit of the impellers or wheels shrunk on the shaft increases rapidly during this speed reduction. (The bore stress varies with the square of the rotative speed.)
3. At the first critical speed range, maximum vibration amplitudes result; the rotor is then in a simple bow configuration, with the maximum deflection at approximately mid-span.
4. The shrunk-on elements thus literally lock the rotor in this bow shape and the clamping action increases below the critical speed. The outermost fibers of the shaft are incapable of sliding axially at the clamp or interference fit areas.
5. At rest, the rotor exhibits a large residual shaft bow, and gross imbalance causes high vibration amplitudes upon restarting.
6. Usually, the shaft is bowed elastically; disassembly of the rotor generally results in restoration of the bare shaft to an acceptable run-out condition.

Multistage centrifugal compressors tend to be particularly susceptible to the foregoing if they incorporate the same interference fit at each end of the impeller bore. It has been said that rotative speed tends to reduce the interference fit to near zero with the first lateral critical speed at 50 percent of the maximum continuous speed. In some cases the impellers will actually move on the shaft. Susceptibility to this problem increases as the ratio between the maximum continuous speed and the lateral critical speed increases. A very successful method of avoiding this problem is:

1. Making the static interference fit at the impeller *toe* equal to approximately *one-half* that at the *heel*, while maintaining the same land axial length.
2. Making the land axial length of the spacer at the end adjacent to the impeller heel equal to approximately 15–20 percent of the length at the opposite end, while maintaining the same interference fit.
3. A 4–6 mil axial gap between components (i.e., impellers, spacer sleeves, etc.) is provided.

Tightness of the impeller and spacer on the shaft at full operating speed is thus assured, while simultaneously providing for controlled axial sliding at the interference fits during deceleration. Some increase in rotor internal friction of course results, with a consequent minor effect on the nonsynchronous whirl threshold speed.

A transient bow condition similar to the above is sometimes produced in built-up rotors which are subject to rapid starting. This occurs if all shrunk-on elements are fitted tightly together axially during cold assembly. In effect, the shrunk-on elements heat up much faster than the bare shaft after startup; the resulting thermal growth of these elements, when combined with a lack of perpendicularity of the vertical mating faces, results in shaft bowing. An axial clearance is usually provided between an impeller or wheel and its adjacent spacer to avoid this. Four to six mils is usually adequate.

This discussion is paraphrased from some of the writings of Roy Greene, who has experience in centrifugal compressor design, manufacture, and operation for Clark, Cooper-Bessemer, and Ingersoll Rand, and who currently is a consultant[9].

Balancing

Since the details of rotor balancing have been fully covered in Chapter 6, we can confine our discussion to a brief recap of this important topic. It is simple to balance a rigid rotor which runs on rigid bearings, so long as neither change shape, or deflect at operating speed. For a high-speed rotor, this is no longer true, because both the rotor and the bearings will deflect as they are exposed to centrifugal forces, and these deflections will result in a very complex unbalance condition that was not there when the rotor was balanced at low speeds in a precision balancing machine. A low-speed balancing machine, no matter how precise and/or sophisticated, cannot detect such conditions: They lie dormant in the machine until high speeds are reached, and *only then* will the beast show its true nature. There is absolutely no way to accurately predict the behavior of an existing high-speed rotor, except to get it up to speed on its own bearings, in its own casing. Even then it is impossible to determine the precise location of unbalance, and to make definitive and 100 percent effective corrections that will eliminate the vibration throughout the speed range. The best one can hope for is to get a correction, essentially by compromise, that permits smooth operation at one speed.

The *only* way to get a predictable rotor is to maintain perfect balance of each individual component during all operations of machining and assembly and *never* to disturb this balance by making indiscriminate corrections on the finished assembly. To do all this properly is exceedingly difficult, and the methods and accuracies required border on (and often exceed) the limits of available technology of manufacturing and measuring. These limits dictate how fast a rotor may run and how many impellers can safely be used at a given speed, both factors being of great economic importance (cost, efficiency).

Once a high-speed rotor is assembled and found out of balance at operating speed, the *only* way to reestablish predictable balance is to completely disassemble the rotor and to start from scratch.

Clean Up and Inspection of Rotor

Compressor rotors must be carefully inspected for any damage. To accomplish this, these guidelines should be followed:

1. The rotor should rest on the packing area that must be protected by soft packing, annealed copper, or lead to avoid any marring of polished surfaces. *Do not use* Teflon® strips since Teflon® impregnation of the metal surfaces can alter the adhesion characteristics of the lubricant in contact with the journal. Lubrication problems could ensue.
2. The rotor should be given an initial inspection for the following:
   a. Impeller hub, cover, and vane pitting or damage.
   b. Are there any rubs or metal transfer on the hub or cover indicating a shifting of rotor position? All foreign metal should be ground off and the area inspected for heat checking as described in Item 8.
   c. Journals
      • Journal diameter—Roundness and taper are the two most critical dimensions associated with a bearing journal. These dimensions are established with a four-point check taken in the vertical and horizontal planes (at 90° to one another) at both the forward and aft edge of the journal. A micrometer is normally used for this purpose. The journal diameter must be subtracted from the liner bore diameter to determine the clearance. If the journal diameter is 0.002 in. or more outside of its drawing tolerance, it is necessary to remachine the journal. Another parameter that must be carefully watched is journal taper. Excessive taper produces an increase in the oil flow out one of the ends of the bearing, thereby starving the other end. This can result in excessive babbitt temperatures. Journal tapers greater than about 0.0015 in. require remachining.

- Journal surface—Surfaces that have been scratched, pitted, or scraped to depths of 0.001 in. or less are acceptable for use. Deeper imperfections in the range of 0.001 to 0.005 in. must be restored by strapping.
- Thrust collar—does it have good finish? Use same guidelines as for journals. Is the locking nut and key tight? If the collar is removed, is its fit proper? It should have 0.001 to 0.0005 in. interference minimum.

3. The journals, coupling fits, overspeed trip, and other highly polished areas should be tightly wrapped and sealed with protective cloth.
4. The rotor should be sandblasted using No. 5 grade, 80/120 mesh, polishing compound, silica sand, or aluminum oxide.
5. When the rotor is clean, it should be again visually inspected.
6. Impellers and shaft sleeve rubs—rubs in excess of 5 mils deep in labyrinth areas require reclaiming of that area.
7. Wheel location—have any wheels shifted out of position? Wheel location should be measured from a thrust collar locating shoulder. There should be a 4–5 mil gap between each component of the rotor; i.e., each impeller, each sleeve, etc.
8. On areas suspected of having heat checking or cracks, a dye penetrant check should be made using standard techniques or "Zyglo":
   a. Preparation
   Cracks in forgings probably have breathed; that is, they have opened and closed during heat cycles, drawing in moist air that has condensed in the cracks, forming oxides and filling cracks with moisture. This prevents penetration by crack detection solutions. To overcome this condition, all areas to be tested should be heated by a gas torch to about 250°F and allowed to cool before application of the penetrant.
   These tests require a smooth surface as any irregularities will trap penetrant and make it difficult to remove, thus giving a false indication or obscuring a real defect.
   b. Application
   The penetrant is applied to the surface and allowed to seep into cracks for 15 to 20 minutes. The surface is then cleaned and a developer applied. The developer acts as a capillary agent (or blotter) and draws the dyed penetrant from surface defects so it is visible, thus indicating the presence of a discontinuity of the surface. In "Zyglo" an ultraviolet light is used to view the surface.
9. A more precise method of checking for a forging defect would require magnetic particle check, "Magnaflux" or "Magnaglow." As

these methods induce a magnetic field in the rotor, care must be taken to ensure that the rotor is degaussed and all residual magnetism removed.

10. The rotor should be indicated with shaft supported at the journals:
    a. Shaft run out (packing areas) 0.002 in. TIR max.
    b. Impeller wobble—0.010 in. TIR—measured near O.D.
    c. Shroud band wobble—0.020 in. TIR.
    d. Thrust collar—0.0005 in. TIR measured on vertical face.
    e. Vibration probe surfaces 0.0005 in. TIR—no chrome plating, metallizing, etc., should be permitted in these areas.
    f. Journal areas—0.0005 in. TIR, 20 micro in. rms or better.
    g. Gaps between all adjacent shrink fit parts—should be 0.004 to 0.005 in.

11. If the shaft has a permanent bow in excess of the limit or if there is evidence of impeller distress, i.e., heavy rubs or wobble, the rotor must be disassembled. Similarly, if the journals or seal surfaces on the shaft are badly scored, disassembly in most cases is indicated as discussed below.

Disassembly of Rotor for Shaft Repair

If disassembly is required the following guidelines will be helpful.

1. The centrifugal rotor assembly is made with uniform shrink fit engagement ($^3/_4$ to $1^1/_2$ mil/in. of shaft diameter), and this requires an impeller heating process or, in extreme cases, a combination process of heating the impeller and cooling the shaft.

2. The shrinks are calculated to be released when the wheel is heated to 600°F *maximum*. To exceed this figure could result in metallurgical changes in the wheel. Tempil® sticks should be used to ensure this is not exceeded. The entire diameter of an impeller must be uniformly heated using "Rosebud" tips—two or more at the same time.

3. Generally a turbine wheel must be heated so that it expands 0.006–0.008 in. more than the shaft so that it is free to move on the shaft.

4. The important thing to remember when removing impellers is that the heat must be applied quickly to the rim section first. After the rim section has been heated, heat is applied to the hub section, starting at the outside. *Never apply heat toward the bore with the remainder of the impeller cool.*

5. To disassemble rotors, naturally the parts should be carefully marked as taken apart so that identical parts can be replaced in the proper

position. A sketch of rotor component position should be made using the thrust collar as a reference point. Measure and record distance from the thrust collar or shoulder to first impeller hub edge. Make and record distance between all impellers.

6. When a multistage compressor is to be disassembled, each impeller should be stencilled. From thrust end, the first impeller should be stencilled T-1, second wheel T-2, and so on. If working from coupling end, stencil first wheel C-1, second wheel C-2, and so on.

7. The rotor should be suspended vertically above a sand box to soften the impact of the impeller as it falls from the shaft. It may be necessary to tap the heated impeller with a lead hammer in order to get it moving. The weight of the impeller should cause it to move when it is hot enough.

Shaft Design

It is not uncommon to design for short-term loads approaching 80 percent of the minimum yield strength at the coupling end of the shaft. The shaft is not exposed to corrosive conditions of the compressed gas at this point. Inside of the casing, the shaft size is fixed by the critical speed rigidity requirements. The internal shaft stress is about 5,000–7,000 psi— very low compared to the impellers or at the coupling area. With drum-type rotors there is no central portion of the shaft, there are only shaft stubs at each end of the rotor. The purpose of the shaft is to carry the impellers, to bridge the space between the bearings and to transmit the torque from the coupling to each impeller. Another function is to provide surfaces for the bearing journals, thrust collars, and seals.

The design of the shaft itself does not present a limiting factor in the turbomachinery design. The main problems are to maintain the shaft straight and in balance, to prevent whipping of overhangs, and to prevent failure which may be caused by lateral or torsional vibration, chafing of shrunk-on parts, or manufacturing inadequacies. The shaft must be accurately made, but the limits of technology are not approached as far as theory or manufacturing techniques are concerned. A thermally unstable shaft develops a bow as a function of temperature. To reduce this bow to acceptable limits requires forgings of a uniformity and quality that can only be obtained by the most careful manufacturing and metallurgical techniques.

Rotors made of annealed material are not adequate, because many materials, for example AISI 4140, have a high ductility transition temperature in the annealed condition. This has caused failures, especially of shaft ends. Therefore, it is very important to make sure that the material has been properly heat-treated.

Most compressor shafts are made from AISI 4140 or 4340. AISI 4340 is preferred because the added nickel increases the ductility of the metal. Most of the time the yield strength is over 90,000 psi and the hardness no greater than 22 Rockwell "C" in order to avoid sulfide stress cracking.

While selection of the material is fairly simple, quality control over the actual piece of stock is complicated. There are several points to consider.

1. *Material Quality:* Forgings of aircraft quality (= "Magnaflux quality") are required for all but the simplest machines. Bar-stock may not have sufficient thermal stability, and therefore must be inspected carefully. Note that shafts—as well as all other critical components—must be stress-relieved after rough machining, which usually leaves $^1/_{16}$ in. of material for finishing.
2. *Testing:* Magnaglow of finished shaft. Ultrasonic test is desirable for large shafts. Heat indication test is required for critical equipment.
3. *Shaft Ends:* Should be designed to take a moderate amount of torsional vibration, not only the steady operating torque.
4. The shaft must be able to withstand the shrink stresses. Any medium strength steel will do this. After some service the impeller hubs coin distinct depressions into such shafts, squeezing the shaft, so to speak. This squeezing process also causes shaft distortion and permanent elongation of the shaft, which can lead to vibration problems or internal rubbing. Since part of the initial shrink fit is lost, this may cause other types of problems, such as looseness of impellers, which then can lead to looseness-excited vibrations such as hysteresis whirl.

Rotor Assembly

1. Remove the balanced shaft from the balancing machine, and position it vertically in a holding fixture providing adequate lateral support; the stacking step on the shaft should be at the bottom.
2. Remove all of the half-keys.
3. Assembly of the impellers and spacers on the shaft requires heating, generally in accordance with the procedure previously outlined for mandrel balancing. The temperature that must be attained to permit assembly is determined by the micrometer measurement of the shaft and bore diameters, and calculation of the temperature differential needed.
4. Due to extreme temperatures, a micrometer cannot be used; therefore, a go-no go gauge, 0.006 in. to 0.008 in. larger than the shaft

diameter at the impeller fit, should be available for checking the impeller bore before any assembly shrinking is attempted.

5. Shrink a ring (0.003 in. to 0.004 in. tight) on the shaft extending about $1/32$ in. past the first impeller location. Machine the ring to the exact distance from the machined surface of the impeller to the thrust shoulder, and record it on a sketch. This gives a perfect location and helps make the impeller run true.

6. Heating the impeller for assembly is a critical step. The important thing to keep in mind is that the hub bore temperature must not get ahead of the rim temperature by more than 10°–15°F. The usual geometry of impellers is such that they will generally be heated so that the rim will expand slightly ahead of the hub section and tend to lift the hub section outward. With long and heavy hub sections, extreme care must be taken to not attempt too rapid a rate of heating because the bore of the hub can heat up ahead of the hub section and result in a permanent inward growth of the bore.

   Heating of the wheel can be accomplished in three different ways:

   a. Horizontal furnace: the preferred method of heating the wheel for assembly because the temperature can be carefully controlled.

   b. Gas ring: The ring should be made with a diameter equal to the mass center of the impeller.

   c. "Rosebuds": The use of two or more large diameter oxyacetylene torches can be used with good results. The impeller should be supported at three or more points. Play the torches over the impeller so that it is heated evenly, remembering the 600°F limitation. Tempil® sticks should be used to monitor the temperature.

7. The wheel fit of the shaft should be lightly coated with high temperature antiseize compound.

8. The heated wheel should be bore checked at about the center of the bore fits. As soon as a suitable go-no go gauge can be inserted freely into the impeller fit bore, the impeller should be quickly moved to the shaft. With the keys in place, the impeller bore should be quickly dropped on the shaft, using the ring added in step 5 as a locating guide.

9. Shim stock, of approximately 0.004–0.006 in. thickness, should be inserted at three equally-spaced radial locations adjacent to the impeller hubs to provide the axial clearance needed between adjacent impellers. This is necessary to avoid transient thermal bowing in service.

10. Artificial cooling of the impeller during assembly must be used in order to accurately locate the impeller at a given fixed axial position. Compressed air cooling must be immediately applied after

the wheel is in place. The side of the impeller where air cooling is applied is nearest to the fixed locating ring and/or support point. The locating ring should be removed after the impeller is cooled.

11. Recheck axial position of the impeller. If an impeller goes on out of position and must be moved, thoroughly cool the entire impeller and shaft *before* starting the second attempt. This may take three to four hours.

12. After the impellers, with their spacers and full-keys, have been assembled and cooled, the shim stock adjacent to the impeller hubs should be removed.

13. If the rotor has no sleeves, another split ring is needed to locate the second impeller. This split ring is machined to equal the distance between the first and second wheels. Then, a split ring is required for the next impeller, etc. Any burrs raised by previously assembled impellers should be carefully removed and the surfaces smoothed out.

14. Check for shaft warpage and impeller runout as each impeller is mounted. It may be necessary to unstack the rotor to correct any deficiencies.

15. The mounting of sleeves and thrust collars requires special attention. Sleeves have a lighter shrink than wheels and because of their lighter cross section can be easily damaged by uneven heating or high temperature. Thrust collars can be easily warped by heat. The temperature of the thrust collar and sleeves should be limited to about 300°–400°F.

16. Mount the rotor, now containing all the impellers, in the balancing machine, and spin it at the highest possible speed for approximately five minutes.

17. Shut down and check the angular position of the high spots and runout at the three previously selected spacer locations between journals. The high spots must be within ±45°, and the radial runouts within $1/2$ mil, of the values recorded during bare shaft checking. If these criteria are not satisfied, it indicates that one or more elements have been cocked during mounting, thus causing the shaft to be locked-up in a bow by the interference fits. It is then necessary to remove the two impellers and spacers from the shaft, and to repeat the vertical assembly process.

18. Install the rotor locknut, being careful not to over tighten it; *shaft bowing can otherwise result*. If the rotor elements are instead positioned by a split ring and sleeve configuration, *an adjacent spacer must be machined to a precise length determined by pin micrometer measurement after all impellers have been mounted.*

19. Many compressors are designed to operate between the 1st and 2nd lateral critical speeds. Most experts agree that routine check

balance of complete rotors with correction on the first and last wheels is wrong for rotors with more than two wheels. The best method is to balance the *assembled* rotor in three planes.

The residual dynamic couple imbalance should be corrected at the ends of the rotor, and the remaining residual static (force) imbalance should be corrected at about the middle of the rotor.

For compressors that operate below the first critical (stiff shaft machines), two plane balance is satisfactory.

20. Install the thrust disc on the rotor; this should require a small amount of heating. *It is most important that cold clearances not exist at the thrust disc bore, since it will permit radial throwout of a relatively large mass at operating speed. Install the thrust-bearing spacer, and lightly tighten the thrust-bearing locknut.*

21. Spin the rotor at the highest possible balancing speed, and identify the correction(s) required at the thrust-bearing location. Generally, a static correction is all that is necessary, and it should be made in the relief groove at the OD of the thrust disc. No correction is permitted at the opposite end of the rotor.

22. Check the radial runout of the shaft end where the coupling hub will mount. This runout must not exceed 0.0005 in. (TIR), as before.

Shaft Balancing

Despite its symmetrical nature the shaft must be balanced. Again, the reader may wish to refer to Chapter 6 for details on the following.

1. Prepare half-keys for the keyways of the bare shaft. These should be carefully taped in position, using high-strength fiber-impregnated tape; several turns are usually required.

   Note: Tape sometimes fails during spinning in the balancing machine. It is therefore important that adequate shields be erected on each side of the balancing machine for the protection of personnel against the hazard of flying half-keys.

2. Mount the bare shaft, with half-keys in place, in the balancing machine with the supports at the journal locations. Spin the bare shaft at a speed of 300–400 rpm for approximately ten minutes. Shut down, and check the radial runout (TIR) at mid-span using a $1/10$ mil dial indicator; record the angular position of the high spot and run out valve. Spin the bare shaft at a speed of 200–300 rpm for an

additional five minutes. Shut down, and again check the radial run-out (TIR) at mid-span; record the angular position of the high spot and runout valve. Compare the results obtained after the ten minute and five minute runs; if they are the same, the bare shaft is ready for further checking and balancing. If the results are not repetitive, additional spinning is required; this should be continued until two consecutive five minute runs produce identical results.

3. Check the radial run-out (TIR) of the bare shaft in at least three spacer locations, approximately equidistant along the bearing span, and near the shaft ends. Record the angular position of the high spots and the runout values at each location. The shaft is generally considered to be satisfactory if both of these conditions are satisfied:
   a. The radial runout (TIR) at the section of the shaft between journals does not exceed 0.001 in.
   b. The radial runout (TIR) outboard of the journals does not exceed 0.0005 in.
4. With the balancing machine operating at its pre-determined rpm, make the required dynamic corrections to the bare shaft using wax. When satisfactory balance is reached, start removing material at the face of the step at each end of the center cylindrical section of the shaft. Under no circumstances should material be removed from the sections of the shaft outboard of the journal bearings.

Rotor Thrust in Centrifugal Compressors

Thrust bearing failure has potentially catastrophic consequences in compressors. Almost invariably, failure is due to overloading because of the following:

1. Improper calculation of thrust in the design of the compressor.
2. Failure to calculate thrust over the entire range of operating conditions.
3. A large increase in thrust resulting from "wiping" of impeller and balance piston labyrinth seals.
4. Surging of machine so that rotor "slams" from one side of thrust bearing to the other, and the oil film is destroyed.
5. Thrust collar mounting design is inadequate.

Rotor Thrust Calculations

Thrust loads in compressors due to aerodynamic forces are affected by impeller geometry, pressure rise through the compressor, and internal leakage due to labyrinth clearances. The impeller thrust is calculated, using correction factors to account for internal leakage and a balance piston size selected to compensate for the impeller thrust load. The common assumptions made in the calculations are as follows:

1. Radial pressure distribution along the outside of disc cover is essentially balanced.
2. Only the "eye" area is effective in producing thrust.
3. Pressure differential applied to "eye" area is equal to the difference between the static pressure at the impeller tip, corrected for the "pumping action" of the disc, and the total pressure at inlet.

These "common assumptions" are grossly erroneous and can be disastrous when applied to high pressure barrel-type compressors where a large part of the impeller-generated thrust is compensated by a balance piston. The actual thrust is about 50 percent more than the calculations indicate. The error is less when the thrust is compensated by opposed impellers, because the mistaken assumptions offset each other.

Magnitude of the thrust is considerably affected by leakage at impeller labyrinth seals. Increased leakage here produces increased thrust independent of balancing piston labyrinth seal clearance or leakage. A very good discussion of thrust action is found in Reference 3.

The thrust errors are further compounded in the design of the balancing piston, labyrinths, and line. API-617, "Centrifugal Compressors," specifies that a separate pressure tap connection shall be provided to indicate the pressure in the balance chamber. It also specifies that the balance line shall be sized to handle balance piston labyrinth gas leakage at twice initial clearance without exceeding the load ratings of the thrust bearing, and that thrust bearings for compressors should be selected at no more than 50 percent of the bearing manufacturer's rating.

Many compressor manufacturers design for a balancing piston leakage rate of about $1\frac{1}{2}$–2 percent of the total compressor flow. Amoco and others feel that the average compressor, regardless of vendor, has a leakage rate of 3–4 percent of the total flow, and the balance line must be sized accordingly. This design philosophy would dictate a larger balance line to take care of the increased flow than normally provided. The balancing chamber in some machines is extremely small and probably highly susceptible to eductor type action inside the chamber which can increase leakage and increase thrust action. The labyrinth's leakage should not be permitted to exceed a velocity of 10 ft per second across the drum. The short balancing piston design of many designs results in a very high leakage velocity rate. Since the thrust-bearing load is represented by the difference between the impeller-generated thrust and the compensating balance piston thrust, small changes can produce overloading, particularly in high-pressure compressors.

Design Solutions

Many of these problems have been handled at Amoco by retrofitting 34 centrifugal compressors (57 percent of the total) with improved bearing designs. Most of the emphasis has been toward increased thrust capacity via adoption of a Kingsbury-type design, but journal bearings are always upgraded as part of the package. Design features include spray-lubed thrust bearings (about a dozen cases), copper alloy shoes, ball and socket tilting pad journals, pioneered by the Centritech Company of Houston, Texas, and many other advanced state-of-the art concepts.

Some of the balancing piston leakage problems have been solved by use of honeycomb labyrinths. The use of honeycomb labyrinths offers better control of leakage rates (up to 60 percent reduction of a straight pass-type labyrinth). Honeycomb seals operate at approximately $^1/_2$ the radial clearance of conventional labyrinth seals. The honeycomb structure is composed of stainless steel foil about 10 mils thick. Hexagonal-shaped cells make a reinforced structure that provides a larger number of effective throttling points.

Compressor shaft failures frequently occur because of loose fit of the thrust collar assembly. With no rotor positioning device left, the rotor shifts downstream and wrecks the machine. The practice of assembling thrust collars with a loose fit (1 to 5 mils) is very widespread because it makes compressor end seal replacements easier. The collar is thin (sometimes less than 1 in. thick) and tends to wobble. The shaft diameter is small in order to maximize thrust bearing area. A nut clamps the thrust collar against a shoulder. Both the shoulder and the nut are points of high stress concentration. With a thrust action of several tons during surging, the collar can come loose. In addition, fretting corrosion between the collar and the shaft can occur.

The minimum thrust capacity of a standard 8-in. (32.0 square in.) Kingsbury-type bearing with flooded lubrication at 10,000 rpm is well in excess of $6^1/_2$ tons. The thrust collar and its attachment method must be designed to accommodate this load. In most designs the inboard bearing has a solid base ring and the thrust collar must be installed after this thrust bearing is in place. The collar can be checked by revolving the assembled rotor in a lathe. The collar is subsequently removed for seal installation and must be checked for true running, i.e., the face is normal to the axis of the bearing housing again after it is finally fitted to the shaft.

This problem has been addressed at Amoco by redesigning the thrust collar to incorporate the spacer sleeve as an integral part and have a light shrink fit (0 to 1 mil tight). A puller is used to remove the collar after a small amount of heat is applied.

Managing Rotor Repairs at Outside Shops

When it becomes necessary to have rotor repairs performed away from your own plant, the outside shop should be required to submit such procedures as are proposed for inspections, disassembly, repair, reassembly, balancing, and even crating and shipping. And, while it is beyond the scope of this text to provide all possible variations of these procedures, two or three good sample procedures are given for the reader's information and review.

In the following section, the procedure proposed by a highly experienced repair shop for work to be performed on centrifugal compressor rotors is shown.

Procedures for Inspection, Disassembly, Stacking, and Balance of Centrifugal Compressor Rotors*

Incoming Inspection

1. *Prepare incoming documentation.* Note any defects or other damage on rotor. Note any components shipped with rotor, such as coupling hub or thrust collar.
2. *Clean rotor.* Protect all bearing, seal, probe, and coupling surfaces. Blast clean with 200 mesh grit. Glass bead, walnut shell, solvent, and aluminum oxide available if requested by equipment owner. After cleaning, coat all surfaces with a light oil.
3. *Perform non-destructive test.* Use applicable NDT procedure to determine existence and location of defects on any components. Record magnitude and location of any defects as indicated in Figure 9-6.
4. *Measure and record all pertinent dimensions of the rotor as shown in Figures 9-7 and 9-8.* Record on a sketch designed for the particular rotor. Record the following dimensions:
   • Impeller diameter and suction eye
   • Seal sleeves, spacers, and shaft

---

*Source: Hickham Industries, Inc., La Porte, Texas 77571. Reprinted by permission.

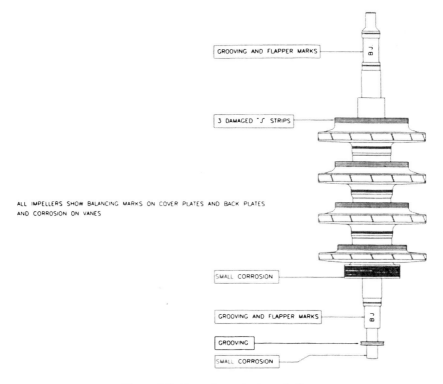

Figure 9-6. Recording rotor imperfections.

- Journal diameters
- Coupling fits and keyways
- Gaps between adjacent shrunk-on parts

5. *Check and record pertinent runouts.* Rotor is supported at the bearing journals on "V" blocks. Runouts should be phase-related using the coupling (driven end) keyway as the 0° phase reference. If the coupling area is double-keyed or has no keyway, the thrust collar keyway should be used as the zero reference. If this is not possible, an arrow should be stamped on the end of the shaft to indicate plane of zero-phase reference.

6. *Check and record electrical runout probe area.* Use an 8-mm diameter eddy probe. Probe should be calibrated to shaft material only. Probe area tolerance should be 0.25 mil maximum.

7. *Check and record all pertinent axial stack-up dimensions.* Referenced from thrust collar shaft shoulder or integral thrust collar.

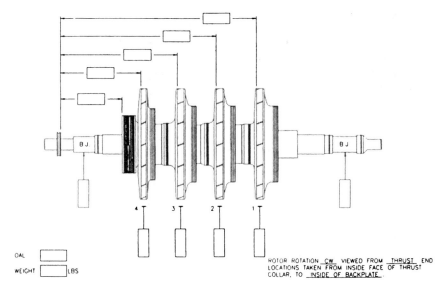

Figure 9-7. Typical record of axial distances for centrifugal compressor rotor.

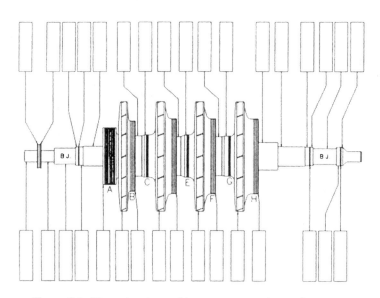

Figure 9-8. Dimensional record for compressor rotor sealing areas.

If Disassembly Is Required

1. *Visually inspect.* Visually inspect each part removed. Measure and record all pertinent shaft and component dimensions as follows:
   - Impeller bore sizes—key size where applicable
   - Shaft sleeve bore sizes
   - Balance piston bore sizes
   - Thrust collar bore size—key size where applicable
2. *Use of applicable non-destructive test procedures.* Use NDT procedures to determine existence and location of any cracks on shaft and component parts. Maximum allowable residual magnetism 2.0 gauss.
3. *Completion of inspection procedures.* Upon completion of inspection procedures, customer is notified and the results evaluated and discussed. The repair scope most advantageous to the customer is confirmed and completed.

Assembly

1. *Check dynamic balance.* Check dynamic balance of shaft. Balance tolerance, unless otherwise specified, is 4 w/n per plane. Correction of unbalance is analyzed and made on an individual basis.
   - Rotors that stack from the center out stack two wheels at a time
   - Rotors that stack from one end stack one wheel at a time
   - After each stacking step, allow components to cool to 120°F or less
   - Runouts should not change more than 0.5 mil between component stacks[†]
   - Maximum allowable runout on shaft is 1.0 mil[†]
   - Maximum allowable eye face runout is 2.0 mil[†]
   - If runouts exceed tolerance, de-stack (to problem point), check shaft runout, and restack
   - Perform 12-point residual per plane after final trim balance is completed

Final Inspection

1. *Document.* Document final runouts and submit to customer.
2. *Probe area.* As required, check and record vibration probe area for electrical/mechanical runouts; correct as required; maximum allowable 0.25 mil peak to peak.

[†]Based on 5,000-lb, 5,000-rpm rotor.
Source: Hickham Industries, Inc., La Porte, Texas 77571. Adapted by permission.

3. *Preserve.* Preserve rotor as follows: coat rotor completely with Cesco 140, wrap rotor, and notify customer with shipping or storage information.

A sample specification or procedure that the responsible (and responsive) repair shop furnishes to its sub-vendors is shown in the following section.

Turbo Specification Chrome Plating and Finish Grinding*

Repair Facility to Provide to Vendor

1. Clear, concise drawing detailing:
   - Areas requiring plating or grinding
   - Finish dimension required and tolerance to hold—unless specified OD tolerance + 0.0005 − 0.0000. Finish 16 RMS maximum
   - Shop contact and job number
   - Desired delivery date
   - Hardness of area to be chromed
2. Proper support cradle for safe transportation (when the repair facility is providing transportation).
3. A calibrated standard-taper ring gauge on taper coupling chroming.
4. No chroming on a sleeve area under any circumstances.

Vendor to Provide to Repair Facility

1. Proper support cradle for safe transportation (when vendor furnishes transportation).
2. Incoming inspection, to note any areas of concern not covered in original scope.
3. Chrome plating. Prepare areas to be chromed by grinding all grooves, pits, scratches, and other blemishes in area to be plated.
4. Blending in of all sharp corners and edges, both internal and external, with adjacent surface. Chrome deposit should thus be blended into adjacent surfaces so as to prevent lack of deposit or build-up of deposit.
5. Anodic cleaning of surfaces to be coated to assure maximum adhesion of chrome.
6. Chrome plating deposited directly on the ground surface without the application of any undercoat.
7. Chrome plating free of any visible defects. It should be smooth, fine grained, and adherent. A dye penetrant inspection qualifies the above.

8. No chrome plating on top of chromium, unless specified by the repair facility.

Finish Grinding

1. Chrome coating should be ground to finish dimensions specified. Tolerance on OD should be + 0.0005 − 0.0000 unless otherwise specified.
2. Grinding should be done with proper coolant and wheel speed to produce proper surface finish.
3. Desired surface finish should be 16 rms maximum unless otherwise specified.
4. Taper shaft fits—appropriate, calibrated, and approved; ring gauge should accompany to ensure standard taper. A blue check should be made prior to shipment.
5. Final inspection—dimensional and dye penetrant.
6. Prepare for shipment by wrapping finished areas with protective cloth to resist damage during handling and shipping, and notify the repair facility representative upon completion for shipping arrangements.

Mounting of Hydraulically Fitted Hubs*

Modern turbomachinery rotors are commonly fitted with coupling hubs. For years, these hubs have incorporated keys. Lately, however, keyless hubs have gained favor.

Hubs that are not provided with a keyway receive (or transmit) the torque from the shaft through friction. Hence, the hubs must grip the shaft tightly. This gripping is accomplished by advancing the hub on the tapered shaft a specified amount. To facilitate this advance one must expand the hub bore. Two methods are used most often: heating or hydraulic pressure.

When hydraulic pressure is used, a few specialized tools are needed. Basically, they are an installation tool, a high pressure oil pump with pressure gauge, and a low pressure pump with pressure gauge.

To ensure satisfactory performance, the following procedure is recommended for proper installation when using the hydraulic pressure method. It assumes that your installation employs O-rings, although experience shows that many modern coupling hubs can be installed *without* the use of O-rings. On these, please disregard any references to O-rings.

---

*Source: Koppers Company, Inc., Power Transmission Division, Baltimore, Maryland 21203. Reprinted by permission.

Check for Proper Contact. After the shaft and hub bore are thoroughly cleaned, spread a thin layer of mechanics blue on the shaft and push the hub snugly. A very slight rotation of the hub is permitted after it is pushed all the way. Remove the hub and check the bore for blue color. At least 80 percent of the bore should have contact.

Improve the Contact. If less than 80 percent contact is found, the shaft and hub should be *independently* lapped using a ring and plug tool set.

Clean the Lapped Surfaces. Remove all traces of lapping compound using a solvent and lint-free towels. Immediately afterward, spread thin oil on the shaft and hub bore to prevent rusting. Recheck the hub to shaft contact.

Determine Zero Clearance (START) Position. Without O-rings in the shaft or hub, push the hub snugly on the shaft. This is the "start" position. With a depth gauge, measure the amount the hub overhangs the shaft end and record this value.

Prepare for Measuring the Hub Draw (Advance). The hub must be advanced on the shaft exactly the amount specified. Too little advance could result in the hub spinning loose; too much advance could result in the hub splitting at or shortly after installation. As the overhang cannot be measured *during* installation, other means to measure the advance must be found. The best way is to install a split collar on the shaft, away from the hub by the amount of the specified advance. Use feeler gauges for accurate spacing. See Figure 9-9.

Install O-Rings and Back-Up Rings. The oil is pumped between the hub and shaft through a shallow circular groove machined either in the hub or in the shaft. Install the O-rings *toward* this groove, the back-up rings *away* from this groove. Do not twist either the O-rings or the back-up rings while

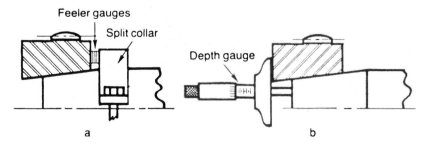

Figure 9-9. Methods of determining and limited hub advance on tapered shaft.

installing. After they are installed, look again! The O-rings must be *between* the back-up rings and the oil groove! Spread a little bit of thin oil on the rubber surfaces.

Mount "Other" Components. Read the coupling installation procedure again. Must other components (such as a sleeve) be mounted on the shaft before hub? Now is the time to do it.

Mount the Hub on the Shaft. Avoid pinching the O-rings during mounting. The O-rings will prevent the hub from advancing to the "start" position. This is acceptable.

Mount the Installation Tool. Wet the threads with thin oil, and rotate the tool until it butts against the shaft shoulder. The last few turns will require the use of a spanner wrench.

Connect the Hydraulic Lines. Connect the installation tool to the *low* pressure oil pump (5,000 psi minimum). Connect the *high* pressure oil pump (40,000 psi minimum) to the hole provided either in the center of the shaft or on the outside diameter of the hub, depending on design. Loosen the pipe plug of the installation tool and pump all the air out; retighten the plug. Both pumps *must* be equipped with pressure gauges. See Figure 9-10.

Advance the Hub to the Start Position, through pumping the low pressure oil pump. Continue pumping until the hub advances 0.005 to 0.010 in. beyond the start position.

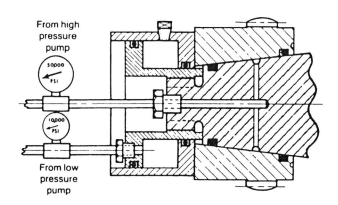

Figure 9-10. Hydraulic system for hub installation.

Expand the Hub. Pump the high pressure pump until you read 15,000 to 17,000 psi on the gauge. As the pressure increases, the hub will tend to move off the shaft. Prevent this movement by occasionally increasing the pressure at the installation tool.

Check for Oil Leaks. The hub *should not* be advanced on the shaft if leaks exist! The pressure at the high pressure oil pump will drop rapidly at first because the air in the system escapes past the O-rings. Continue pumping until pressure stabilizes. A pressure loss of no more than 1,000 psi *per minute* is acceptable. If the pressure drops faster than that, remove the hub and replace the O-rings. However, before removing the hub make sure that the leaks do not occur at the hydraulic connections.

Advance the Hub. Increase the pressure at the installation tool and the hub will advance on the shaft. If all the previous steps were observed, the pressure at the high pressure gauge will gradually *increase* as the hub advances. If the pressure *does not* increase, then *stop.* Remove the hub and check O-rings. If the pressure increases, keep advancing the hub until it touches the split collar or until the specified advance is reached. *Do not* allow the pressure to exceed 30,000 psi. If it does, open the pressure valve slowly and release some oil. If in doing this the pressure drops below 25,000 psi, pump the high pressure pump to 25,000 psi, and continue the hub advance.

Seat the Hub. *Very slowly* release *all* the pressure at the high pressure pump. Do not work on that hub for $^1/_2$ hour, or one hour in cold weather. After that, release all the pressure at the installation tool and remove it.

Verify the Advance. Measure and then record the new overhang of the hub over the shaft. Subtract from the overhang measured in the start position and the result must be the specified advance.

Secure the Hub. Remove the split collar from the shaft and install the retaining nut, but do not overtighten. Secure the nut with the setscrews provided.

Dismounting of Hydraulically Fitted Hubs

In current practice, when a hydraulically fitted hub is removed, it comes off the shaft with sudden movement. Lead washers or other damping means are used to absorb the energy of the moving hub.

Koppers Company, Inc., engineers have developed a dismounting procedure that eliminates the sudden movement of the hub. Without this sudden movement the dangers related to removing hydraulically fitted hubs are greatly reduced. However, normal safety procedures should continue to be used.

Koppers' dismounting method requires the use of the same tools used when mounting the hub. The following procedure is recommended:

1. *Remove the shaft nut.*
2. *Mount the installation tool.* Wet the shaft threads with thin oil and rotate the tool until it butts against the shaft shoulder. There should be a gap between the tool and the hub equal to or larger than the amount of advance when the hub was installed (check the records). If the gap is less than required, the wrong installation tool is being used.
3. *Connect the hydraulic lines.* Connect the installation tool to the *low* pressure oil pump (5,000 psi minimum). Connect the *high* pressure oil pump (40,000 psi minimum) to the hole provided either in the center of the shaft or on the outside diameter of the hub, depending on the design. Loosen the pipe plug of the installation tool and pump all air out; retighten the plug. Both pumps *must* be equipped with pressure gauges.
4. *Activate the installation tool.* Pump oil into the installation tool. The piston will advance until it contacts the hub. Continue pumping until the pressure is between 100 to 200 psi. Check for leaks.
5. *Expand the hub.* Pump oil between the hub and the shaft by using the high pressure pump. While pumping watch both pressure gauges. When the high pressure gauge reads about 20,000 psi the pressure at the low pressure gauge should start increasing rapidly. This pressure increase is caused by the force that the hub exerts on the installation tool, and is an indication that the hub is free to move. Continue to pump until pressure reaches 25,000 psi. In case the low pressure at the installation tool does not increase even if the high pressure reaches 30,000 psi, wait for about $1/2$ hour while maintaining the pressure. It takes time for the oil to penetrate in the very narrow space between the hub and the shaft. Do not exceed 30,000 psi.
6. *Allow the hub to move. Very slowly* open valve at the low pressure pump. The oil from the installation tool will flow into the pump and allow the hub to move. The pressure at the high pressure gauge will also drop. Do not allow it to fall below 5,000 psi. If it does, close the valve and pump more oil at the high pressure pump. Continue the process until the valve at the low pressure pump is completely open and the pressure is zero.

7. *Remove the hub.* Release the high pressure and back off the installation tool until only two or three threads are still engaged. Pump the high pressure pump and the hub will slide off the shaft. When the hub contacts the installation tool, release all the pressure and remove the tool. The hub should now come off the shaft by hand. *Do not remove the installation tool unless the pressure is zero.*

8. *Inspect O-rings.* Reusing even slightly damaged rings invites trouble. The safest procedure is to always use new seals and discard the old ones.

Chapter 10
Protecting Machinery Parts Against Loss of Surface

Many repairs of worn machinery surfaces can be achieved by hard surfacing. By definition, hard surfacing is the process of applying, by specialized welding techniques, a material with properties superior to the basis metal*.

Perhaps the chief factor limiting wide acceptance of this process today is the aura of mystery surrounding the properties of the various hard-surfacing alloys. There are literally hundreds of hard-surfacing alloys commercially available, each with a strange sounding name and a vendor's claim that it is the ultimate material for this or that application. Rather than sift through the chaff to determine which should be used on an urgent problem, many machinery maintenance people drop the idea of hard surfacing and rely on more familiar techniques. This section will discuss the various forms of wear, and show how a few hard-surfacing materials can solve most wear problems. The following information is intended to simplify the field of hard surfacing so that maintenance and design engineers can effectively use the process to reduce maintenance and fabrication costs.

Basic Wear Mechanisms

The first step in solving any wear problem is to determine the mode or modes of wear present[1]. This is of the utmost importance. The same

---
*From "Guide to Hard Surfacing," by K.G. Budinski, Eastman Kodak Co., in *Plant Engineering*, 1974. By permission.

approach must be taken in new designs, except that then the question to be asked is not, "How did this part fail?" but "How might wear occur on this part?" By way of a quick review, there are four basic types of wear: adhesive, abrasive, corrosive, and surface fatigue. Each basic type can also be further categorized, as in Figure 10-1.

Adhesive Wear

The mechanism of adhesive wear is the removal of material from one or both mating surfaces by the action of particles from one of the surfaces bonding to the other. With repeated relative motion between the surfaces, the transferred particles may fracture from the new surface and take on the form of wear debris. Adhesive wear is thus analogous to friction, and is present in all sliding systems. It can never be eliminated—only reduced.

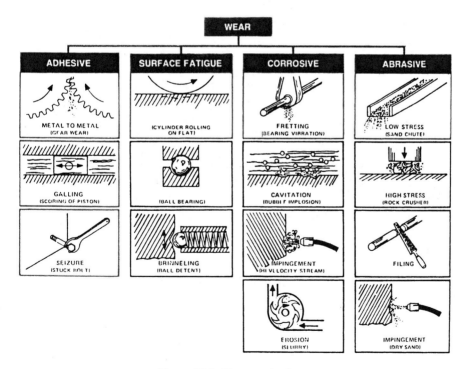

Figure 10-1. Wear mechanisms.

Abrasive Wear

Abrasive wear occurs when hard, sharp particles, or hard, rough surfaces, contact soft surfaces and remove material by shearing it from the softer surface. The amount of metal removed is a function of the nature of the abrading substance, and of the loading. For this reason, it is common to subdivide abrasive wear into high stress, low stress, high-velocity impingement, and filing.

Corrosive Wear

There is no one mechanism to describe corrosive wear. Fretting, the first form of corrosive wear shown in Figure 10-1, occurs in systems that are not supposed to move. One of the most common instances of fretting occurs on shafts, under the press-fitted inner race of a rolling element bearing. Vibration provides a slight relative motion between the shaft and the race. This oscillatory motion causes small fragments of one surface to adhere to the other (adhesive wear), and with repeated vibration or oscillation, the fragments oxidize (corrosive wear) and form abrasive oxides (abrasive wear) which amplify the surface damage.

Cavitation occurs in highly agitated liquids where turbulence and the implosion of bubbles cause removal of the protective oxide film on the metal surface, followed by corrosive attack on the base metal. If the implosion of the bubbles is particularly energetic, as is the case in ultrasonic devices, the material removal can be quite rapid.

Impingement by high velocity liquids causes removal of protective oxide films. Metal removal then occurs by corrosion of the active metal surface.

Erosion involves the same mechanism as impingement. However, the liquid in this instance contains abrasive particles that enhance the removal of surface films.

Surface Fatigue

This last form of wear to be discussed results from high compressive stress because of point or line-contact loading. These high stresses, with repeated rolling, produce subsurface cracks that eventually propagate and cause particles to be removed from the surface. Once this occurs, the deterioration of the entire rolling surface starts. This is a result of the additional compressive stresses that are generated when the first fragments detached are rerolled into the surface. Surface fatigue of this nature occurs in rolling-element bearings, rails, and other surfaces subjected to point or line-contact loads.

Hard-Surfacing Techniques

Almost every welding technique can be used to apply a hard-surfacing material. Referring to the definition of hard-surfacing—applying by welding or spraying techniques a material with properties superior to those of the basis metal—it can readily be seen that this can be accomplished in many ways. Figure 10-2 illustrates most of the methods used. Each has advantages and disadvantages. Shielded metal-arc welding is the most common and versatile welding technique, but many of the hard-surfacing alloys have not been available in a coated electrode form.

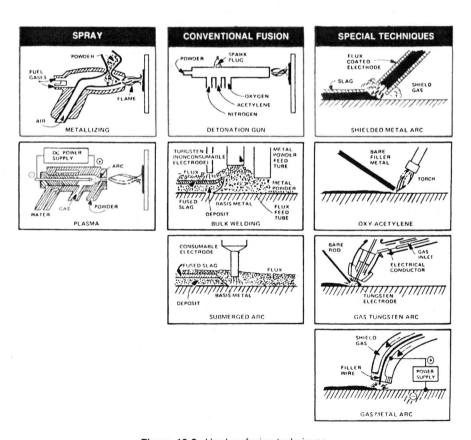

Figure 10-2. Hard-surfacing techniques.

Oxyacetylene welding is the preferred method for applying bare filler metals, in that it minimizes "dilution" or mixing of the filler metal with the basis metal, so that the hardest deposit is achieved with this mode of deposition. Gas welding is slow, and it is difficult to control the deposit profile. Gas tungsten-arc welding is also slow, but provides the most accurate deposit profile of any of the fusion processes; it has the major disadvantage of significant dilution, with a corresponding loss in deposit hardness. Gas metal-arc welding is one of the fastest processes for applying hard-surfacing; however (once again) not all surfacing alloys are available as wire that can be roll-driven through the welding gun. (When used for hard-surfacing, the gas metal-arc process is often used without a shielding gas, and then is referred to as the "open arc" process.)

Under the heading of spray surfacing techniques, there are three primary processes: metallizing, plasma spray, and detonation gun. Metallizing is commonly done by spraying a powder at the surface with air pressure. The powder is heated to a highly plastic state in an oxyacetylene flame, coalesces, and mechanically bonds to the substrate. Preparation of the substrate often involves knurling or abrasive blasting. In some systems, a wire, instead of a powder, is fed into the flame, and the molten droplets are sprayed on the surface with a gas assist. In another modification of this process, heating of the wire is accomplished by two carbon electrodes. In the powder system, it is also possible to spray ceramic materials; as in all spray systems, the deposit does not fuse to the basis metal, and there is a high degree of porosity. Some sprayed alloys may be heated with a larger torch after spraying, and the deposit will thus be fused to the substrate.

The plasma system is used only with powders, and its 15,000°F flame provides a denser deposit with improved bonding to the surface. Ceramics are commonly sprayed with plasma.

Special techniques used in hard-surfacing include detonation gun, bulk welding, and submerged arc. The detonation gun system is a proprietary process which provides even greater densities and better bonds than can be achieved by plasma spraying. This process is described later.

Bulk welding is a process combining tungsten arc and submerged arc. It is designed for surfacing large areas quickly and cheaply. The mechanization involved in this process makes it more economical than most other fusion processes for hard-surfacing large areas. Submerged arc welding, the last on the list of special techniques, is also used for surfacing large areas. However, it has the limitation that it is best suited to hard-surfacings that are available in a coiled wire form. This rules out use on many of the hard cobalt and nickel-based alloys.

One of the principal factors that limit acceptance of hard-surfacing is the confusion surrounding selection of an appropriate alloy for a given service. There are literally hundreds of alloys on the market. The

American Welding Society (AWS) has issued a specification on surfacing materials (AWS A5. 13), detailing 21 classes of electrodes and 19 classes of rods. Each class may contain rods from a dozen or so manufacturers, each slightly different. And many proprietary alloys are not included in this specification.

Each vendor of surfacing materials has his own selection system; most systems are based on application. If the machinery maintenance man's application is not included in the list, there is no way for him to know which material to choose. One thing that seems to be common to most vendors is a reluctance to supply information on the chemistry of their products. Many claim that this is proprietary and that the user does not need the information. This is like buying a "pig in a poke." No surfacing material should be used unless the plant engineer knows the composition of the alloy, the basic structure of the deposit, and the one and two layer as-deposited hardness. Other considerations of importance are cracking tendencies, bond, all position capability (for electrodes), and slag removal (again, electrodes).

Hard-surfacing alloys derive their wear characteristics from hard phases or intermetallic compounds in their structure. For example, due to their microstructure, two hardened steels may each have a Rockwell hardness of RC 60. However, the steel with the higher chromium content will have much greater wear resistance than the plain carbon steel. This is because of the formation of hard intermetallic compounds (chromium carbides) within the RC 60 martensitic matrix. Thus, in selecting tool steels and surfacing alloys, one must consider not only the macroscopic Rockwell-type of hardness, but the hardness and volume percentage of the micronconstituents.

If the machinery maintenance person wishes to solve an abrasive wear problem involving titanium dioxide, he or she must select a surfacing that is harder than TiO_2. Since many abrasive materials are harder than the hardest metals, this requires thinking in terms of absolute hardness. As indicated in Figure 10-3, TiO_2 has an absolute hardness of approximately $1,100 \, kg/mm^2$. The hardest steel is only 900 on this scale.

To solve this wear problem, machinery maintenance personnel must select an alloy that has a significant volume concentration of chromium carbide, vanadium carbide, or some other intermetallic compound which is even harder than the TiO_2. Most surfacing alloys have significant amounts of these hard intermetallic compounds in their structure, and this is why they are effective.

In an attempt to simplify the subject of surfacing alloys without going into detail on microstructures, nine general classifications based on alloy composition have been established. These are illustrated in Figure 10-4. To effectively use hard-surfacing, it is imperative that the engineer become familiar with the characteristics of each class.

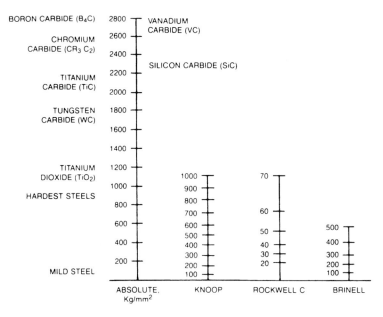

Figure 10-3. Comparison of commercial hardness tester scales.

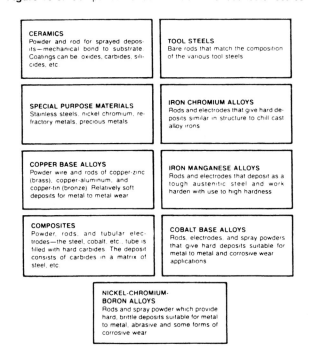

Figure 10-4. Classification of hard-surfacing materials.

Tool Steels

By definition, hard-surfacing is applying a material with properties superior to the basis metal. In repair welding of tool steels, a rod is normally selected with a composition matching that of the basis metal. When this is done, repair welding of tool steels is not really rod surfacing. However, tool steel rods are available in compositions to match hot-work, air-hardening, oil-hardening, water-hardening, high-speed, and shock-resisting steels, and these rods can be applied to basis metals of differing composition. If the expected hardness is achieved, the surface deposit will have the service characteristics of the corresponding tool steel. Tool steel rods are normally only available as bare rod.

Iron-Chromium Alloys

The iron-chromium alloys are essentially "white irons." For many years, the foundry industry has known that most cast irons will become very hard in the chilled areas if rapidly cooled after casting. If chromium, nickel, or some other alloying element is added to the cast iron, the casting may harden throughout its thickness. These alloy additions also provide increased wear resistance in the form of alloy carbides. Iron-chromium hard-surfacings are based upon the metallurgy of these white irons. Hardnesses of deposit can range from RC 40 to RC 60. Some manufacturers use boron as the hardening agent instead of carbon, but the metallurgy of the deposit is still similar to white iron.

Iron-Manganese Alloys

These alloys are similar to the "Hadfield Steels." They are steels with manganese contents in the 10 to 16 percent range. The manganese causes the steel to have a tough austenitic structure in the annealed condition. With cold working in service, surface hardnesses as high as RC 55 can be obtained.

Cobalt-Base Alloys

These materials contain varying amounts of carbon, tungsten, and chromium in addition to cobalt, and provide hardnesses ranging from RC 35 to RC 60. Their wear resistance is derived from complex carbides in a cobalt-chromium matrix. The size, distribution and the types of carbides vary with the alloy content. The matrix can be harder than the austenitic matrix of some of the iron-chromium alloys.

Nickel-Chromium-Boron Alloys

This family of alloys forms deposits consisting of hard carbides and borides in a nickel eutectic matrix. The macrohardness can be as low as RC 35 and as high as RC 60, but at all hardness levels these alloys will provide good metal-to-metal wear resistance when compared with an alloy steel of the same hardness. These alloys are normally applied by oxy-acetylene or gas tungsten-arc welding deposition of bare rod, or by powder spraying.

Composites

A composite, in hard surfacing, is a metal filler material containing substantial amounts of nonmetals. Typically, these are intermetallic compounds such as tungsten carbide, tantalum carbide, boron carbide, titanium carbide, and others. All of these intermetallics are harder than the hardest metal. Thus, they are extremely effective in solving abrasive wear problems. Composite electrodes usually consist of a steel, or soft alloy, tube filled with particles of the desired compound.

During deposition, some of these particles dissolve and harden the matrix, while the undissolved particles are mechanically included in the deposit of welding techniques. Oxyacetylene deposition is the preferred technique for application since fewer particles dissolve. The hard particles are available in various mesh sizes and can be so large that they can readily be seen on the surface. Composites are not recommended for metal-to-metal wear problems since these large, hard particles may enhance this type of wear. However, composites of smaller particle size can be applied by thermal spraying techniques, such as plasma and detonation gun.

Copper-Base Alloys

Brasses (copper and zinc) or bronzes (copper and aluminium, tin or silicon) can be deposited by most of the fusion welding techniques, or by powder spraying. Oxyacetylene deposition is the most common method. These alloys are primarily used for metal-to-metal wear systems with the copper alloy surfacing being the perishable component. These alloys should run against hardened steel for optimum performance.

Ceramics

Ceramics can be applied as surfacings by plasma, detonation gun spraying or with some types of metallizing equipment. Coating thicknesses are normally in the range of 0.002 to 0.040 in. Commonly sprayed ceramics include carbides, oxides, nitrides, and silicides. These coatings are only mechanically bonded to the surface, and should not be used where impact is involved.

Special Purpose Materials

Many times metals are surfaced with austenitic stainless steels or soft nickel-chromium alloys for the sole purpose of corrosion resistance. For some applications, costly metals such as tantalum, silver, or gold are used as surfacings. If a particular application requires a very special material, a surfacing technique probably can be used to put this special metal on only the functional surfaces, with a reduction in cost.

In an effort to come up with a viable hard surfacing selection system, a series of wear tests was conducted on fusion surfacing materials from each of the classifications detailed in the preceding pages. Several vendors' products in each classification were tested, and the welding characteristics of each material determined. Ceramics, tool steels, and special purpose materials were not tested.

The specific procedure for evaluating the fusion surfacings was to make multilayer test coupons, determine the welding characteristics, and run metal-to-metal and abrasive wear tests on the materials that performed satisfactorily in the welding tests. The compositions of the hard surfacing alloys tested are shown in Table 10-1.

Test Results

As shown in Figure 10-5, the abrasive wear resistances of certain compositions, such as FeCr-5 and Composites 2 and 3 were superior. (The three mentioned are notable for ease of application.) Harder nickel and cobalt-based alloys with macrohardnesses of approximately RC 60 did not perform as well. The manganese steels (FeMn-1 and 2), the low chromium iron alloys (Fe-1 and 2), and the copper-based alloy Cu-1 all had poor abrasive wear resistance.

Adhesive wear test results are shown in Figure 10-6. Co-2 had the lowest net wear. The composite surfaces, Com-1, 2, and 3 performed very well, but produced more wear on the mating tool steel than did the

Table 10-1
Compositions of Some Hard-Surfacing Alloys

Alloy**	Nominal Composition*													Hardness Rockwell C
	C	Si	Cr	W	Co	Al	Mn	B	Fe	Ni	Cu	Other	Other	
Fe-1 Δ	0.18	0.30	2.90				1.02		BAL					35
Fe-2 Δ	0.68	0.47	6.80				1.48		BAL					20
FeCr-1 Δ	2.20	0.90	30.00				1.30		BAL			3.8 Mo	1.50 Ti	50
FeCr-2 Δ	0.44	1.20	29.60				1.70		BAL			4.0 Mo		60
FeCr-3 Δ	4.00	1.00	23.00						BAL			0.5 Mo	7.50 Cb	60
FeCr-4 Δ	1.00	3.00	13.00				0.70	3.00	BAL					60
FeCr-5 Δ	6.00	1.00	13.00				2.70		BAL			5.2 Ti		62
FeMn-1 Δ	0.44	0.52	14.10				16.60		BAL					20
FeMn-2 Δ	0.64	0.26	0.47				13.60		BAL					20
Co-1 +	1.20	2.06	30.00	4.50	BAL		2.00		3.00	3.00		1.5 Mo		40
Co-2 +	1.80		29.0	9.00	BAL									50
Co-3 +	2.00	0.85	30.90	13.80	BAL				2.30					55
Co-4 +	2.50		32.50	17.50	BAL									58
Co-1A +	0.95		27.40	5.00	BAL				1.90					30
Co-1C Δ	1.10	1.20	29.00	4.50	BAL					3.00				40
Co-7 Δ														36
Cu-1 +						14.00					BAL			20
NiCr-4 +	0.45	2.25	10.00					2.00	4.00	BAL				35
NiCr-5 +	0.65	3.75	11.50					2.60	2.50	BAL				50
NiCr-6 +	0.75	4.25	13.50					3.00	4.25	BAL				56
NiCr-C +	0.04	0.86	15.50	4.10				0.24	5.70	BAL		16 Mo	0.34 V	30
COM-1 +									BAL			60 WC		62
COM-2 Δ									BAL			60 WC		62
COM-3 +	0.48	0.80	12.00	1.80	10.00		0.80		1.20	1.20		60 WC		60

*Fe—Iron Base; FeMn—Iron manganese; FeCr—Iron Chromlum; Co—Cobalt Base; NiCr—Nickel Chromlum; Cu—Copper Base; COM—Composite.

**+ Rod; Δ Electrode.

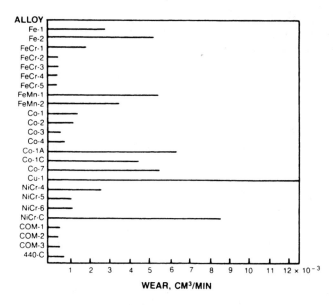

Figure 10-5. Performance of hard-surfacing materials subjected to low-stress abrasive wear. Numbers indicate formulations shown in Table 10-1.

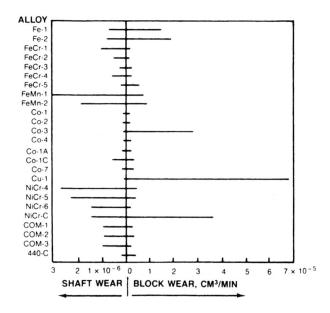

Figure 10-6. Adhesive wear graph showing results of running test blocks in contact with 440-C stainless steel shaft of HRC 58. In most applications, neither shaft wear nor block wear is desired; several cobalt alloys gave superior results.

cobalt-based alloys. This result was also experienced with the nickel-based alloys. The copper-based alloy Cu-1 showed the highest surfacing wear rate, but one of the lowest shaft wear rates.

Discussion

The results of the abrasive wear tests indicated that the theoretical prediction that the abrasive wear rate is inversely proportional to the hardness of the material subjected to wear held true. However, as was mentioned earlier, this result does not refer to the macrohardness, but to a combination of macrohardness and microconstituent hardness. The surfacings that performed best in the abrasive wear tests—Com-2 and 3, and FeCr-5—all had large volume percentages of intermetallic compounds with hardnesses greater than the abrading substance, which in this case was silicon dioxide.

Another significant observation was that the iron chromium alloy FeCr-5 with high carbon (6 percent) and titanium (5.2 percent) concentrations outperformed the arc-welded tungsten carbide deposit Com-1. Thus it was shown that a coated electrode (FeCr-5) could be used to get abrasive wear resistance almost as good as that of gas-deposited tungsten-carbide composite. All of the very hard alloys exhibited cracking after welding, making them unsatisfactory for some applications, such as knife edges. The cobalt-based alloy Co-2 had the best abrasive wear resistance of those alloys that did not crack after welding. Cracking and checking do not mean a loss of bond; and thus, in many surfacing applications, cracking tendencies can be neglected.

In explanation of the results of the adhesive wear tests, it can be hypothesized that the hard microconstituents present in many of the surfacing alloys tested promoted wear of the mating metal surface. The cobalt-based alloys that performed best in this test do not have a large volume fraction of hard microconstituents. In fact, there are few particles large enough to allow a hardness determination. This may account for the low wear of the cobalt-based alloys on the mating tool steel. In any case, adhesive wear, because it is a complex interaction between metal surfaces, cannot be predicted by simple property measurements—a wear test is required.

Selecting a Surfacing Method

The first step is to determine the specific form of wear that is predominant in the system. Once this has been done, the next step will be to select a process for application. The final step will be to select the surfacing material. Here are some guidelines for process selection.

- If a large area has to be surfaced, consider the use of open arc, submerged arc, or bulk welding
- If distortion cannot be tolerated in a surfacing operation, consider use of spray surfacing by plasma arc, metallizing, or detonation gun
- If optimum wear resistance is required, use oxyacetylene to minimize dilution, or use a spray technique
- If accurate deposit profiles are required, use gas tungsten-arc welding
- If surfacing must be done out of position, use shielded, metal-arc welding

The process of application will limit alloy selection to some extent. For example, if spray surfacing is required because of distortion, many of the iron chromium, iron manganese, or tool steel surfacings cannot be employed because they are not available as powders.

Selecting a Surfacing Material

Here are some guidelines for choosing the right alloy:

- Tool steels should be used for small gas tungsten-arc welding deposits where accurate weld profiles are required
- Iron-chromium alloys are well-suited to abrasive wear systems that do not require finishing after welding
- The composite alloys should be used where extreme ·abrasion is encountered, and when finishing after welding is not necessary
- Iron-manganese alloys should be used where impact and surface fatigue are present. Deformation in service must occur to get work hardening. These alloys are not well suited for metal-to-metal wear applications
- Cobalt-based alloys are preferred for adhesive wear systems. They have the additional benefit of resistance to many corrosive and abrasive environments
- Nickel-chromium-boron alloys are suitable for metal-to-metal and abrasive wear systems, and they are preferred where finishing of a surfacing deposit is necessary
- Copper-based surfacing alloys are suitable only to adhesive wear systems. They are resistant to seizure when run against ferrous metal, but may be subject to significant wear

- Ceramics are the preferred surfacings for packing sleeves, seals, pump impellers, and similar systems involving no shock, but with severe low-stress abrasion

These surfacings should not be run against themselves without prior compatibility testing.

Table 10-2 lists specific alloys likely to give exceptionally good performance, based on the tests summarized in Figures 10-5 and 10-6.

Table 10-2
Hard-Surfacing Selection Guide (Typical Only)

Surfacing	Form	Deposition Process	Characteristics	Uses
Chromium Oxide	Powder	Plasma Spray	Excellent resistance to very low stress abrasion. Thickness 5–40 mils. Can be ground to very good finish. No welding distortion. (>HRC 70)	Low Stress Erosion
AISI 431 Stainless Steel	Powder	Metallize	Good adhesive wear resistance when lubricated. Poor abrasion resistance. Can be ground to good finish. No welding distortion. (HRC 35)	Fretting, Galling
NiCr-4	Powder	Metallize and Fuse	Good adhesive wear resistance; corrosion resistant. Coating thickness to 0.125 in. with fusion bond. Distortion may occur in fusing, but application is faster than oxyacetylene rod surfacing.	Metal-to-Metal Wear, Galling, Seizure, Cavitation, Erosion, Impingement, Brinelling
FeCr-1 (Iron-Chromium)	Electrode	Shielded Metal-Arc Welding	Moderate resistance to low stress abrasion and adhesive wear. Can be easily finished by grinding. Low cost. (HRC 50)	Low Stress, High Stress, Cylinder and Ball Rolling

Table 10-2
Hard-Surfacing Selection Guide (Typical Only)—cont'd

Surfacing	Form	Deposition Process	Characteristics	Uses
FeCr-5 (Iron-Chromium + TiC)	Electrode	Shielded Metal-. Arc Welding	Very good resistance to low stress abrasive wear. Easy to apply (HRC 60)	Low Stress, Filing, Impingement
Co-1 (Cobalt-Chromium)	Rod	Oxyacetylene	Very good resistance to adhesive wear. Moderate resistance to low stress abrasion. Corrosion resistant. The alloy is expensive and application is slow. (HRC 43)	Metal-to-Metal Wear, Galling, Seizure, Fretting, Cavitation, High Velocity Liquid, Erosion, Brinelling
Co-1C (Cobalt-Chromium)	Electrode	Shielded Metal-Arc Welding	Good resistance to adhesive wear. Easy to apply, Suitable for all position welding. The alloy is expensive. (HRC 43)	Metal-to-Metal Wear, Galling, Seizure, Fretting, Cavitation
NiCr-4	Rod	Oxyacetylene	Good resistance to metal wear. Machinable. Will not rust. Costly to apply. (HRC35)	Metal-to-Metal Wear, Galling, Seizure
COM-1 (Tungsten Carbide-Steel Matrix)	Rod	Oxyacetylene	Excellent resistance to low stress abrasion. Use as deposited. Costly to apply. (>HRC 60)	Low Stress, Filing, Impingement, Erosion
COM-3 (Tungsten Carbide-Cobalt-Chromium Matrix)	Rod	Oxyacetylene	Excellent resistance to low stress abrasive wear. Use as deposited. Corrosion resistant. Costly to apply. (>HRC 60)	Low Stress, Filing, Impingement, High Velocity Liquid

The Detonation Gun Process*

If we look at the repair of rotating machinery shaft bearings, journals, seal surfaces, and other critical areas in the context of hard-surfacing, it becomes apparent that there are numerous methods available. As we saw, one of these methods is by the use of detonation gun coatings. In review, the detonation gun is a device that can deposit a variety of metallic and ceramic coating materials at supersonic velocities onto a workpiece by controlled detonation of oxygen-acetylene gas mixtures. Coatings applied by this method are characterized by high bond strength, low porosity, and high modulus of rupture. Table 10-3 shows some of the physical properties of detonation gun coatings. This section describes the equipment used to apply D-Gun coatings and provides data on coating thicknesses used, surface finish available, and physical properties of some popular D-Gun coatings used in machinery repair. Examples are cited showing increases in operating life that can be achieved on various pieces of equipment by properly selected and applied coatings.

Shaft repairs on turbomachinery and other equipment can be accomplished in many ways. Repair methods include weld deposit, sleeving, electroplated hard chromium, flame spraying, plasma arc spraying, and detonation gun coatings. Each of these methods has its own advantages and disadvantages. Again, factors such as time needed to make the repair, cost, machinability, surface hardness, wear resistance, corrosion resistance, material compatibility, friction factor, minimum or maximum allowable coating thickness, surface finish attainable, bond strength, coefficient of thermal expansion, coating porosity and the amount of thermal distortion from the repair; all have varying degrees of importance depending on the particular application. In some cases, the repair method to be used is simply based on the availability of a shop in the area that can make the repair within the desired schedule. Sometimes compromise coatings or repair methods are selected. In other cases, a planned, scheduled and engineered solution is used to effect a repair that provides service life that is far superior to the original equipment.

A properly chosen method of repair can provide improved durability of the repaired part over that of the original part with properties such as higher hardness, better surface finish, improved wear resistance and improved corrosion resistance. Properly chosen coatings can combine the favorable attributes of several materials, thus lessening the compromises that would have to be made if a single material was used. Equipment users have frequently found that repaired components have withstood service

---

* A proprietary process of Praxair Surface Technologies.

Table 10-3
Physical Properties of Some Detonation Gun
Coatings (UCAR D-Gun)

COMMERCIAL DESIGNATION	LW-1N30	LW-15	LW-5	LC-1C
Nominal Composition (Weight %) (a)	87WC, 13Co	86WC, 10Co, 4Cr	73WC, 20Cr, 7Ni	800r$_3$C$_2$, 16Ni, 4Cr
Tensile Bond Strength (psi) (b)	>10,000	>10,000	>10,000	>10,000
Modulus of Rupture (psi)	90,000		40,000	70,000
Modulus of Elasticity (psi)	31 × 16^6		17 × 10^6	18 × 10^6
Metallographic Apparent Porosity (Vol. %)	≤1	≤1.5	<1	≤1
Nominal Vickers Hardness (kg/mm^2, 300 g load)	1,150 HV	1,100 HV	1,100 HV	775 HV
Rockwell "C" Hardness—Approx.	71	70	70	63
Max. Rec. Operating Temp. (°F)	1,000	1,000	1,400	1,400
Avg. Coef. of Thermal Expansion (in/in/°F)	4.5 × 10^{-6} (70 to 1,000°F)	4.2 × 10^{-6} (70 to 1,832°F)	4.6 × 10^{-6} (70 to 1,400°F)	6.1 × 10^{-6} (70 to 1,475°F)
Characteristics	Extreme Wear Resistance	Good Wear Resistance to Approx. 1,000°F. Greater corrosion Resistance than WC-Co.	Good Wear Resistance to Approx. 1,400°F. Greater oxidation and corrosion Resistance than WC-Co.	Excellent Wear Resistance at Elevated Temp.

(a) The composition shown represents the total chemical composition, but not the complex microstructural phases present.
(b) Measured per ASTM C633-69 modified to use a reduced coating thickness of 10 mils.
® UCAR is a registered trademark of the Union Carbide Corporation.

better than the original equipment manufacturer's components. This has led many users to specify specialized coatings on key components of new equipment being purchased. In some cases the use of coatings has led to reduced first cost of components since the special properties of coatings allow the use of lower-cost, less exotic base materials.

Comparing repair prices to the purchase price of new parts, assuming that the new parts are available when needed, shows that the price of repaired parts may be only $\frac{1}{5}$ to $\frac{1}{2}$ that of new OEM parts. If the repair method eliminates the need for expensive disassembly such as rotor unstacking, the savings become even more dramatic. Coupling these savings with the frequently extended service life of the repaired parts over the original ones, which in turn extends periods between inspections and repairs, the coating repair of parts is extremely attractive from an economic standpoint.

Process Details. In the following we will concentrate on the detonation gun process of coating which is often referred to as the D-Gun process. The system is shown again schematically in Figure 10-7. It consists of a water-cooled gun barrel, approximately three feet long, that is fed with oxygen, acetylene and coating powder. Ignition of the oxygen-acetylene mixture is accomplished by means of a spark plug. The detonation wave in the gun barrel, resulting from the ignition of the gas mixture, travels at ten times the speed of sound through the barrel, and temperatures reach or exceed 6,000°F inside the gun. Noise levels generated by the D-Gun require isolating the process in a noise-attenuating enclosure. The equipment operator monitors the coating operation from a control console while observing the operation through a view port. Detonation is cyclic, and subsequent to each detonation the barrel is purged with nitrogen before a fresh

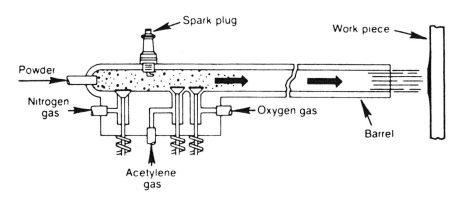

Figure 10-7. Detonation gun schematic.

charge of oxygen, acetylene, and coating powder is admitted. The particles of coating powder are heated to plasticity and are ejected at supersonic speeds averaging approximately 2,500 ft per second. Kinetic energy of the D-Gun particles is approximately ten times the kinetic energy per unit mass of particles in a conventional plasma arc gun and 25 times the energy of particles in an oxyacetylene spray gun. The high temperature, high velocity coating particles attach and conform to the part being coated, giving a very strong coating bond at the interface and low porosity in the coating. This coating does not depend on a severely roughened surface to provide mechanical interlocking to obtain a bond. Surface preparation for hardened steel consists of grinding to the desired undersize plus, in some instances, grit blasting. Titanium parts do not need grit blasting before coating.

In spite of the high temperatures generated in the barrel of the D-Gun, the part being coated remains below 300°F, so there is little chance of part warpage and the base material metallurgy is not affected.

Coating Details. The D-Gun deposits a very thin coating of material per detonation, so multiple passes are used to build up to the final coating thickness. Figure 10-8 shows the pattern formed by the overlapping circular deposits being built up on the surface of a piston rod. Finished

Figure 10-8. Detonation gun coated piston rod.

coating thickness may be as low as 1.5 to 2 mils for some high pressure applications such as injection pump plungers or polyethylene compressor piston rods but many typical applications use finished thicknesses of three to five mils. Greater thicknesses may be used for repair jobs. Finished thicknesses greater than is practical for a given cermet or ceramic coating may require prior build-up with metallic coatings such as nickel.

A number of ceramic and metallic coatings are available for application with the D-Gun. These include mixtures or alloys of aluminum oxide, chromium oxide, titanium dioxide, tungsten carbide, chromium carbide, titanium carbide, cobalt, nickel, and chromium. Table 10-3 lists some of the more popular coatings with their compositions and some key physical properties. Tungsten carbide and cobalt alloys are frequently used for coating journal areas and seal areas of shafts. In cases where additional corrosion resistance is required, the tungsten carbide and cobalt alloys have chromium added. Such a powder is often used on the seal areas of rotors. Greater oxidation and corrosion resistance at elevated temperatures is accomplished by using powder with chromium and nickel in conjunction with either tungsten carbide or chromium carbide.

Carbide coatings exhibit excellent wear resistance by virtue of their high hardnesses. Chromium carbide coatings have a cross-sectional Vickers hardness number (HV) in the range of 650 to 900 kg/mm^2 based on a 300 g-load which is approximately equal to 58 to 67 Rockwell "C." The tungsten carbide coatings are in the range of 1,000 to 1,400 HV or approximately 69 to 74 Rockwell "C."

Coatings applied by the D-Gun have high bond strengths. Bond strengths, as measured per ASTM C633-69 modified to use a reduced coating thickness of 10 mils, are in excess of 10,000 psi, which is the limit of the epoxy used in the test. Special laboratory methods of testing bond strengths of D-Gun coatings by a brazing technique have given values in excess of 25,000 psi. This type of test, however, may change the coating structure. Porosity is less than 2 percent by volume for these coatings. Figure 10-9 shows a photomicrograph of a tungsten carbide coating applied to steel. The original photo was taken through a 200 power microscope. The markers in the margin denote from top to bottom: the coating surface, tungsten carbide and cobalt coating, bond interface and base metal. The tight bond and low porosity are clearly evident. Low porosity is an important factor in corrosion resistance and it enhances the ability of a coating to take a fine surface finish.

The as-deposited surface finishes of carbide coatings are in the range of 120 to 150 microinches rms when deposited on a smooth base material. Finishing of low tolerance parts, such as bearing journals, is usually accomplished by diamond grinding. Parts that do not require extremely close dimensional control such as hot gas expander blades can be left as

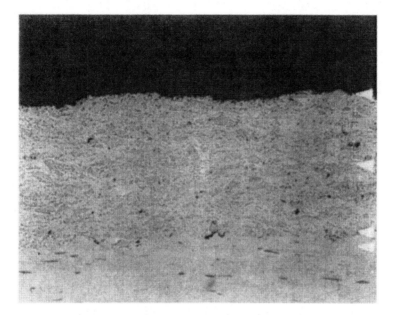

Figure 10-9. Photomicrograph of tungsten carbide-cobalt coating.

coated or, if a smoother finish is desired, they can be given a nondimensional finishing by means of abrasive belts or wet brushing with an abrasive slurry.

A combination of grinding, honing, and polishing is routinely used to finish tungsten carbide coatings to eight microinches, and finishes as fine as two microinches or better are attainable with these coatings.

For many applications however, plasma and D-Gun coatings can be used as coated. In fact, in at least one application, a D-Gun tungsten carbide-cobalt coating is grit blasted to further roughen the surface for better gripping action. Probably in the majority of applications, the coatings are finished before being placed in service. Finishing techniques vary from brush finishing to produce a nodular surface, to machining, honing, grinding, and lapping to produce surfaces with surface roughness down to less than microinches rms. Machining can be used on some metallic coatings, but most coatings are ground with silicon carbide or diamond (diamond is usually required for D-Gun coatings). The best surface finish that can be obtained is a function not only of the finishing technique, but also of the coating type and the deposition parameters. Finishing of D-Gun coatings is usually done by the coating vendor, since great care must be exercised to avoid damaging the coatings.

A typical check list for grinding of most hard surface coatings follows:

1. Check diamond wheel specifications.
   a. Use only 100 concentration.
   b. Use only resinoid bond.
2. Make sure your equipment is in good mechanical condition.
   a. Machine spindle must run true.
   b. Backup plate must be square to the spindle.
   c. Gibs and ways must be tight and true.
3. Balance and true the diamond wheel on its own mount—0.0002 in. maximum runout.
4. Check peripheral wheel speed—5,000 to 6,500 surface feet per minute (SFPM).
5. Use a flood coolant—water plus 1–2 percent water soluble oil of neutral pH.
   a. Direct coolant toward point of contact of the wheel and the workplace.
   b. Filter the coolant.
6. Before grinding each part, clean wheel with minimum use of a silicon carbide stick.
7. Maintain proper infeeds and crossfeeds.
   a. Do not exceed 0.0005 in. infeed per pass.
   b. Do not exceed 0.080 in. crossfeed per pass or revolution.
8. Never spark out—stop grinding after last pass.
9. Maintain a free-cutting wheel by frequent cleaning with a silicon carbide stick.
10. Clean parts after grinding.
    a. Rinse in clean water—then dry.
    b. Apply a neutral pH rust inhibitor to prevent atmospheric corrosion.
11. Visually compare the part at 50X with a known quality control sample.

Similarly, a typical check list for lapping is:

1. Use a hard lap such as GA Meehanite or equivalent.
2. Use a serrated lap.
3. Use recommended diamond abrasives—Bureau of Standards Nos. 1, 3, 6, and 9.
4. Imbed the diamond firmly into the lap.
5. Use a thin lubricant such as mineral spirits.
6. Maintain lapping pressures of 20–25 psi when possible.
7. Maintain low lapping speeds of 100–300 SFPM.

8. Recharge the lap only when lapping time increases 50 percent or more.
9. Clean parts after grinding and between changes to different grade diamond laps—use ultrasonic cleaning if possible.
10. Visually compare the part at 50X with a known quality control sample.

Limitations. All thermal spray-applied coatings have restrictions in their application since a line of sight is needed between the gun and the surface to be coated. The barrel of a D-Gun is positioned several inches away from the surface to be coated, and the angle of impingement can be varied from about 45° to the optimum of 90°. Coating of outside surfaces generally presents no problem, but small diameter, deep or blind holes may be a problem. It is possible to coat into holes when the length is no more than the diameter. The structure and properties of the coating may vary somewhat as a function of the geometry of the part, because of variations in angle of impingement, stand-off, etc. Portions of a part in close proximity to the area being plated may require masking with metal.

Applications. Detonation gun coatings have been used in a large number of applications for rotating and reciprocating machinery as well as for special tools, cutters and measuring instruments. References 2 and 3 attest to the success of such coatings. Table 10-4 shows typical applications in a petrochemical plant utilizing tungsten carbide based coatings.

The tungsten carbide family of coatings is used principally for its wear resistance. Tungsten carbide is combined with up to 15 percent cobalt by weight. Decreasing the amount of cobalt increases wear resistance, while

Table 10-4
Cobalt Alloy Applications in a Petrochemical Refinery[2]

Alloy	Specification	Substrate	Deposition	Component	Wear type
91WC-9Co (powdered form)	AMS 2435A	SAE 4140 SAE 1040 SAE 60304 SAE 60310	D-Gun D-Gun D-Gun D-Gun	Compressor rod Pump sleeve Pump sleeve Pump plunger	Sliding Rubbing Fretting Sliding
85WC-15Co (powdered form)	P & W A* Spec. No. 46	SAE 4140	D-Gun	Turbine shaft	Rubbing
55Co-29Cr-4.5W-1.25Si-1C (E CoCr-C)	AWS A5-13-70	11-13%Cr steel	Arc welding	Slide valve gate	Erosion
52Co-30Cr-13W-2.4C (solid castings)	AMS 5788	—	Cast	Pump sleeve	Rubbing
66Co-26Cr-5W-1C (solid castings)	AMS 5373, 5387, 5788	— — —	Cast Cast Cast	Pump impeller Control valve plug Control valve seat	Erosion Erosion Erosion
66Co-26Cr-5W-1C (coated electrode)	—	ASTM A216	Arc welding	Pump casing	Erosion

Pratt & Whitney Aircraft (this coating was developed for use in jet engines).

adding cobalt increases thermal and mechanical shock resistance. Coatings of this type are frequently used to coat bearing journals and seal areas on compressors, steam turbines, and gas turbines. These coatings have a high resistance to fretting and they have been used on midspan stiffeners of blades for axial flow compressors. Their fretting resistance and ability to carry high compressive loads make them suitable to correct loose interference fits on impellers and coupling hubs. Addition of chromium to the tungsten carbide and cobalt mixtures adds corrosion resistance and improves wear resistance at high temperature levels. In general, this family of coatings is most frequently used in neutral chemical environments but can be used with many oxidizing acids. Cobalt mixture coatings are usually not used in strongly alkaline environments.

Coatings that combine tungsten carbide with chromium and nickel exhibit greater oxidation and alkaline corrosion resistance than the tungsten carbide-cobalt coatings. Their wear resistance capabilities are good up to about 1,200°F, which is about 200°F higher than that of the tungsten carbide-cobalt coatings. This higher temperature capability makes these coatings useful for applications such as coating rotor blades on hot gas expanders used for power recovery from catalytic crackers. Coated blades resist the wear from catalyst fines and have extended life from just a few months, as experienced with uncoated blades, to a life of three to five years. This type of coating is suitable for use in many alkaline environments.

Chromium carbide combined with nickel and chromium provides excellent wear resistance at elevated temperatures and is recommended for temperatures up to about 1,600°F. These coatings do not have the wear resistance of tungsten carbide coatings at low temperatures, but they do perform well at high temperatures. Such coatings have found numerous applications in hot sections of gas turbines. Cobalt base alloys with excellent wear resistance to temperatures over 1,800°F are also available.

In applications where hydrogen sulfide is present, ferrous base materials should not exceed a hardness of 22 Rockwell "C," per recommendation of the National Association of Corrosion Engineers, in order to avoid sulfide stress cracking. The tungsten carbide-cobalt-chromium coatings and tungsten carbide-chromium-nickel coatings have imparted wear resistance to parts in such service while the base material retains a low hardness to avoid sulfide stress cracking.

Application of D-Gun coatings on reciprocating machinery has resulted in extended parts life. Uncoated hardened Monel piston rods in oxygen booster compressors that previously required rod resurfacing in one to two years of service have shown virtually no measurable wear in five to six years of service when coated with tungsten carbide-cobalt coatings. In addition, average life of the gas pressure packings was more than doubled.

The high bond strength of the D-Gun coatings has also proven useful on polyethylene hypercompressors. There have been examples of tungsten carbide-cobalt coated plunger pistons that have operated 16,000 hours at 20,000 psi with a wear of only one mil and without any coating peeling problems.

In summary, we find that detonation gun coatings are useful to both designers and machinery maintenance personnel as a means of providing dependable wear and corrosive resistant surfaces on machine components operating under difficult service conditions. Properly selected coatings used within their intended limits are significantly capable of extending wear life of parts. The extended wear life reduces the ratio of parts cost per operating hour, justifying the expenditure for coatings on both new and refurbished equipment.

Industrial Plating. Another process that will restore worn or corroded machinery surfaces is industrial plating, usually electroplating. This process is not normally applied on-site but parts in need of restoration have to be shipped to a company specializing in this type of work.

Surface preparation for plating is usually achieved by smooth machining or grinding. In some cases, shot or grit blasting may be suitable. A very rough surface before plating is neither necessary nor desirable. Unless a greater thickness of deposit is required for wear, corrosion allowance, or for bearing material compatibility, there is no need to remove more metal than required to clean up the surface. Sharp corners and edges should be given as large a radius or diameter as possible. Areas not requiring resurfacing will be protected by the plating shop.

Materials that can be repaired belong to the majority of metals used in normal design practice. It is, however, very important that the plating company be informed of the composition or specification.

The properties of steel can be adversely affected by plating unless precautions are taken. Such effects become increasingly important with high strength materials, which may become brittle or lose fatigue strength. Heat treatment or shot peening can help to reduce these effects[4].

Plating metals normally used for machinery component salvage are chromium and nickel, either singly or in combination. If needed, other metals may be specified, for example, copper, in cases where heat or electrical conductivity is of importance. In the following we would like to concentrate on chromium as the preferred plating metal for machinery wear parts.

Industrial chrome-plating has been applied successfully whenever metal slides and rubs. The excellent wear characteristics of chromium make it well suited for use on liners of power engines, reciprocating compressors and, in some cases, on piston rods.

The process offers two major approaches: *Restorative plating*, to salvage worn parts, and *preventive plating*, to condition wear parts for service. The following advantages are usually stressed[5]:

- Chromium is extremely hard and therefore gives longer life to plated parts
- Chromium withstands acid contamination and corrosive vapors found in engine crankcase oils and fuels
- Chromium-plated parts possess a very low friction factor coupled with high thermal conductivity while permitting the parts to operate at more efficient temperatures
- Chrome-plating extends life of engine parts. It is generally accepted that chromium is four to five times harder than the original cast iron wearing surface
- Electroetching can provide porosity in a chrome-plated surface where this is necessary to hold a lubricant

These characteristics of chrome-plating are further explained in the next section.

Chrome-Plating of Cylinder Liners.* In reciprocating engine and compressor cylinders surface finish of the liner is accomplished through smooth turning, grinding, or honing. It is important that the wearing surfaces get a finish that gives the material a maximum of resisting power against the strains to which it is subjected, and also offers a low coefficient of friction and the best possible conditions for the retention of the lubricating film.

It is desirable to develop a "glaze" on the wearing surface of the liner, under actual conditions of service. This glaze is produced by subtle structural and chemical changes in the surface of the liner and is not easily achieved; it is more a product of chance than design, due to the wear-promoting factors mentioned earlier.

Other surface finishing approaches include chemical and metallic coatings, with substantial reductions in ring wear, but uniform coating of cylinders has not been fully documented as of this writing.

Just how good a step toward the solution of cylinder liner wear is chromium?

Bonding. First of all, proper chrome-plating is deposited one ion at a time, assuring a molecular bond that approaches the integrity of fusion between the basis metal and the chrome-plated surface. The chromium literally

---

* Courtesy of Van Der Horst Corporation of America, Olean, New York.

grows roots into the basis metal to make its bond the strongest in the industry. With a bond of that quality, good chromium actually adds to the structural strength of a restandard-sized liner, or even a new one, to make it stronger and tougher than before . . . and eligible for resalvaging time and time again. The importance of bond in chrome-plating includes its value in strengthening the liner wall, even after successive reborings and replatings. It assures that the chrome layer is "locked" to the basis metal, even under heavy wear conditions that would strip and spall conventional chrome-plating.

Low Coefficient of Friction

Chromium is known to have the lowest coefficient of friction of any of the commonly used structural metals for engine cylinder liners. The value of the sliding coefficient for chromium on chromium has been given as 0.12, for chromium on steel as 0.16, while steel on steel is 0.20[6]. For rotating shafts, chromium was also found to have the lowest friction of any of the metals tested in a study conducted by Tichvinsky and Fisher[7].

Hardness

The hardness of chromium has definite advantages over cast iron for long wear characteristics. The hardness is maintained throughout the thickness of the chrome-plating, while the hardness of metals treated with processes like nitriding and carburizing decreases with depth. The value of hardness in a chrome-plated surface lies in its ability to resist abrasion and scoring. Contaminants in lube oil and fuel and their deposits cannot be eliminated, but their abrasive action on the liner wear surfaces has a negligible effect on chromium due to its extreme hardness. Table 10-5 shows some typical hardness tests, in Brinell notation.

Because of the relative immunity of chromium to scuffing, which often occurs in "green" tests, or the initial runs of an engine, ring scuffing and piston seizure are eliminated, and engine production is accelerated. This is an example of the use of *preventive plating*, through processing of a new liner before it is installed for the first time.

Corrosion Resistance

The corrosion resistance of chromium is high, partly due to the dense packing of the chromium molecules during electrodeposition. This resistance makes chromium especially adaptable to cylinder liners.

Table 10-5
Hardness Table

	Brinell hardness (load 3,000 kg: 10-mm ball) or equivalent
Chromium	800–1,000
Cast-iron cylinder compositions	150–275
Cast-iron cylinder compositions, heat-treated	Max. 400
Steel 4140 heat-treated	250–375
Steel, carburized	625
Nitralloy in cylinders after removal of surface stock	650–750

Sulfuric acid, which attacks cast iron vigorously, has little effect on chromium. Hydrogen sulfide is another corrosive agent found in diesel engines running on sour gas. In a controlled test of the effects of hydrogen sulfide on steel and chromium, a steel rod partially chrome-plated was subjected to moist hydrogen sulfide at temperatures ranging from 120° to 200°F for 252 hours. After this exposure, the unplated portion of the rod was blackened and badly corroded, but the chrome-plated portion was not disturbed[8].

Chromium is also unaffected by nitric acid and saturated solutions of ammonia, but it is susceptible to hydrochloric acid. However, as discussed earlier, modern fresh-water cooling prevents the introduction of HCl-producing chlorides into diesel engines now being used. The possibility of the occurrence of sufficient quantities of HCl to attack the chrome-plated surface in a given application is negligible, thus even in marine engines.

Atmospheric corrosion also has no effect on chromium. Porous chromed cylinders may be stored indefinitely with little or no protection, without detrimental results to the plated area.

Lubrication

One chrome plating company* has developed a proprietary process for etching channels and pockets in the surface of the chrome layer, to provide tiny reservoirs for lube oil. This porous surface provides high oil retention even under the high temperatures of the combustion area in the

---

* Van Der Horst.

cylinder, and under the constant sliding of adjacent metal parts, like piston rings. This phenomenon of oil retention is termed "wettability" and describes the dispersive characteristics of oil on a microscopically uneven metal surface.

The oil collects in the recesses of the metal surface and disperses outward in an enveloping movement. In contrast, the surface tension of oil will cause an oil drop on a smooth plane surface to exhibit a tendency to reach a state of equilibrium where it will neither spread nor recede, and thus does not provide a lubricating coating for that surface. Such a surface inside a cylinder liner will not be adequately covered with an oil film and will require greater volumes of lube oil to achieve adequate protection from friction, temperatures, corrosion, and abrasion.

This proprietary porous chrome surface prevents the action of the oil's molecular cohesion in trying to achieve a perfect sphere, and forming into a drop. The configuration of the chrome surface disrupts this tendency.

This porous chrome presents such a varied surface that a portion of the area $\frac{1}{4}$ in. in diameter may contain from 50 to several hundred pores or crevices depending on the porosity pattern applied to the chrome surface. Even one drop of oil, encountering such a surface, tends to disperse itself indefinitely over the flats, downslopes of pores, and the upsloping side.

The importance of lubrication has been discussed in numerous books, and a direct correlation between a successfully maintained oil film and wear on piston rings and liner surfaces can be shown. Thus, the ability of porous chrome surfaces to provide an unbroken oil film indicates its desirability in preventing many types of liner wear, including gas erosion, which is due to a leaky piston ring seal; friction and frictional oxidation, by protecting the surface from oxygen in the combustion area; preventing metal stresses resulting in abrasion from excessive loading, which does not break the oil film maintained by the chrome; and by protecting the surface from corrosive agents produced by lube oil breakdown and combustion products.

The load capacity of porous chrome involves a condition known as boundary-layer lubrication. This term refers to an oil film thickness that is so thin it approaches the characteristics of dry lubrication. It has lost its mobility as a fluid, but reduces the mutual attraction of adjacent, sliding metallic surfaces, and thereby the friction. Fluid lubrication, e.g., thicker layers of lube oil, are not desired under the high-temperature conditions of the combustion area of the cylinder, because of the susceptibility of lube oil to flash point combustion, breakdown into deposits, unnecessarily high lube oil consumption, and the production of air pollutants.

Thermal Conductivity

Thermal conductivity in chromium is higher than for cast iron and commonly used steels, by approximately 40 percent, as shown in Table 10-6. Maximum metal surface temperatures in the cylinder are at the liner surface, especially in the combustion zone, and any improvement in heat transfer provides a lower wall temperature and will improve piston and ring lubrication. The heat reflection qualities of chromium add to the combustion and exhaust temperatures, helping to reduce incomplete combustion and its products.

While the coefficient of expansion, also shown in Table 10-6, of chromium is lower than that of cast iron or steel, there is a decided advantage in the difference. The surface of the cylinder liner has a much higher temperature than the underlying basis metal, because of that sharp temperature gradient through the wall. The effect of this gradient on a homogeneous metal, e.g., distortion, is eliminated with chromium plating, because it is desirable to have a variable coefficient of expansion ranging from a lower value at the inner wall surface to a higher value in the outer wall, where the coolants are operating. Tests run on air-cooled airplane engines for 700 to 1,000 hours showed no tendency of the chromium surfaces to loosen due to differential expansion between the chrome and the basis metal. This is significant, considering that these engines normally run at higher cylinder wall temperatures than engines in stationary installations.

Table 10-6
Expansion Coefficients and Thermal Conductivity

Metal	Linear thermal expansion in./in. at 68°F ($\times 10^{-6}$)	Thermal conductivity cal/cm²/cm/ deg C/sec at 20°C	Melting point, F°
Chromium, electrolytic	4.5	0.165	3,325+
Cast iron	6.6	0.12	2,500
Steels	6.2–6.6	0.11	2,700–2,800
Aluminum	12–13	0.52	1,216
Copper	9.1	0.92	1,981

Table 10-7
Cylinder Wear

| Location | Diametral wear, in. | | Wear ratio, cast-iron cyl./ porous-chrome cyl. |
	Cast-iron cylinder	Porous-chrome cylinder	
⅛ in. below head end	0.0077	0.0018	4.3–1
½ in. below head end	0.0071	0.0021	3.4–1
2 in. below head end	0.0083	0 0022	3.8–1

Table 10-8
Piston and Ring Wear

| | In cast iron cylinder | | In porous-chrome cylinder | | Wear ratio, cast-iron cyl./ porous-chrome cylinder |
	Grams	Percent of original weight	Grams	Percent of original weight	
Aluminum piston loss in weight	0.589	0.90	0.316	0.45	1.9 : 1
Top ring lost in weight	2.506	56.0	0.849	18.9	3.0 : 1
Second ring loss in weight	2.586	57.8	0.972	21.8	2.6 : 1
Oil ring loss in weight	1.260	46.5	0.865	31.9	1.5 : 1

Abrasion Resistance

Wear rates for chromium are substantially lower than for cast iron, as the data in Tables 10-7 and 10-8 show. The data were developed in a test of two engines run for nine hours under load, with one a cast iron cylinder and the other a porous-chrome surfaced cylinder. Allis-Chalmers Type B abrasive dust (85 percent below five microns, 100 percent below 15 microns) was added to the initial charge of crankcase oil, with a proportion of abrasive dust of 8.6 percent by weight of the lube oil, to accelerate wear.

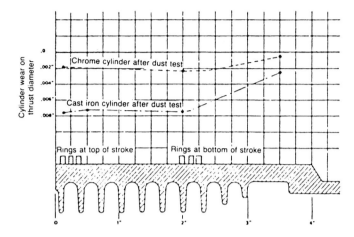

Figure 10-10. Cylinder wear on chrome and cast-iron cylinders.

The constrast in wear ratios between the cast iron and chromium in this test is substantial, reaching as much as four to one in the cylinder and three to one for the pistons and rings. Figure 10-10 shows a plot of these data for the cylinder wear comparison[9].

Often, the boring out of worn cylinders requires the deposition of extra-thick layers of chrome to bring the surface back to standard size. This is not advisable, because on the next resalvaging, it may not be possible to bore down further into the basis metal and still retain enough structural strength in the liner wall to justify salvaging.

With another proprietary process, 99.9 percent pure iron is electro-deposited on the basis metal with a special bond, to build up the basis metal, to a thickness where normal chrome layer thicknesses are practical.

Chromium in Turbocharged Engines

The operation of turbocharged engines involves the exaggeration of all the wear factors described in this section because the temperatures are higher, fuel and lube oil consumption are higher, the engine runs faster, and corrosive agents seem to be more active and destructive. Turbocharging, however, increases the horsepower of an engine from 10 to 25 percent. During the last decade, many stationary engines were retrofitted for turbocharging, and engines with liners not surfaced with chrome have had the chance to be upgraded.

Just as an example, the high heat of turbocharged engines creates a lubrication problem with cast iron liner surfaces. Even microporosity in iron casting will not retain oil under such high temperatures. The corresponding increase in wear factor effects will accelerate liner and piston ring wear and increase downtime.

Special chromium, with variable porosity tailored to the operating characteristics of the engine, can make the difference between a productive engine installation and a liability.

Operating Verification

In a detailed study assessing the conditions and circumstances influencing machinery maintenance on motor ships, Vacca[10] plots the operating performance of several marine engine liners and arrives at a documented conclusion that chrome-plated liners show a wear rate that is less than half that of nonchrome-plated liners. The indirect result is considerable improvement in fuel economy and ship speed. Figure 10-11 shows these data plots.

Another application study emphasized the benefits of chrome-plating engine liners and was seen to have a direct effect on labor requirements and the workloading of engine room staffs.

For more documented low wear rates, a study on engine liner performance by Dansk-Franske Dampskibsselskab of Copenhagen on one of their ships, the "Holland," produced some interesting statistics. All

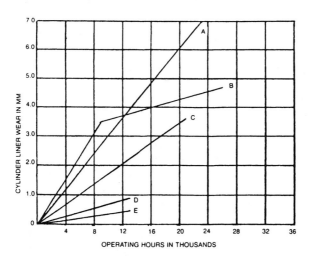

Figure 10-11. Graphs of cylinder liner wear. Curves A and C refer to opposed piston engines and curves B, D, and E are for poppet valve engines. Curves D and E show results using chromium plated liners.

cylinder liners were preventive plated with chromium before they were installed. The results well repaid the effort, in less overhaul, reduced ring wear, and extremely low cylinder wear. The highest wear rates on the six cylinder liners were 0.20 mm/10,000 hours, as shown on the chart in Figure 10-12. This negligible wear led to the conclusion that the liners:

". . . will still have a life of more than 10,000 hours . . . In fact, it means that this ship will never need any liner replacement."[11]

Even though these studies represented only a fraction of the operating and test data that supports this contention, they indicate the considerable benefits in terms of cost-savings and long-lived performance that the use of chrome-plating can provide. The fact that the studies cited were performed on motor ships, in salt-water environments, where corrosion agents are more active than in stationary facilities, adds further emphasis to this position.

The question of chrome plating economy has been raised and can be answered by an example. Chrome-plating offers a twofold economy. First, in the cost of restandardizing with chrome over the cost of a new liner, and second, the extended length of operating life of the plated liner, whether new or reconditioned.

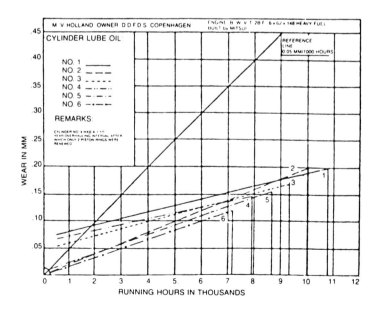

Figure 10-12. Cylinder liner wear—with chrome plating.

As stated in the *Diesel Engineering Handbook*:

"A (chrome) plating . . . will cost 65 to 75 percent of the price of a new unplated cast iron liner, or 50 to 60 percent of the price of a new chrome-plated liner. It must be remembered that the plated liner will have three to five times the life of a new unplated liner."[12]

The significance of the last sentence in the quote is often overlooked. Even if the chrome-plating restandardsizing of the worn liner were *100 percent* of the cost of a new unplated liner, a cost savings will be achieved because the replated liner will still last three to five times as long. At 100 percent, the replated liner is thus still only about 30 percent of the cost of all the new replacement liners that would be required to match its normal operating life.

Conclusion

Our principal conclusions can be summarized as follows. Directly or indirectly, all of the effects of the wear factors described in this section can be mitigated or eliminated completely with the use of special chromium-plating on cylinder liners, crankshafts, and piston rods.

Whether the method of liner salvage is restandardsizing or oversize boring with oversize piston rings, or even with new liners and parts to be conditioned for long wear before going into service, proprietary chromium plating processes can add years of useful operating life in a continuing, cost effective solution to the problems of wear.

On-Site Electroplating Techniques. Where parts cannot be moved to a plating work station, deposition of metal by the *brush electroplating technique* may be considered.* This process serves the same varied functions that bath electroplating serves. Brush electroplating of machinery components is used for corrosion protection, wear resistance, improved solderability or brazing characteristics and the salvaging of worn or mismatched parts. Housed in a clean room, the equipment needed for the process is:

1. The power pack.
2. A lathe.
3. Plating tools.

---

* Dalic Plating Process.

4. Masking equipment and plating solutions.
5. Drip retrieval tray.
6. Pump to return solutions through a filter to the storage bath.
7. Trained operator.
8. Supply of clean water for rinsing parts between plating operations.

Brush electroplating thickness in excess of 0.070 in. is generally more economic if done in a plating bath.

Electrochemical metallizing, another form of electroplating, is a hybrid between electric arc welding and bath electroplating. It is a portable system for adding metal to metal. As a special type of metallizing, the process is claimed to offer better adhesion, less porosity, and more precise thickness control than conventional flame spray or plasma types of metallizing. Unlike conventional metallizing or bulk welding, the base metal is not heated to high temperatures, thus avoiding thermal stresses.

In the rebuilding of main bearing saddle caps—a typical application—one flexible lead is connected to a working tool or "stylus" of appropriate size and shape. The stylus serves as an anode, and is wrapped in an absorbent material. The absorbent is a vehicle for the aqueous metallic plating solution. Metal deposits rapidly onto the cathodic—negative charged—workpiece surface. Deposit rates of 0.002 in. per minute are typical. One repair shop uses multistep processes in which the prepared metal surface initially is built to approximate dimensions with a heavy-build alkaline copper alloy solution. Then a hardened outer surface is created by depositing a tungsten alloy from a second solution.

Not only engine saddle caps, but cylinder heads, crankcases, manifolds, engine blocks, crankshafts, and other machinery castings have been successfully repaired using the electrochemical metallizing process. The process has replaced conventional oxyacetylene high-heat bronze welding that was used to build new metal onto worn saddle caps. The high-heat welding associated with oxyacetylene spraying had disadvantages in terms of excessive machining time, metal waste, lost time in cool-down, and high temperature distortion of the workpiece.

In field use, the hardness and durability of electrochemically metallized material appears to equal the original casting. In contrast to other metal rebuilding methods, flaking or cracking of parts rebuilt with the process has not been experienced[13].

The following equipment is required for an electrochemical plating process[14]:

1. The power pack and flexible leads.
2. Turning heads and assorted stylus tools. The turning head is a low speed reversible, variable speed rotational device for use in electro-

chemically plating cylindrical components. It enables rotation of shafts, bearings and housings, so that either inside or outside diameters can be uniformly plated.

3. Handles and selected anodes.
4. Accessories such as cotton batting, wrapping material, stylus holders, evaporating dishes, solution pump, and tubing.
5. Selection of plating solutions from some 100 different primary metals or alloy solutions.
6. A trained operator.

Hardening of Machinery Components. In trying to achieve improved wear resistance it would be well not to neglect proven traditional steel-hardening methods. In surface hardening of alloy steels the core of a machinery part may be treated to produce a desired structure for machinability or a strength level of service, whereas the surface may be subsequently hardened for high strength and wear resistance.

Flame hardening involves very rapid surface heating with a direct high temperature flame, followed by cooling at a suitable rate for hardening. The process utilizes a fuel gas plus air or oxygen for heating.

Steels commonly flame hardened are of the medium, 0.30 to 0.60 percent carbon range with alloy suitable for the application. The quenching medium may be caustic, brine, water, oil, or air, as required. Normally quenchants are sprayed, but immersion quenching is used in some instances.

To maintain uniformity of hardening, it is necessary to use mechanical equipment to locate and time the application of heat, and to control the quench.

As with conventional hardening, residual stresses may cause cracking if they are not immediately relieved by tempering. In some instances residual heat after quenching may be sufficient to satisfactorily relieve hardening stresses. As size dictates, either conventional furnace tempering or flame tempering may be used. With flame tempering, the heat is applied in a manner similar to that used for hardening but utilizing smaller flame heads with less heat output[15].

Carburizing is one of the oldest heat treating processes. Evidence exists that in ancient times sword blades and primitive tools were made by carburization of low carbon wrought irons. Today, the process is a science whereby carbon is added to steel within desired limitations to a controlled amount and depth. Carburizing is usually, but not necessarily, performed on steels initially low in carbon.

If selective or local case hardening of a part is desired, it may be done in one of three ways:

1. Carburize only the areas to have a hardened case.
2. Remove the case from the areas desired to be soft, either before or after hardening.
3. Case carburize the entire surface, but harden only the desired areas.

The first method is the most popular and can be applied to the greatest variety of work.

Restricting the carburizing action to selective areas is usually done by means of a coating that the carburizing gas or liquid will not penetrate. A copper plate deposited electrolytically, or certain commercial pastes generally prove satisfactory. The several methods employed in adding carbon come under the general classification of park carburizing, gas carburizing, and liquid carburizing[16].

Nitriding is a process for the case hardening of alloy steel in an atmosphere of ammonia gas and dissociated ammonia mixed in suitable proportions. The steel used is of special composition, as seen in Table 10-9. The process is carried out at a temperature below the transformation range for steel and no quenching operation is involved unless optimum core properties are desired. Nitrided parts evidence desirable dimensional stability and are, therefore, adaptable to some types of close tolerance elevated temperature applications[17].

The parts to be nitrided are placed in an airtight container and the nitriding atmosphere is supplied continuously while the temperature is raised and held at 900° to 1,150°F. A temperature range of 900° to 1,000°F is generally considered optimum to produce the best combination of hardness and penetration. The hardening reaction takes place when nitrogen from the ammonia diffuses into the steel and reacts with the nitride

Table 10-9
Composition of Various Nitriding Steels[17]

	AISI 7140 AMS 6470E	AMS 6425	135 Type G	N	EZ
Carbon	0.38–0.43	0.21–0.26	0.30–0.40	0.20–0.27	0.30–0.40
Manganese	0.50–0.70	0.50–0.70	0.40–0.70	0.40–0.70	0.50–1.10
Silicon	0.20–0.40	0.20–0.40	0.20–0.40	0.20–0.40	0.20–0.40
Chromium	1.40–1.80	1.00–1.25	0.90–1.40	1.00–1.30	1.00–1.50
Aluminum	0.95–1.30	1.10–1.40	0.85–1.20	0.85–1.20	0.85–1.20
Molybdenum	0.30–0.40	0.20–0.30	0.15–0.30	0.20–0.30	0.15–0.25
Nickel	—	3.25–3.75	—	—	—
Selenium	—	—	—	—	0.15–0.25

formers (aluminum, chromium, molybdenum, vanadium, and tungsten) to produce precipitates of alloy nitrides.

Nitrogen is absorbed by the steel only in the atomic state, and therefore, it is necessary to keep fresh ammonia surrounding the steel surfaces. This is accomplished by adequate flow rates and circulating the gases effectively within the container.

The nitriding cycle is quite long depending upon the depth of case required. A 50 hour cycle will give approximately 0.021 in. case of which 0.005 to 0.007 in. exceeds 900 Vickers Diamond Pyramid hardness. The handling of nitrided steels in general is similar to that of any other alloy steel. However, due to their high aluminum content, these steels do not flow as readily in forging as other alloy steels and, therefore, require somewhat greater pressures. Where large sections are encountered, normalizing prior to nitriding is recommended.

To develop optimum core properties, nitriding steels must be quenched and tempered before nitriding. If the part is not properly heat treated and all traces of decarburization removed from the surface, nitrogen will penetrate along the ferrite grain boundaries and thereby produce a brittle case that has a tendency to fail by spalling.

In tempering, the temperature must exceed the nitriding temperature; otherwise, significant distortion may result during the nitriding cycle.

If a large amount of machining is to be done, it is sometimes advisable to anneal, rough machine, heat treat, and finish machine. In very large parts, it is advisable to stress relieve before final machining if the parts were rough machined in the heat treated condition. In all instances where machining is done after heat treatment, it is important that sufficient surface be removed to ensure freedom from decarburization.

Nitrided surfaces can be ground, but whenever possible this should be avoided. In nitriding, some growth does occur due to the increase in volume of the case. However, this is constant and predictable for a given part and cycle. Therefore, in most instances, parts are machined very close to final dimensions before nitriding. When necessary, lapping or honing is preferred to grinding because the extremely hard surface is shallow. If threads and fillets are to be protected or areas are to be machined after nitriding, an effective means of doing so is to tin plate those locations which are to remain soft. A 1 : 1 mixture of tin and lead is commonly used when electroplating is not possible. Since the nitriding temperatures exceed the melting point of the tin and tin alloys, it is essential that an extremely thin coat be applied to prevent the coating from flowing onto surfaces other than those to be protected.

Nitrided parts have a combination of properties that are desirable in many engineering applications. These properties include:

1. An exceptionally high surface hardness which is retained after heating to as high as 1,100°F.
2. Very superior wear resistance particularly for applications involving metal-to-metal wear.
3. Low tendency to gall and seize.
4. Minimum warpage or distortion and reduced finishing costs.
5. High resistance to fatigue.
6. Improved corrosion resistance.

Here is a list of typical machinery applications:

Bushings	Piston Rods
Cams	Plungers
Camshafts	Pump Sleeves
Connecting Rods	Pump Shafts
Crankshafts	Push Rods
Cylinder Barrels	Racks and Pinions
Cylinder Liners	Ratchets
Diesel Engine Fuel	Retaining Rings
Injector Pump Parts	Seats and Valves
Gears	Shafts
Guides	Splines
King Pins	Sprockets
Knuckle Pins	Studs
Needle Valves and Seats	Thrust Washers
Nozzles	Timing Gears
Pinions	Thrust Washers
Pinion Shafts	Water Pump Shafts
Piston Pins	Wear Plates
Pistons	Wrenches

Diffusion Alloys.* Since carburizing dealt with earlier is, by definition, a diffusion alloying system, the primal history of diffusion alloys is quite lost in antiquity. But, we can state that the modern systems began during World War II in Germany when precious chromium was diffused into steel parts to form a stainless surface. Until recently, almost the sole beneficiaries of this work were gas turbine and rocket engine manufacturers. These engines make use of diffusion alloys resistant to high temperature oxidation and sulfidation. Now we are able to produce diffusion alloys tailored to specific industrial needs: Hardness, corrosion resistance, erosion

---

*Courtesy Turbine Metal Technology, Inc., Tujunga, California 91042.

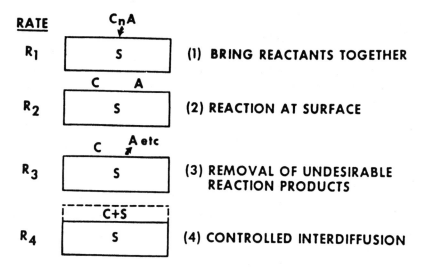

Figure 10-13. Principles of the diffusion alloy process.

resistance, and oxidation resistance, including combinations of these properties. Diffusion alloys can be produced on a wide spectrum of alloys, allowing interesting combinations of substrate properties and alloys optimized for cost, strength, or other considerations.

Diffusion alloys are alloys and/or intermetallic compounds formed by the high temperature reaction of atoms at the surface of the part to be alloyed with atoms brought to that surface by a suitable process such as chemical vapor deposition (CVD). This is illustrated schematically in Figure 10-13. Since diffusion alloy deposition is conducted at fairly high temperatures there is significant atom mobility for both alloy and substrate elements, i.e., diffusion of all atom species will occur.

Properties of diffusion alloys are quite different from metals in many respects. In general they are single phase, but if multiple phases should exist, these are not intermingled but occur in layers.

There are no grain boundaries, and grain boundaries that exist in the substrate disappear in the alloying. Although the ductility of the alloys is limited, they are not glass-brittle and will allow some plastic deformation of the substrate without cracking. Unlike overlay coatings such as plasma spray, there is no weak interface between the alloy and the substrate to sometimes fail under thermal shock or differential thermal expansion; diffusion alloys are an integral part of the system. In reality then, they are not a "coating," but a conversion of the surface.

How can a thin diffusion alloy prevent erosion? Nothing totally prevents erosion, but erosion can be slowed by a diffusion alloy. As pointed out previously, this is a single phase system. In a hardened metal, the hard precipitate is slowly eroded, but the soft matrix in which it is held erodes very quickly. As soon as the support for the hard particle is worn away, the particle simply drops off. By producing a hard, single phase system on the surface, there is no soft matrix to erode, and a much slower erosion rate results. This rate is low enough so that increases in life of 3 to 30 times are common.

Tungsten carbide and diffusion alloying. There are a number of advantages to diffusion alloying tungsten carbide. Like hardened metals, tungsten carbide is a two-phase system and the matrix is readily eroded. Technology has developed a system that not only hardens the matrix, but reacts with the tungsten carbide particles to form an even harder material. Another advantage is realized by using carbides with higher binder content. The more erosion-resistant grades of tungsten carbide contain very little binder. This results, however, in an extremely brittle material, having low resistance to both thermal and mechanical shock. By utilizing a diffusion alloy with the higher binder carbides, the properties of the alloy are not impaired and a better structural part is produced.

How does a diffusion alloy prevent wear? As described before, wear can be divided into two basic types, adhesive wear and rubbing wear. Adhesive wear usually occurs when two metals rub against each other under either very heavy pressure or extremely de-oxidizing conditions. In both cases metal migrates across the interface of the parts, resulting in an actual weld. Further movement tears a piece of the material from one or the other of the two parts. This is usually called seizing or galling. Again, the high bond strength between the atoms of an intermetallic compound prevents their migration across the interface with the mating part. When there is no migration there is no welding.

How corrosion resistant are diffusion alloys? Different combinations of metals in the part and elements introduced by the process give differing results in corrosion. Generally, diffusion alloys are acid resistant, and various combinations will yield resistance to hydrochloric acid, sulfuric acid, nitric acid, and hydrofluoric acid. Oxidation resistance can be imparted to over 2,000°F. Most of the diffusion alloys are resistant to hydrogen sulfide and mercaptans. Diffusion alloys can be tailored for specific properties. An intermetallic compound behaves, chemically, very differently from those elements of which it is composed.

Application. Diffusion alloys build with remarkable uniformity, following each asperity of the original surface. Total alloy thickness variation on a part is normally 0.0001 to 0.0002 in. If the original surface is eight microinches rms (root mean square), then the surface of the

Table 10-10
Hardness of Diffusion Alloys

Surface Treatment	Hardness (Vickers)
Nitriding	600–950
Carbonitriding	700–820
Carburizing	700–820
Hard Chrome Plating	950–1,100
TMT-5 Steel	1,600–2,000
TMT-5 WC	2,200–2,350
TMT-5 Molybdenum	2,900–3,100

diffusion alloyed part will also be eight rms. Finer finishes require slight lapping.

Hardnesses of diffusion alloys are shown in Table 10-10. How brittle is such a hard material? Although the hard diffusion alloys cannot stand extensive elongation, they are sufficiently ductile, for example, to allow straightening of shafts which have been heat treated following diffusion alloying. As long as the plastic deformation is below about five percent, the alloy will not crack. Equally, thermal and mechanical shock do not have any effect. Unlike the "stuck on" coatings, thermal differentials do not load an interface in shear. There is no true interface to load.

What metals can be diffusion alloyed? Almost any alloy of iron, nickel, or cobalt can be diffusion alloyed. Naturally, some alloys are preferred for specific systems, but the general rule holds. Aluminum, copper, zinc, and cadmium cannot be diffusion alloyed. Tungsten molybdenum, niobium, and titanium can be diffusion alloyed.

Is a high strength alloy affected by the high temperature process? In general, when a part is diffusion alloyed it is in the annealed state. If high strength is required, the part is heat treated following alloying. With some simple precautions, the heat treating can be carried out in a normal manner.

Can a diffusion alloy be formed in any shape? There are virtually no configuration restraints. Internal passages and blind holes pose no problems. The elements added are transported in a gaseous phase. Spray patterns or "line of sight" are not a part of the system.

The following *specific* process machinery applications of diffusion alloys have been successfully implemented:

1. Pump impellers and casings in fluid catalytic cracking units suffering from erosion by catalytic fines.

2. Pump impellers and casings used in coking service.
3. Pump impellers in lime slurry service.
4. Steam turbine nozzles and blades.
5. Expander turbines in contaminated gas streams.

Table 10-11 represents a more comprehensive overview of diffusion alloy applications.

Table 10-11
Characteristics and Applications of Diffusion Alloys

System	Substrates	Characteristics	Applications
TMT-56 Boron Carbide	Some Chrome Stainless Steels; Low-Alloy Steels	1,800–2,200 KHN; Erosion and wear resistance.	Centrifugal pumps; Screw pumps; Piping
TMT-601 Complex Boride	Alloy or Carbon Steel	1,700–1,900 KHN; Erosion and wear resistance.	Centrifugal pumps; Valves; Piping
TMT-601	Chromium Stainless Steel	1,900–2,000 KHN; Erosion and wear resistance.	Centrifugal pumps; Valves
TMT-601	Chrome-Nickel Stainless Steel	1,700–1,800 KHN; Erosion and wear resistance.	Valves; Pump plungers; Shafting
TMT-601	Nickel Alloys	1,900–2,200 KHN; Erosion and wear resistance.	Centrifugal pumps; Valves; Pump plungers
TMT-601	Cobalt Alloys	2,000–2,200 KHN; Erosion and wear resistance.	Valves; Pump plungers; Seal rings
TMT-745 Titanium Diboride	Cobalt-bonded Tungsten Carbide	4,000–4,500 KHN; Erosion and wear resistance.	Valves; Chokes; Seal rings; Dies; Blast tubes & joints; Orifice plates; Wear plates
TMT 5 A modified complex Cobalt Boride	Cobalt-bonded Tungsten Carbide	3,800–4,200 KHN; Erosion and wear resistance.	Valves: Chokes; Seal rings; Dies; Blast tubes and joints; Orifice plates; Wear plates
TMT-2413 Aluminide	Co or Ni bonded WC; Ni, Co, and Cr-Ni Stainless	Corrosion and galling resistance.	Pump plungers; Piston rings; Shafts; Seal rings; Bearings; Threads

Table 10-11
Characteristics and Applications of Diffusion Alloys—cont'd

System	Substrates	Characteristics	Applications
TMT-2813 Nickel Aluminide	Carbon, Low-Alloy, Chrome and Cr-Ni steel	Corrosion and galling resistance.	Piston rings; Bearings; Piping
TMT-213 Chrome Aluminide	Carbon and Low-Alloy Steels	Galling resistance.	Bearings
EC-114 Complex Aluminide	Nickel and Iron-Base Alloys	Friction and oxidation resistance.	Turbine hot section; Oil tools
KS-138 Dispersed Phase Aluminide	Nickel Alloys	Corrosion and erosion resistance.	High temperature fans; Valves; Power recovery turbines
PS-138 Disp. Phase Platinum Aluminide	Nickel Alloys	Hot gas corrosion resistance.	Turbine hot section; Oil tools
RS-138 Disp. Phase Rhodium Aluminide	Cobalt Alloys	Hot gas corrosion resistance.	Turbine hot section; Oil tools

Electro-Spark Deposition Coatings. Electro-spark deposition involves the transfer of minute molten droplets of the desired coating material from a contacting electrode to the surface of the part. At the completion of the spark-induced transfer, the droplet welds to the part. By careful control, usually by computer, these microwelds will overlap, yielding a complete new surface. Heat input to the part is extremely low, and the maximum temperature rise is just a few degrees above ambient.

Some of the remarkable things that are being done by this process include carbide coatings on aluminum, carbide coatings on titanium, nickel or gold on aluminum, and nickel aluminide on steels, as well as the seemingly simple coating of stainless steel with stainless steel.

Bond strengths equal those of the base components, rather than being limited to 10,000 or 12,000 psi. Thus, the ESD coatings will withstand bend tests, thermal shock, and mechanical shock that no other coating system can match.

This process is now in use in critical nuclear reactor components and may well be the answer to some of the most difficult wear and corrosion

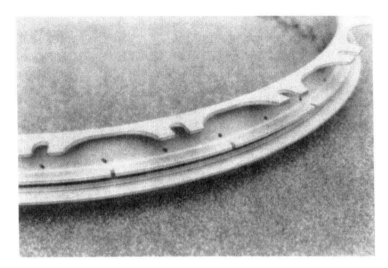

Figure 10-14. Triboloy 700 applied by Electro-Spark Deposition to seal ring surfaces to prevent fretting.

problems facing us today. Figure 10-14 illustrates Triboloy 700 applied by electro-spark deposition to seal ring surfaces to prevent fretting.

High-Velocity Thermal Spray Coatings. These coatings are available for steam-turbine blading and other components. In a number of applications, high-velocity thermal spray systems have produced coatings that are equal to or better than D-Gun and high-energy plasma spray deposits when evaluated for bond strength, density, and oxide content. More specifically, the bond strength of tungsten carbide/cobalt coatings produced with the HV system has been measured at more than 12,000 psi on a grit-blasted surface.

High-velocity thermal spray systems use high-velocity combustion exhaust gases to heat and propel metallic powder onto a workpiece, thereby producing a coating. The exhaust is produced by internal combustion of oxygen and a fuel gas. Propylene, MAPP, and hydrogen have all been used as fuels with propylene being the recommended fuel. The combustion temperature is approximately 5,000°F, with exhaust velocities of 4,500 ft/sec, more than four times the speed of sound.

The powder particles are introduced axially into the center of the exhaust jet. When the powder particles (hot and possessing high kinetic energy) hit a solid workpiece, they are deformed and quenched. The resulting coatings exhibit high bond strength and density and are exceedingly smooth. Table 10-12 highlights the characteristics and principal applications for high-velocity thermal sprays.

Table 10-12
Characteristics and Applications of High-Velocity Thermal Sprays

Description	Characteristics	Application
TMT proprietary high density fused.	Very high wear and impact resistance. Operating temperature to 1,200°F. Hardness >1,200 DPH. Bond strength >25,000 psi.	Gate valves and mill rolls.
80 Cr3C2, 20 Ni-Cr	Excellent wear resistance to temperatures approaching 1,600°F (1,400°F continuous). Not recommended in corrosive environments.	Gas turbine hot section components.
ESD prepared substrate. May be used with all TMT WC systems listed.	Forms metallurgical bond with substrate for all TMTHV applied tungsten carbide systems.	Same applications for all WC systems listed.
73 WC, 20 CR, 7 NI	Good wear resistance with improved oxidation and corrosion resistance at temperatures approaching 1,400°F. Hardness >1,100 DPH.	Oilfield machinery and chemical processing equipment. Gas turbine components.
83 (W, Ti)C 17 Ni	Smooth as-coated resistant surface. Resistant to alkaline solutions. Hardness 1,200 DPH. Operating temperature below 1,000°F.	Plastics industry.
85 WC, 15 Co	Similar to TMTHV-387 with greater impact resistance.	Gas turbine compressor components.
87 WC. 13 Co	Excellent resistance to wear. Particle erosion resistance approx. 50% improvement over weld deposited STELLITE 6. Hardness >1,100 DPH. Operating temperature to 1,100°F.	Industrial machinery. Replacement for cemented carbide. Gas turbine components.
91 WC, 9 Co	A hard, erosion and wear resistant surface. Impact resistance fair. Hardness 1,200 DPH. Operating temperature to 1,000°F.	High wear components in aerospace and industrial applications.

(Table continued on next page)

Table 10-12
Characteristics and Applications of High-Velocity
Thermal Sprays—cont'd

Description	Characteristics	Application
Triboloy 400	Very high strength & good wear resistance. Hardness 800 DPH. Operating temperature to 1,200°F.	Gas turbine bleed air components.
Triboloy 800	Excellent resistance to metal-to-metal wear, galling. and corrosion to 1,500°F. Hardness 600 DPH.	Gas turbine components. Extrusion dies. Piston rings.
Haynes STELLITE 6	High resistance to particle erosion, abrasive wear, and fretting to 1,500°F. Hardness 490 DPH.	Valve and pump components. Exhaust valves and seats, conveyor screws, hot crushing rolls.
Hastelloy C	Excellent corrosion resistance. Good metal-to-metal wear and abrasion resistance to 1,900°F. Hardness 470 DPH.	Valve and pump components for chemical industry. Boiler tubes, digesters, guide rolls, fan blades.

*Comparable in material composition only. TMT HV Systems offer improved density, bond strength and oxide control.

Triboloy, STELLITE, and Hastelloy are tradenames of Stellite Div. Cabot Corporation. ES is a patented process of Intermetallics, a joint venture of Turbine Metal Technology Inc., and Westinghouse Electric Corp.

Source: Turbine Metal Technology, Inc., Tujunga, California 91402.

Other Coatings for Machinery Components. There are many proprietary coating processes that can be applied to machinery components either in a restorative or preventive manner. These coatings may be used for services in moderate wear and corrosion environments but also in applications where metal to metal contact is made and the danger of galling of the two surfaces exists. The *iron-manganese-phosphate bath process** is a typical example. The use of this process is especially indicated for cams, rollers, and gears.

*Parko Lubriting Process.

The iron-manganese-phosphate process adds from 0.0002 to 0.0003 in. to the surface of the workpiece. The process specification calls for the cleaning of the workpiece to be coated, preheating in a water bath to 200°F and immersion into the iron-manganese-phosphate bath until all reaction stops. The piece is then rinsed and immersed in a hot solution of soluble oil and colloidal graphite. It is finally wiped and dried thoroughly.

Hard coating treatment of aluminum alloys (anodizing) is a process that increases surface hardness and abrasion and corrosion resistance of aluminum and aluminum alloys. This is accomplished by formation of a dense aluminum oxide in a suitable electrolyte. Coating thicknesses range from 0.0015 to 0.0025 in. Typical applications are the coating of reciprocating compressor pistons and centrifugal compressor labyrinths.

Application of thin films of Teflon® to metals. These films often have the advantage over other dry lubricants by producing very low coefficients of friction. The operable temperature range of thin Teflon® lubricating coatings is from −80°F to 550°F. These coatings have also been successfully applied for corrosion protection of machinery parts.

Teflon® coatings are formed by baking an air-dried coating deposited from an aqueous dispersion. Several aqueous suspensions are available. Application methods are well defined[18]. Film thickness ranges between 0.0002 and 0.0003 in. for most lubrication applications. For corrosion protection, multiple coatings are applied for a final thickness of 0.0015 to 0.003 in.

Fluoropolymer (Teflon®) Infusion Process*. This process entails the infusion of a friction-reducing fluoropolymer into the surface of machinery parts. The process does not result in a surface coating, although an initial coating of 0.005 in. is provided after treatment. Since the infusion process is not a rebuilding process, the workpiece must be serviceable before treatment. The process has been applied to steam turbine trip valve stems, pump plungers, compressor sliding parts, shafts, and bushings. The treatment adds built-in lubrication and corrosion resistance but does not harden the original surface[19,20].

Concluding Comments on Coatings and Procedures

Recall that the coatings and compositions given in this text are representative of typical industry practices and availabilities. There are hundreds of variations and proprietary formulations. Users are encouraged

---

* Impreglon® Process.

Table 10-13
Elastomer Preference by Application

DYNAMIC	STATIC
NUTRILE (B, C, OR D)	NITRILE (B, C, OR D)
ETHYLENE-PROPYLENE (E)	ETHYLENE-PROPYLENE (E)
SBR (G)	NEOPRENE (N)
FLUOROCARBON (V)	SBR (G)
NEOPRENE (N)	SILICONE (S)
PHOSPHONITRILIC	FLUOROCARBON (V)
FLUOROELASTOMER (Q)	
POLYURETHANE (U)	POLYACRVLATE (L)
POLYACRYLATE (L)	FLUOROSILICONE (F)
BUTYL (J)	POLYURETHANE (U)
EPICHLOROHYDRIN (Z)	BUTYL (J)
	PHOSPHONITRILIC FLUORDELASTOMER (Q)
	EPICHLOROHYDRIN (Z)
	POLYSULFIDE (K)
	CHLOROSULFONATED POLYETHYLENE (H)

to seek out experienced vendors, i.e., vendors that use proven materials and application processes. Refer to Appendix 10-A for examples.

Selection and Application of O-Rings[‡]

In hydrocarbon processing plants, mechanical seals for pumps and compressors, tube fittings and pipe flanges often use O-rings to prevent fluid flow or leakage. According to application, O-rings can be categorized as static (seal between flange facings) and dynamic (subjected to movement or wobble). Table 10-13 lists the commonly available O-ring materials in decreasing order of preference based on an overall desirability for O-ring sealing service, with cost and availability considered secondary. When following the design steps results in several candidate elastomers for a specific application, this table may be used for final selection. (Letter suffixes identify elastomers compound designations.)

Next, the user has to consider temperature limitations of the elastomers. Here Table 10-14 will be helpful.

Chemical compatibility of O-rings with a process fluid and temperature limits will define the method of O-ring production, using full-circle

---

[‡] Source: National® O-Rings Division of Federal Mogul, 11634 Patton Road, Downey, California 90241. Adapted by permission.

Table 10-14
Elastomer Temperature Ranges

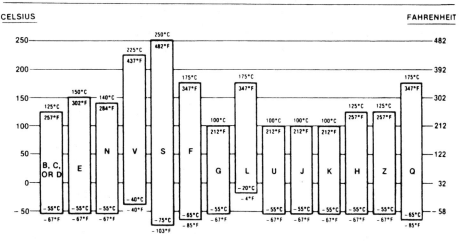

B, C, or D — NITRILE	L — POLYACRYLATE
E — ETHYLENE-PROPYLENE	U — POLYURETHANE
N — NEOPRENE	J — BUTYL
V — FLUOROCARBON	K — POLYSULFIDE
S — SILICONE	H — CHLOROSULFONATED POLYETHYLENE
F — FLUOROSILICONE	Z — EPICHLOROHYDRIN
G — SBR	Q — PHOSPHONITRILIC FLUOROELASTOMER

molding, ambient adhesive bonding and hot bonding or vulcanizing. Having no joint and hence no weak point, full-circle molded O-rings are the most common for reliability in operation. Available in a wide range of stock sizes and materials, O-rings of this type also can be custom-molded. Ambient adhesive-bonded O-rings of any diameter can be quickly and easily made, using cord stock of most materials except silicone rubber. A simple jig used for cutting square ends and aligning them for bonding gives a smooth joint, which can sometimes be made in place without machine disassembly. Vulcanizing is considered to be an intermediate method in terms of nonstock O-ring delivery, chemical, and temperature resistance. Thermal and chemical resistance of the hot-bonded O-rings is superior to the adhesive-bonded, but inferior to the molded ones.

O-ring failure analysis can be instructive to prevent machine failure. There are several common causes of O-ring failure.

Deterioration in Storage. Some synthetic rubbers such as neoprene and Buna N are sensitive to ultraviolet radiation, others to heat and ozone. O-rings should therefore not be exposed to temperatures above 120°F (49°C) or air, light, ozone, and radiation generating electrical devices. Generally, storing O-rings in polyethylene bags inside larger cardboard boxes under normal warehouse conditions will ensure maximum storage life.

Temperature. Exceeding an allowable temperature is a common cause of O-ring failure. Many O-rings fail from overheating because they are deprived of lubricant and/or coolant. Others are unable to recover from compression set at a high temperature, and they remain "flattened out," so to speak. Yet others fail because of overheating or chemical attack. Table 10-14 lists temperature limits for common O-ring materials. These limits can often be exceeded for short periods. Process fluid temperature may not be equal to O-ring temperature. A cool flush may reduce O-ring temperature, as may heat dissipation through a barrier structure. On the other hand, localized frictional heating may increase the O-ring temperature. If high or low temperatures are suspected of causing failure, it may be practical to change the environmental temperature or the O-ring material. In extreme cases, an all-metal bellows seal may permit elimination of O-rings.

Mechanical Damage. Tearing, pinching, foreign matter embedment, dry rubbing and various other mechanical damage can occur during installation, operation, and removal. A sharp steel tool used for O-ring removal can scratch the groove or sealing surface and cause leakage. Brass, wood or plastic tools can be used without risk of scratching. Removal and installation instructions are shown in Figure 10-15.

Chemical Attack. Tables 10-15 and 10-16 are useful in selecting O-ring materials compatible with various fluids. Experimental verification is sometimes worthwhile. Complications may occur because of different properties of supposedly identical O-rings produced by various manufacturers, Also, a loss of identification in storage and handling is possible. To minimize unpleasant surprises for critical services, consider discarding even new O-rings which come with new and rebuilt seal assemblies, and replace them with O-rings of known material from a single manufacturer whom you consider reliable. To check whether an unknown, used O-ring is Viton or something else, you can immerse it in carbon tetrachloride. If it sinks, it is probably Viton; if it floats, it is not. Kalrez® perfluoroelastomer would also sink, but this material is uncommon and usually identified by tagging or other means.

O-ring Removal and Installation for Specific Applications

The diagram below illustrates three different O-ring sealing applications: External Static, Internal Static and External Dynamic.

The replacement O-rings in each of these situations will require its own set of procedures.

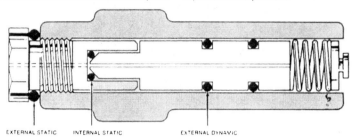

EXTERNAL STATIC   INTERNAL STATIC   EXTERNAL DYNAMIC

Removal and Installation in External Static Applications

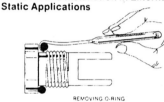

REMOVING O-RING

When removing an O-ring in this type of application, using a pull-type hook tool is very useful. Slip the hook under the O-ring and pull it over the threads and off.

As you would with any O-ring installation, make sure the area is clean and free of all foreign substances to prohibit deformation of sealing surfaces.

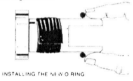

INSTALLING THE NEW O-RING

After you have cleaned the surfaces, put tape over the threads to prevent damage to the O-ring. Lubricating the ring with a compatible material with system fluid will help the O-ring slide easily into position without damage.

Removal and Installation in Static Internal Applications

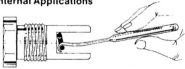

REMOVING THE OLD O-RING

INSTALLING THE NEW O-RING

Most internal static applications require the use of tools because O-rings are hard to get at. For removal, use the pull-type hook. For installation of the new O-ring, use a push-type hook. It may be necessary during removal to loosen the O-ring with a spoon type tool.

Removal and Installation in Dynamic External Applications

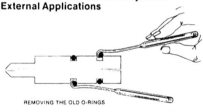

REMOVING THE OLD O-RINGS

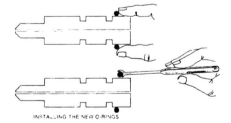

INSTALLING THE NEW O-RINGS

Figure 10-15. O-ring removal and installation instructions.

Table 10-15
Elastomer Capabilities Guide

NOMENCLATURE	NITRILE (BUNA N)	ETHYLENE PROPYLENE	CHLOROPRENE (NEOPRENE)	FLUOROCARBON (VITON, FLUOREL)	SILICONE	FLUOROSILICONE	STYRENE BUTADIENE (SBR)
NATIONAL COMPOUND PREFIX	B, C, D	E	N	V	S	F	G
ASTM D2000 PREFIX	BG, BK, CH	CA	BC, BE	HK	FC, FE, GE	FK	AA, BA
ASTM D1418 DESIGNATION	NBR	EPDM EPM	CR	FKM	PVMQ, VMQ	FVMQ	SBR
GENERAL							
HARDNESS RANGE A SCALE	40-90	50-90	40-80	70-90	40-80	60-80	40-80
RELATIVE O-RING COST	LOW	LOW	LOW/MOD	MOD/HIGH	MODERATE	HIGH	LOW
CONTINUOUS HIGH TEMP LIMIT	257°F,125°C	302°F,150°C	284°F,140°C	437°F,225°C	482°F,250°C	347°F,175°C	212°F,100°C
LOW TEMPERATURE CAPABILITY	67°F,55°C	67°F,55°C	67°F,55°C	40°F,40°C	103°F,75°C	85°F,65°C	67°F,55°C
DYNAMIC SERVICE/ABRASION RESISTANCE	EXCELLENT	VERY GOOD	VERY GOOD	VERY GOOD	POOR	POOR	EXCELLENT
COMPRESSION SET RESISTANCE	VERY GOOD	VERY GOOD	GOOD	VERY GOOD	EXCELLENT	VERY GOOD	GOOD
FLUID COMPATIBILITY SUMMARY							
ACID, INORGANIC	FAIR	GOOD	FAIR/GOOD	EXCELLENT	GOOD	GOOD	FAIR/GOOD
ACID, ORGANIC	GOOD	VERY GOOD	GOOD	GOOD	EXCELLENT	GOOD	GOOD
AGING (OXYGEN, OZONE, WEATHER)	FAIR/POOR	VERY GOOD	GOOD	VERY GOOD	EXCELLENT	EXCELLENT	POOR
AIR	FAIR	VERY GOOD	GOOD	VERY GOOD	EXCELLENT	VERY GOOD	FAIR
ALCOHOLS	VERY GOOD	EXCELLENT	VERY GOOD	FAIR	VERY GOOD	VERY GOOD	VERY GOOD
ALDEHYDES	FAIR/POOR	VERY GOOD	FAIR/POOR	POOR	GOOD	POOR	FAIR/POOR
ALKALIS	FAIR/GOOD	EXCELLENT	GOOD	GOOD	VERY GOOD	GOOD	FAIR/GOOD
AMINES	POOR	VERY GOOD	VERY GOOD	POOR	GOOD	POOR	FAIR
ANIMAL OILS	EXCELLENT	GOOD	GOOD	VERY GOOD	GOOD	EXCELLENT	POOR
ESTERS, ALKYL PHOSPHATE (SKYDROL)	POOR	EXCELLENT	POOR	POOR	GOOD	FAIR/POOR	POOR
ESTERS, ARYL PHOSPHATE	FAIR/POOR	EXCELLENT	FAIR/POOR	EXCELLENT	GOOD	VERY GOOD	POOR
ESTERS, SILICATE	GOOD	POOR	FAIR	EXCELLENT	POOR	VERY GOOD	POOR
ETHERS	POOR	FAIR	POOR	POOR	POOR	FAIR	POOR
HYDROCARBON FUELS, ALIPHATIC	EXCELLENT	POOR	FAIR	EXCELLENT	FAIR	EXCELLENT	POOR
HYDROCARBON FUELS, AROMATIC	GOOD	POOR	FAIR/POOR	EXCELLENT	POOR	VERY GOOD	POOR
HYDROCARBONS, HALOGENATED	FAIR/POOR	POOR	POOR	EXCELLENT	POOR	VERY GOOD	POOR
HYDROCARBON OILS, HIGH ANILINE	EXCELLENT	POOR	GOOD	EXCELLENT	VERY GOOD	EXCELLENT	POOR
HYDROCARBON OILS, LOW ANILINE	VERY GOOD	POOR	FAIR/POOR	EXCELLENT	FAIR	VERY GOOD	POOR
IMPERMEABILITY TO GASES	GOOD	GOOD	GOOD	VERY GOOD	POOR	POOR	FAIR/GOOD
KETONES	POOR	EXCELLENT	POOR	POOR	POOR	FAIR/POOR	POOR
SILICONE OILS	EXCELLENT	EXCELLENT	EXCELLENT	EXCELLENT	GOOD	EXCELLENT	EXCELLENT
VEGETABLE OILS	EXCELLENT	GOOD	GOOD	EXCELLENT	EXCELLENT	EXCELLENT	POOR
WATER/STEAM	GOOD	EXCELLENT	FAIR	FAIR	FAIR	FAIR	FAIR

What Makes an O-ring

O-rings are manufactured from a variety of elastomers which are blended to form compounds. These compounds exhibit unique properties such as resistance to certain fluids, temperature extremes, and life. The following section describes the most prominent elastomers and their inherent properties.

Nitrite, Buna N, or NBR. Nitrile is the most widely used elastomer in the seal industry. The popularity of nitrile is due to its excellent resistance to petroleum products and its ability to be compounded for service over a temperature range of −67° to 257°F (−55°C to 125°C).

Nitrile is a copolymer of butadiene and acrylonitrile. Variation in proportions of these polymers is possible to accommodate specific requirements. An increase in acrylonitrile content increases resistance to heat plus petroleum base oils and fuels but decreases low temperature flexibility. Military AN and MS O-ring specifications require nitrile compounds with low acrylonitrile content to ensure low temperature performance. Nitrile

Table 10-16
Elastomer Capabilities Guide

NOMENCLATURE	POLYACRYLATE	POLYURETHANE	BUTYL	POLYSULFIDE (THIOKOL)	CHLOROSULFONATED POLYETHYLENE (HYPALON)	EPICHLOROHYDRIN (HYDRIN)	PHOSPHONITRILIC FLUOROELASTOMER (PNF)
NATIONAL COMPOUND PREFIX	L	U	J	K	H	Z	Q
ASTM D2000 PREFIX	DF DH	BG	AA BA	AK BK	CE	DK DJ	NOT ASSIGNED
ASTM D1418 DESIGNATION	ACM	EU	IIR	T	CSM	ECO	PZ
GENERAL							
HARDNESS RANGE A SCALE	70 90	60 90	50 70	50 80	50 90	50 90	50 90
RELATIVE O-RING COST	MODERATE	MODERATE	MODERATE	MODERATE	MODERATE	MODERATE	HIGH
CONTINUOUS HIGH TEMP LIMIT	347°F 175°C	212°F 100°C	212°F 100°C	212°F 100°C	257°F 125°C	257°F 125°C	347°F 175°C
LOW TEMPERATURE CAPABILITY	4°F 20°C	67°F 55°C	67°F 55°C	67°F 55°C	67°F 55°C	67°F 55°C	85°F 65°C
DYNAMIC SERVICE/ABRASION RESISTANCE	GOOD	EXCELLENT	GOOD	FAIR/POOR	POOR	FAIR	VERY GOOD
COMPRESSION SET RESISTANCE	FAIR	FAIR	FAIR/GOOD	FAIR	FAIR/POOR	FAIR/GOOD	GOOD
FLUID COMPATIBILITY SUMMARY							
ACID INORGANIC	POOR	POOR	GOOD	POOR	EXCELLENT	FAIR	POOR
ACID ORGANIC	POOR	POOR	VERY GOOD	GOOD	GOOD	FAIR	FAIR
AGING (OXYGEN OZONE WEATHER)	EXCELLENT	EXCELLENT	VERY GOOD	EXCELLENT	VERY GOOD	VERY GOOD	EXCELLENT
AIR	VERY GOOD	GOOD	GOOD	GOOD	EXCELLENT	GOOD	EXCELLENT
ALCOHOLS	POOR	POOR	VERY GOOD	FAIR/GOOD	VERY GOOD	GOOD	FAIR
ALDEHYDES	POOR	POOR	GOOD	FAIR/GOOD	FAIR/GOOD	POOR	POOR
ALKALIS	POOR	FAIR/GOOD	EXCELLENT	POOR	EXCELLENT	FAIR	GOOD
AMINES	POOR	POOR	GOOD	POOR	POOR	POOR	GOOD
ANIMAL OILS	EXCELLENT	GOOD	GOOD	POOR	GOOD	GOOD	FAIR
ESTERS ALKYL PHOSPHATE (SKYDROL)	POOR	POOR	VERY GOOD	POOR	POOR	POOR	POOR
ESTERS ARYL PHOSPHATE	POOR	POOR	EXCELLENT	GOOD	FAIR	POOR	EXCELLENT
ESTERS SILICATE	FAIR/POOR	POOR	POOR	FAIR/POOR	FAIR	GOOD	EXCELLENT
ETHERS	FAIR/POOR	FAIR	FAIR/POOR	GOOD	POOR	GOOD	POOR
HYDROCARBON FUELS ALIPHATIC	VERY GOOD	GOOD	POOR	EXCELLENT	FAIR	VERY GOOD	EXCELLENT
HYDROCARBON FUELS AROMATIC	POOR	FAIR/POOR	POOR	GOOD	FAIR/POOR	VERY GOOD	EXCELLENT
HYDROCARBONS HALOGENATED	FAIR/GOOD	FAIR	POOR	GOOD	FAIR	EXCELLENT	FAIR
HYDROCARBON OILS HIGH ANILINE	EXCELLENT	EXCELLENT	POOR	VERY GOOD	EXCELLENT	EXCELLENT	EXCELLENT
HYDROCARBON OILS LOW ANILINE	EXCELLENT	VERY GOOD	POOR	GOOD	VERY GOOD	EXCELLENT	EXCELLENT
IMPERMEABILITY TO GASES	VERY GOOD	FAIR	EXCELLENT	VERY GOOD	VERY GOOD	EXCELLENT	FAIR
KETONES	POOR	POOR	EXCELLENT	EXCELLENT	GOOD	FAIR	POOR
SILICONE OILS	EXCELLENT	EXCELLENT	EXCELLENT	EXCELLENT	EXCELLENT	EXCELLENT	EXCELLENT
VEGETABLE OILS	GOOD	FAIR	GOOD	POOR	GOOD	EXCELLENT	FAIR
WATER/STEAM	POOR	POOR	EXCELLENT	FAIR	FAIR	GOOD	FAIR

provides excellent compression set, tear, and abrasion resistance. The major limiting properties of nitrile are its poor ozone and weather resistance and moderate heat resistance.

Advantages:

- Good balance of desirable properties
- Excellent oil and fuel resistance
- Good water resistance

Disadvantages:

- Poor weather resistance
- Moderate heat resistance

Ethylene-Propylene, EP, EPT, or EPDM. Ethylene-propylene compounds are used frequently to seal phosphate ester fire resistant hydraulic fluids such as Skydrol. They are also effective in brake systems, and for sealing hot water and steam. Ethylene-propylene compounds have good resistance to mild acids, alkalis, silicone oils and greases, ketones, and alcohols. They are not recommended for petroleum oils or diester lubricants. Ethylene-propylene has a temperature range of −67°F to 302°F (−55°C to 150°C). It is compatible with polar fluids that adversely affect other elastomers.

Advantages:

- Excellent weather resistance
- Good low temperature flexibility
- Excellent chemical resistance
- Good heat resistance

Disadvantage:

- Poor petroleum oil and solvent resistance

Chloroprene, Neoprene, or CR. Neoprene is a polymer of chlorobutadiene and is unusual in that it is moderately resistant to both petroleum oils and weather (ozone, sunlight, oxygen). This qualifies neoprene for O-ring service where many other elastomers would not be satisfactory. It is also used extensively for sealing refrigeration fluids. Neoprene has good compression set characteristics and a temperature range of −57°F to 284°F (−55°C to 140°C).

Advantages:

- Moderate weather resistance
- Moderate oil resistance
- Versatile

Disadvantage:

- Moderate solvent and water resistance

Fluorocarbon, Viton, Fluorel, or FKM. Fluorocarbon combines more resistance to a broader range of chemicals than any of the other elastomers. It constitutes the closest available approach to the universal O-ring elastomer. Although most fluorocarbon compounds become quite hard at temperatures below −4°F (−20°C), they do not easily fracture, and are thus serviceable at much lower temperatures. Fluorocarbon compounds provide a continuous 437°F (225°C) high temperature capability.

Advantages:

- Excellent chemical resistance
- Excellent heat resistance
- Good mechanical properties
- Good compression set resistance

Disadvantage:

- Fair low temperature resistance

Silicone or PVMQ. Silicone is a semi-organic elastomer with outstanding resistance to extremes of temperature. Specially compounded, it can provide reliable service at temperatures as low as −175°F (−115°C) to as high as 482°F (250°C) continuously. Silicone also has good resistance to compression set.

Low physical strength and abrasion resistance combined with high friction limit silicone to static seals. Silicone is used primarily for dry heat static seals. Although it swells considerably in petroleum lubricants, this is not detrimental in most static sealing applications.

Advantages:

- Excellent at temperature extremes
- Excellent compression set resistance

Disadvantages:

- Poor physical strength

Fluorosilicone or FVMQ. Fluorosilicones combine most of the attributes of silicone with resistance to petroleum oils and hydrocarbon fuels. Low physical strength and abrasion resistance combined with high friction limit fluorosilicone to static seals. Fluorosilicones are used primarily in aircraft fuel systems over a temperature range of −85°F to 347°F (−65°C to 175°C).

Advantages:

- Excellent at temperature extremes
- Good resistance to petroleum oils and fuels
- Good compression set resistance

Disadvantage:

- Poor physical strength

Styrene-Butadiene or SBR. Styrene-butadiene compounds have properties similar to those of natural rubber and are primarily used in the manufacture of tires. Their use in O-rings has been mostly in automobile brake systems and plumbing. Ethylene-propylene, a more recent development, is gradually replacing styrene-butadiene in brake service. Temperature range is −67°F to 212°F (−55°C to 100°C).

Advantages:

- Good resistance to brake fluids
- Good resistance to water

Disadvantages:

- Poor weather resistance
- Poor petroleum oil and solvent resistance

Polyacrylate or ACM. Polyacrylate compounds retain their properties when sealing petroleum oils at continuous temperatures as high as 347°F (175°C). Polyacrylate O-rings are used extensively in automotive transmissions and other automotive applications. They provide some of the attributes of fluorocarbon O-rings. A recent variation, ethylene-acrylate, provides improved low temperature characteristics with some sacrifice in hot oil resistance.

Advantages:

- Excellent resistance to petroleum oils
- Excellent weather resistance

Disadvantages:

- Fair low temperature properties
- Fair to poor water resistance
- Fair compression set resistance

Polyurethane, AU, or EU. Polyurethane compounds exhibit outstanding tensile strength and abrasion resistance in comparison with other elastomers. Fluid compatibility is similar to that of nitrile at temperatures up to 158°F (70°C). At higher temperatures, polyurethane has a tendency to soften and lose both strength and fluid resistance advantages over other elastomers. Some types are readily damaged by water, even high humidity. Polyurethane seals offer outstanding performance in high pressure hydraulic systems with abrasive contamination, high shock loads, and related adverse conditions provided temperature is below 158°F (70°C).

Advantages:

- Excellent strength and abrasion resistance
- Good resistance to petroleum oils
- Good weather resistance

Disadvantages:

- Poor resistance to water
- Poor high temperature capabilities

Butyl or IIR. Butyl is a copolymer of isobutylene and isoprene. It has largely been replaced by ethylene-propylene for O-ring usage. Butyl is resistant to the same fluid types as ethylene-propylene and, except for resistance to gas permeation, it is somewhat inferior to ethylene-propylene for O-ring service. Temperature range is −67°F to 212°F (−55°C to 100°C).

Advantages:

- Excellent weather resistance
- Excellent gas permeation resistance

Disadvantage:

- Poor petroleum oil and fuel resistance

Polysulfide, Thiokol, or T. Polysulfide was one of the first commercial synthetic elastomers. Although polysulfide compounds have limited O-ring usage, they are essential for applications involving combinations of ethers, ketones, and petroleum solvents used by the paint and insecticide industries. Temperature range is −67°F to 212°F (−55°C to 100°C).

Disadvantages:

- Poor high temperature capabilities
- Poor mechanical strength
- Poor resistance to compression set

Chlorosulfonated Polyethylene, Hypalon, or CSM. Chlorosulfonated polyethylene compounds demonstrate excellent resistance to oxygen, ozone, heat, and weathering. But their mechanical properties and compression set are inferior to most other elastomers, and they are seldom used to advantage as O-rings. Temperature range is −65°F to 257°F (−55°C to 125°C).

Advantages:

- Excellent resistance to weather
- Good resistance to heat

Disadvantages:

- Poor tear and abrasion resistance
- Poor resistance to compression set

Epichlorohydrin, Hydrin, or ECO. Epichlorohydrin is a relatively recent development. Compounds of this elastomer provide excellent resistance to fuels and oils plus a broader temperature range, −65°F to 275°F (−55°C to 135°C), than nitrile. Initial usage has been in military aircraft where the particular advantages of epichlorohydrin over nitrile are of immediate benefit.

Advantages:

- Excellent oil and fuel resistance
- Excellent weather resistance
- Good low temperature resistance

Disadvantage:

- Fair resistance to compression set

Phosphonitrilic Fluoroelastomer, Polyphosphazene, PNF, or PZ. This is another new elastomer family. O-rings of phosphonitrilic fluoroelastomer are rapidly accommodating aircraft sealing requirements where the physical strength of fluorosilicone is inadequate. In other regards, the functional characteristics of phosphonitrilic fluoroelastomer and fluorosilicone are similar. Temperature range is –85°F to 347°F (–65°C to 175°C).

Advantages:

- Excellent oil and fuel resistance
- Wide temperature range
- Good compression set resistance

Disadvantage:

- Poor water resistance

UTEX HTCR® Fluororubber. Typical of many recent elastomeric compounds, this copolymer of tetrafluoroethylene and propylene is too new to be on most charts. In application range, it fits somewhere between fluorocarbon (Viton) and Kalrez®.

HTRC is thermally stable for continuous use in temperatures of 450°F, and depending on the specific application, has serviceability in environments up to 550°F. The US manufacturer, UTEX, claims excellent resistance to a wide variety of chemical environments. Table 10-17 provides an indication of its chemical resistance. Since temperature, concentration, mixtures and elastomer compound selection can affect performance, this chart provides guidelines only.

Table 10-17

HTCR GENERAL MEDIA RESISTANCE GUIDELINES	
Acids...E	Amines..............................G to E
Animal and Vegetable Oils............E	Oils and Lubricants (incl. synthetics, SF CD, etc.).................................G to E
Bases..E	
Brake Fluids................................E	
Hydraulic fluids (incl. phosphate esters, MiL-H -5606, water/glycol, etc.)...............E	Oxidizing Agents....................G to E
	Sour (H$_2$S) Oil and Gas with Corrosion Inhibitors...........G to E
Steam/Water/Brine......................E	Benzene, Xylene, etc...........F to G
Radiation.......................................E	Fuels...................................F to G
Weathering/Ozone.......................E	Ketones......................................F
Alcohol............................G to E	Chloroform...................................P

Perfluoroelastomer (Kalrez®). Kalrez® O-rings have mechanical properties similar to other fluorinated elastomers but exhibit greater heat resistance and chemical inertness. They have thermal, chemical resistance, and electrical properties similar to Teflon® fluorocarbon resins but, made from a true elastomer, possess excellent resistance to creep and set.

Generally, Kalrez® O-rings are capable of providing continuous service at temperatures of 500°–550°F (260°–288°C) and can operate at 600°F (316°C) for shorter periods as long as they are in static service. For long-term dynamic sealing duties, an operating temperature of 450°F (232°C) would be a reasonable limit.

The chemical resistance of Kalrez® O-rings is outstanding. When using specially formulated compositions, little or no measurable effect is found in almost all chemicals, excepting fluorinated solvents which induce moderate swelling. The parts have excellent resistance to permeation by most chemicals.

Resistance to attack is especially advantageous in hot, corrosive environments such as:

- Polar solvents (ketones, esters, ethers)
- Strong commercial solvents (tetrahydrofuran, dimethyl formamide, benzene)
- Inorganic and organic acids (hydrochloric, nitric, sulfuric, trichloroacetic) and bases (hot caustic soda)
- Strong oxidizing agents (dinitrogen tetroxide, fuming nitric acid)
- Metal halogen compounds (titanium tetrachloride, diethylaluminum chloride)
- Hot mercury/caustic soda
- Chlorine, wet and dry
- Inorganic salt solutions
- Fuels (aviation gas, kerosene, JP-5, Jet Fuel, ASTM Reference Fuel C)
- Hydraulic fluids, synthetics and transmission fluids
- Heat transfer fluids
- Oil well sour gas (methane/hydrogen sulfide/carbon dioxide/steam)
- Steam

Back-Up Rings. Back-up rings, as shown in Figure 10-16, are often used to prevent extrusion in high pressure applications, or to correct problems such as spiral failure or nibbling. They are sometimes used in normal pressure range applications to provide an added measure of protection or to prolong O-ring life. These devices also permit the use of a wider clearance gap when close tolerances are impossible to maintain.

A back-up ring is simply a ring made from a material harder than the O-ring, designed to fit in the downstream side of the groove and close to

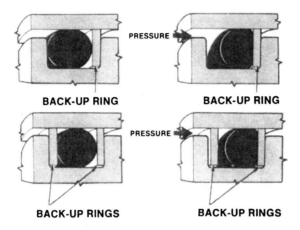

Figure 10-16. Back-up rings used with O-rings.

Table 10-18
Gland Design Guide

	INCHES					MILLIMETERS				
O-Ring Section Diameter	.070	.103	.139	.210	.275	1.78	2.62	3.53	5.33	6.99
STATIC SEALING										
A Gland Depth	.048	.077	.109	.168	.222	1.22	1.96	2.77	4.27	5.64
	.054	.083	.115	.176	.232	1.37	2.11	2.92	4.47	5.89
B Groove Width	.090	.140	.180	.280	.370	2.29	3.56	4.57	7.11	9.40
	.100	.150	.190	.290	.380	2.54	3.81	4.83	7.37	9.65
R Groove Radius (Max.)	.015	.020	.025	.035	.050	.38	.51	.64	.89	1.27
DYNAMIC SEALING										
A Gland Depth	.055	.088	.120	.184	.234	1.40	2.24	3.05	4.67	5.94
	.057	.090	.124	.188	.240	1.45	2.29	3.15	4.76	6.10
B Groove Width	.090	.140	.180	.280	.370	2.29	3.56	4.57	7.11	9.40
	.100	.150	.190	.290	.380	2.54	3.81	4.83	7.37	9.65
R Groove Radius (Max.)	.015	.020	.025	.035	.050	.38	.51	.64	.89	1.27

the clearance gap to provide support for the O-ring. Quite often, O-rings are used as back-up rings, even though back-up rings do not perform any sealing function.

O-Ring, Back-Up Ring, and Gland Dimensions. O-ring sizes have been standardized and range in size from an inside diameter of 0.029 in. and a cross section of 0.040 in. to O-rings with an inside diameter of 16 or more in. and a cross section of 0.210 or more in. Installation dimensions vary with duty and application and the user may find it easy to consult manufacturers' catalogs, which are typically configured as shown in Figure 10-17. Note the small differences in gland dimensions. They depend on whether the O-ring will be axially squeezed, radially squeezed, or will perform dynamic piston and rod sealing duty.

To calculate your own gland design, refer to Table 10-18, "Gland Design Guide."

O-RING SIZE AND GLAND DIMENSIONS

AS 568 MS 28775 DASH	INSIDE DIAMETER	CROSS SECTION	STATIC AXIAL SQUEEZE FACE SEAL GLAND DIMENSIONS (SAE ARP 1234) DIAMETER A (+.005 −.000)	DIAMETER B (+.000 −.005)	GLAND WIDTH C (+.010 −.006)	GLAND DEPTH D (+.000 −.000)	STATIC RADIAL SQUEEZE GLAND DIMENSIONS (SAE ARP 1232) GLAND DIA A (+.001 −.000)	GLAND WIDTH B (+.010 −.000)	ROD DIA C (+.000 −.001)	DYNAMIC PISTON AND ROD SEAL GLAND DIMENSIONS (SAE ARP 1233) GLAND DIA A (+.001 −.000)	GLAND WIDTH B (+.010 −.000)	ROD DIA C (+.000 −.001)
-001	.029 ±.004	.040 ±.003					.083	.101	.035			
-002	.042 ±.004	.050 ±.003					.116	.101	.046			
-003	.056 ±.004	.060 ±.003					.149	.106	.062			
-004	.070 ±.005	.070 ±.003		.075	.125	.049	.162	.114	.077			
-005	.101 ±.005	.070 ±.003		.106	.125	.049	.216	.114	.109			
-006	.114 ±.005	.070 ±.003		.119	.125	.049	.230	.110	.122	.230	.100	.121
-007	.145 ±.005	.070 ±.003		.150	.125	.049	.263	.110	.154	.262	.100	.152
-008	.176 ±.005	.070 ±.003		.181	.125	.049	.296	.105	.186	.294	.100	.183
-009	.208 ±.005	.070 ±.003		.213	.125	.049	.327	.105	.218	.326	.100	
-010	.239 ±.005	.070 ±.003		.244	.125	.049	.350	.105	.250			
-011	.301 ±.005	.070 ±.003	.436			.049	.422	.105				.304
-012		.070 ±.003	.694	.426	.210	.107	.487	.195	.438	.668	.175	.429
-206	.484 ±.005	.139 ±.004	.757	.489	.210	.107	.723	.195	.502	.732	.175	.492
-207	.546 ±.007	.139 ±.004	.819	.551	.210	.107	.769	.195	.567	.795	.175	.556
-208	.609 ±.009	.139 ±.004	.882	.614	.210	.107	.855	.195	.633	.859	.175	.621
-209	.671 ±.009	.139 ±.004	.944	.676	.210	.107	.918	.195	.697	.922	.175	.683
-210	.734 ±.010	.139 ±.004	1.007	.739	.210	.107	.983	.195	.762	.966	.175	.747
-211	.796 ±.010	.139 ±.004	1.069	.80	.210	.107	1.047	.195	.825	1.049	.175	.810
-212	.859 ±.010	.139 ±.004	1.132	.864	.210	.107	1.111	.195	.889	1.112	.175	.873
-213	.921 ±.010	.139 ±.004	1.194	.926	.210	.107	1.174	.195	.953	1.175	.175	.935
-214	.984 ±.010	.139 ±.004	1.257	.989	.210	.107	1.239	.195	1.017	1.238	.175	.998
-215	1.046 ±.010	.139 ±.004	1.319	1.051	.210	.107	1.302	.195	1.080	1.299	.175	1.060
-216	1.109 ±.012	.139 ±.004	1.382	1.114	.210	.107	1.367	.195	1.146	1.365	.175	1.125
-217	1.171 ±.012	.139 ±.004	1.444	1.176	.210	.107	1.432	.195	1.210	1.427	.175	1.187
-218	1.234 ±.012	.139 ±.004	1.507	1.239	.210	.107	1.496	.195	1.274	1.489	.175	1.250
-219	1.296 ±.012	.139 ±.004	1.569	1.301	.210	.107	1.559	.195	1.337	1.552	.175	1.313
-220	1.359 ±.012	.139 ±.004	1.632	1.364	.210	.107	1.623	.195	1.401	1.616	.175	1.376
-221	1.421 ±.012	.139 ±.004	1.694	1.426	.210	.107	1.687	.195	1.465	1.676	.175	1.438
-222	1.484 ±.015	.139 ±.004	1.757	1.489	.210	.107	1.754	.195	1.532	1.744	.175	
-223	1.609 ±.015	.139 ±.004			.210	.107	1.881	.195		1.868		
-224	1.734 ±.015		.952			.107	2.009					.547
-225	±.009	.210 ±.005	1.015	.605	.300				.624			.612
			1.077	.667	.300	.169		.280	.687	1.044	.250	.674
-314	.725 ±.010	.210 ±.005	1.140	.730	.300	.169	1.098	.280	.753	1.108	.250	.738
-315	.787 ±.010	.210 ±.005	1.202	.792	.300	.169	1.160	.280	.816	1.170	.250	.801
-316	.850 ±.010	.210 ±.005	1.265	.855	.300	.169	1.225	.280	.880	1.234	.250	.864
-317	.912 ±.010	.210 ±.005	1.327	.917	.300	.169	1.287	.280	.943	1.295	.250	.926
-318	.975 ±.010	.210 ±.005	1.390	.980	.300	.169	1.353	.280	1.008	1.359	.250	.989
-319	1.037 ±.010	.210 ±.005	1.452	1.042	.300	.169	1.416	.280	1.071	1.420	.250	1.051
-320	1.100 ±.012	.210 ±.005	1.515	1.105	.300	.169	1.482	.280	1.137	1.485	.250	1.116
-321	1.162 ±.012	.210 ±.005	1.577	1.167	.300	.169	1.545	.280	1.200	1.548	.250	1.178
-322	1.225 ±.012	.210 ±.005	1.640	1.230	.300	.169	1.610	.280	1.265	1.610	.250	1.241
-323	1.287 ±.012	.210 ±.005	1.702	1.292	.300	.169	1.673	.280	1.328	1.673	.250	1.304
-324	1.350 ±.015	.210 ±.005	1.765	1.355	.300	.169	1.736	.280	1.392	1.737	.250	1.367
-325	1.475 ±.015	.210 ±.005	1.890	1.480	.300	.169	1.868	.280	1.523	1.865	.250	1.495
-326	1.600 ±.015	.210 ±.005	2.015	1.605	.300	.169	1.995	.280	1.650	1.990	.250	1.620
-327	1.725 ±.015	.210 ±.005	2.140	1.730	.300	.169	2.122	.280	1.778	2.114	.250	1.745
-328	1.850 ±.015	.210 ±.005	2.265	1.855	.300	.169	2.250	.280	1.905	2.240	.250	1.871
-329	1.975 ±.018	.210 ±.005	2.390	1.980	.300	.169	2.381	.280	2.036	2.369	.250	1.999
-330	2.100 ±.018	.210 ±.005	2.515	2.105	.300	.169	2.508	.280	2.163	2.493	.250	2.124

Figure 10-17. A sampling of O-ring and gland dimensions.

References

1. Bloch, H. P. and Geitner, F. K., *Machinery Failure Analysis and Troubleshooting*, Gulf Publishing Company, Houston, Texas, Third Edition, 1997, Pages 42–57.
2. Locke, J. J., "Cobalt Alloy Overlays in a Petro-Chemical Refinery," *Cobalt*, 1974, Vol. 2, Pages 25–31.
3. Mendenhall, M. D., "Shaft Overlays Proven Effective," *Hydrocarbon Processing*, May 1980, Pages 191–192.
4. *Tribology Handbook*, edited by M. J. Neale, John Wiley & Sons, New York-Toronto, 1973, Page E13.
5. *Chrome Plating*, sales brochure by Exline, Inc., Salina, Kansas 67401.
6. Pyles, R., "Porous Chromium in Engine Cylinders," Transactions of the ASME, April 1944, Pages 205–214.
7. Tichvinsky, L. M. and Fischer, E. G., "Boundary Friction in Bearings at Low Loads," Transactions of the ASME, Vol. 61, 1939.
8. Reference 6, Page 206.
9. Reference 6, Page 210.
10. Vacca, A. P., "Extended Periods of Overhaul of Diesel Machinery," *The Motor Ship*, January 1964.
11. Mollerus, A. P. H. J., "Wear Data of Cylinder Liners," study submitted to Ingenieursbureau Lemet Chromium H., Van Der Horst N. V., November 1964.
12. Stinson, K. W., *Diesel Engineering Handbook,* Diesel Publications Inc., Stamford, Connecticut, 1966.
13. Eichenour, C. and Edwards, V. H., "Electromechanical Metallizing Saves Time in Rebuilding Engine Parts," *Plant Services*, December 1982, Pages 49–50.
14. Correspondence with engineers for the Selectron Process, Metal Surface Technology, Addison, Illinois 60101.
15. Republic Steel Corporation, "Heat Treatment of Steel," Advertising Booklet #1302Ra-10M-266, Cleveland, Ohio 44101, Pages 22–23.
16. Reference 15, Pages 24–25.
17. Reference 15, Pages 29–34.
18. Technical Bulletin by E.I. Du Pont de Nemours Co., Finishes Division, Wilmington, Delaware.
19. Technical Bulletin by SR Metal Impregnation Co., Edmonton, Alberta, Canada.
20. Moffat, J. D., "New Metal Impregnation Technology Solves Friction and Corrosion Problems," ASME, New York, New York, publication No. 75-PEM-17, 1975, Pages 2–7.

Appendix 10-A

Part Documentation Record

Table 10-A-1
Typical Part Documentation Record Sheet

THE INTENT OF THIS FORM IS TO RECORD SPECIFIC PART CHARACTERISTICS
THAT WILL BE USED FOR FUTURE EVALUATION.

PART IDENTIFICATION _____ UNIQUE ID _____

CUSTOMER _____ CUSTOMER P.O. _____

INSPECTOR _____ DATE _____

PST JOB NO. _____ DRAWING NO. _____

HARDNESS _____ COATING TYPE _____

OAL _____ MAJOR DIAMETER _____

LENGTH OF COATING FROM THE CROSSHEAD END OF THE SHAFT _____

LENGTH OF COATING _____

MAGNAFLUX—ACCEPTED/REJECTED

COMMENTS _____

_____

_____

_____

Table 10-A-1
Typical Part Documentation Record Sheet—cont'd

LOC	DIMENSION	TIR	RMS
A			
B			
C			
D			
E			
F			
G			

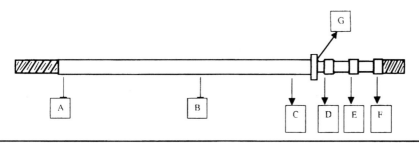

Source: Praxair Surface Technologies, Houston, Texas.

Table 10-A-2
Coating Designations and Physical Properties of Materials Typically Used by Praxair Surface Technologies, Houston, Texas

| PST Coating Name | Coating Designation | | Physical Properties | | | Metallographic Evaluation | | | Typical Coating Thickness Min/Max (Note 8) |
	Nominal Chemistry	AMS 2447B Designation (Note 1)	Average Diamond Pyramid Hardness (Note 2)	Tensile Bond Strength (psi) ASTM C633 (Note 3)	Strain to Fracture (in/in) (Note 4)	Max. Apparent Porosity (Note 5)	Cracks (Note 6)	Interface Separation (Note 7)	
LC-117	75CrC-25NiCr	AMS 2447-3	775 (HV300)	>10,000	0.0020	1.0%	None	None	0.005/0.020
LW-102	83WC-17Co	AMS 2447-7	1,095 (HV300)	>10,000	0.0021	1.0%	None	None	0.005/0.020
LW-103	86WC-10Co-4Cr	AMS 2447-9	950 (HV300)	>10,000	0.0023	0.75%	None	None	0.005/0.020
LW-104	90WC-10Ni	AMS 2447-10	1,075 (HV300)	>10,000	0.0032	0.5%	None	None	0.005/0.020
LN-120	IN 718: 53Ni-20Cr-19Fe-5(Nb + Ta)-3Ti	AMS 2447-6	400 (HV300)	>10,000	0.0066	0.5%	None	None	0.015/0.050

Table 10-A-2
Coating Designations and Physical Properties of Materials Typically Used by Praxair Surface Technologies, Houston, Texas—cont'd

Coating Designation			Physical Properties			Metallographic Evaluation			Typical Coating Thickness
PST Coating Name	Nominal Chemistry	AMS 2447B Designation (Note 1)	Average Diamond Pyramid Hardness (Note 2)	Tensile Bond Strength (psi) ASTM C633 (Note 3)	Strain to Fracture (in/in) (Note 4)	Max. Apparent Porosity (Note 5)	Cracks (Note 6)	Interface Separation (Note 7)	Min/Max (Note 8)

The properties for the following coating are based on limited data.

LN-121	72Ni-16Cr-4Si-3.5B-4Fe-.8C	N/A	500 (HV300)	>9,000	Not available	2.0%	None	None	0.005/0.040

Note 1: The AMS 2447B designation is to be used as a cross reference for chemical composition only. Additional testing for specific coating applications would be required to meet the requirements of AMS 2447B.

Note 2: Average diamond pyramid hardness is based on coating at an optimum spray angle and distance. Actual hardness on a particular part will vary based on part geometry.

Note 3: Bond strength results are based on 1,018 steel specimens coated and tested per ASTM C633. This test is limited by strength of the epoxy used. Bond strength can also vary depending on the specimen material used.

Note 4: Strain to fracture is a measure of ductility using a four-point bend test developed by Praxair Surface Technologies.

Note 5: Maximum apparent porosity is determined by examining a coupon coated at an optimum spray and distance. A cross section of the coupon is examined at 200× and compared to photographic standards. Actual porosity on a part will be influenced by part geometry.

Note 6: Presence of cracking is evaluated by examining a coupon coated at an optimum spray and distance. A cross section of the coupon is examined at 200×.

Note 7: Interface separation is evaluated by examining a coupon coated at an optimum spray and distance. A cross section of the coupon is examined at 200×.

Note 8: Typical coating thickness should be used as a reference only. The coating thickness for a specific part must be evaluated based on factors such as part material type, part geometry, and type of service.

Table 10-A-3
Common Repair Methods and Preferred Selection Sequence

	Common Repair Methods Preferred Selection 1-2-3-4					
	LC-117	LW-102	LW-103	LW-104	LN-120	LN-121
Bearing Journal	1	3		2		4
Thrust Collar Fits	1	3		2		4
Keyed Coupling Fits					2	1
HP-LP Seal Areas	1			2	4	3
Barrel Keyed Wheel Fits					2	1
Inner Stage Seal Areas	3			4	2	1
Impeller Eyes	2			4	3	1
Balance Piston and Impeller Bores					1	2
Balance Piston Diameter	1			2	4	3
Laby Hi-Lo Sections					1	
Hydraulic Tapered Coupling	1				2	

LW-103 should be considered where additional corrosion and abrasion resistance is needed.

Source: Praxair Surface Technologies, Houston, Texas.

Table 10-A-4
Fusion High Velocity Oxygen Fuel (HVOF) Coating Procedure for Repair of Industrial Crankshafts

Scope
Listed on this document are approved standards to be used in the repair of a crankshaft when a customer specification, or no other specification, exists. A customer drawing or specification will always be used in place of these standards. The HVOF repair method will provide a much harder surface than original hardness of substrate, while offering the bond strength and optimum density of compressive coatings. This will allow resistance to wear and corrosion. It is not intended to restore tensile or torsional strength.

Diameter Tolerance:
Rod journal diameter: +0/−0.001″
Main journal diameter: +0/−0.001″
Gear fits: +0/−0.001″
Seal areas: +0/−0.001″

Circular Runout Limits:
Main journals: 0.002″ total indicator reading (T.I.R.); crankshaft supported at each end.

Table 10-A-4
Fusion HVOF Coating Procedure for Repair
of Industrial Crankshafts—cont'd

Roundness Limits:
All diameters: 50 percent of the available diameter tolerance. For example; if the diameter tolerance is 0.001″, roundness limit will be 0.0005″.

Taper Limits (Parallelism):
All diameters: 50% of the available diameter tolerance. For example; if the diameter tolerance is 0.001″, taper limit will be 0.0005″.

Surface Finish (Average):

Regrinds:

Rod and main journals: 16 rms
Seal areas: 16 rms
All other diameters: 16–32 rms

With HVOF Coating—Chrome Carbide or Nickel-Chromium Self-Fluxing Alloy:

Rod and main journals: 16 rms
Seal areas: 16 rms
Gear fits: 16 rms: minimal grinding marks are acceptable
Tapered snouts: 16 rms and 85 percent + blue contact
All other diameters: 16–32 rms: minimal grinding marks are acceptable

1.0   Scope
1.1   This document describes a process for rebuilding worn crankshaft journals and other fits with HVOF coating(s). It will apply to crankshafts that require ABS (American Bureau of Shipping) certification.

1.2   This HVOF procedure will be qualified by testing in accordance with bond tests per ASTM C633.

2.0   Associated Documents
• Bond Test Results (ASTM C633)
• Crankshaft Inspection Reports (FI-007 and FI-008)
• ASTM B499-88 Standard Test Method for measurement of coating/plating thickness by the magnetic method
• Grit Blasting Parameter (FI-7-10 Procedure)
• HVOF Coating Parameter (FI-3-110 Procedure) Nickel-Chromium Self-Fluxing Alloy
• HVOF Coating Parameter (FI-3-60 Procedure) Chrome Carbide

2.1   Copies of these documents are available from the Q.A. Manager or Operations Manager.

2.2   Drawings provided by manufacturer or sketch of crankshaft with size, tolerance, etc. provided by customer.

(Text continued on next page)

Table 10-A-4
Fusion HVOF Coating Procedure for Repair
of Industrial Crankshafts—cont'd

3.0 Cleaning

3.1 Remove all parts that are needed to completely clean and inspect the crankshaft. Mark these parts with the fusion job number.

3.2 Remove all plugs and check all oil holes for blockage; clear oil passages as necessary.

3.3 In a location not to cause damage, stamp the crankshaft with the issued job number.

3.4 Place the crankshaft in a hot caustic solution until oil and grease are no longer present.

3.5 Remove the crankshaft from caustic tank, then steam clean, flushing oil ports, and rinse the entire shaft.

4.0 Initial Inspection (use initial inspection report FI-007)

4.1 Inspect and record the "as received" dimensions of the crankshaft; this includes the bearing journals, thrust width(s), seal fits, gear fits, and coupling fits. If tapered, small end will be measured and any fretting noted.

4.2 Visually inspect and record the overall condition of the above-mentioned areas, plus any threaded holes and keyways.

4.3 Check and record hardness of crankshaft. This to be performed on at least (1) main and (1) rod journal. If any rub or "hot spot" is noticeable, document hardness of area(s) before and after pregrind. If hardness goes beyond 50 Rc, contact customer for alternate or extended repair required.

4.4 If the crankshaft has been previously repaired, check and record the depth of the coating or chrome on each journal. If shaft has been welded, note on space provided.

4.5 Inspect and record the total indicator run-out of each main bearing journal. This can be performed in V-blocks, or, for larger shafts, in crank grinding machine with proper supports in place.

4.6 Using the magnetic particle method, inspect the crankshaft for cracks or other indications.

4.7 Shaft will be degaussed to a residual level of 2 or less.

Author's Note: Good procedures ensure good workmanship!

Table 10-A-4
Fusion HVOF Coating Procedure for Repair
of Industrial Crankshafts—cont'd

Note: When required, an inspector may elect to be present during any or all of the preinspection steps. Their purpose will be as follows:
- To witness magnetic particle examination of the crankshaft to ensure inspection is performed by a certified level II/III inspector.
- To view the initial inspection reports, work order recording of the operations that have been completed, information recorded, and initialed.
- To visually inspect the crankshaft for any previous stamps, markings, or signs of damage.

Option: An ultrasonic inspection can be performed to detect subsurface cracks or flaws in the substrate. This will be performed by outside level III inspector at our facility.

5.0   *Straightening*

5.1   Straightening may be done on crankshafts that are bent more than 0.010″. This can be done by peening or by straightening in a hydraulic press. No heat will be applied to a crankshaft for straightening purposes.

6.0   *Pregrind Operation*

6.1   Grinding wheels used to undercut diameters are to be dressed with a corner radius that conforms with the journal's radius.

6.2   The journal diameters are to be undercut to a diameter that will leave a final minimum coating thickness of between 0.005″ and 0.010″ after finish grinding, or to the undersize limits specified by the OEM (original equipment manufacturer). A maximum of 0.050″ on diameter will be removed if required to remove wear or damage. Alternate repair methods are available if undercut is beyond these limits.

6.3   If there are any "hot spots" or rubs that have discolored the shaft, these areas will, after pregrind, be checked for hard spots.

7.0   *Secondary Magnaflux Inspection*

7.1   After undercutting a magnetic examination will be performed by a certified level II or III inspector.

7.2   Shaft will be degaussed to a residual level of 2 or less.

8.0   *HVOF Coating*

8.1   The crankshaft will be masked and taped off on all areas not to be coated. Oil holes will be plugged to protect from overspray or damage.

(Text continued on next page)

Table 10-A-4
Fusion HVOF Coating Procedure for Repair
of Industrial Crankshafts—cont'd

8.2   Grit blast with aluminum oxide grit, using only new grit, to attain desired anchor profile on areas to be coated. Use Fusion FI-7-10 Parameter Procedure for this purpose.

8.3   Journals will be sprayed using either HVOF Chrome Carbide coating or HVOF Nickel-Chromium Self-Fluxing Alloy coating. Keyed fits, such as coupling areas, will be sprayed with HVOF Nickel-Chromium Self-Fluxing Alloy coating. The coating, as sprayed, will allow for finish grind stock. Use FI-3-60 or FI-3-110 Parameter Procedure for this purpose.

8.4   Document temperature of area(s) coated at point of contact using an infrared gun. This temperature will not exceed 350°F maximum temperature.

8.5   After coating, allow the crankshaft to cool in still air to ambient temperature.

8.6   Record the lot number and the type of coating powder used.

Note: A sample coating coupon. for metallurgical evaluation, can be provided upon request.

9.0   Finish Grinding

9.1   Grind journal diameters to specified OEM dimensions. A diamond grinding wheel will be used to grind journals within 0.001" of finish size.

9.2   Dye penetrant inspect coating.

9.3   All journals will be diamond honed to size and RMS requirement.

10.0   Polishing

10.1   De-burr and polish all oil ports and journal radii to be smooth of any sharp edges or scratches. Radii to be free of any blemishes.

11.0   Final Inspection

11.1   Visually inspect all repaired areas for signs of blemishes and defects.

11.2   Inspect and record the dimensions of all repaired areas on a crankshaft final inspection report (FI-008).

11.3   Record final T.I.R. of each main journal.

11.4   Inspect and record the coating thickness. Micrometers are to be used to measure coating thickness.

Table 10-A-4
Fusion HVOF Coating Procedure for Repair
of Industrial Crankshafts—cont'd

11.5   Document journal(s) rms using profilometer.

Note: ABS to witness final inspection, when required.

12.0   ***Shipment***

12.1   Clear all oil passages and reinstall counterweights if needed.

12.2   Locate and install any other loose components, or parts that were removed from crankshaft, before packaging.

12.3   Review work order to ensure all operations and inspections were completed.

12.4   Apply a rust preventative and prepare for shipment per customer requirement.

Author's Note: Insist on reviewing written repair procedures. Ask the repair specialists for explanation of steps needed to achieve high-quality results!

Table 10-A-5
Documentation (Typical Only) Identifying Procedure Changes
from a Previous Revision

Fusion High Velocity Oxygen Fuel (HVOF) Coating Procedure for
Repair of Industrial Crankshafts: Changes from Revision 5 to
Revision 6

Header:

From: Revision No.: 5
To: Revision No.: 6

From: Effective date: 10/30/98
To: Effective date: 07/09/99

Surface Finish (Average):

From: With HVOF Coating—Chrome Carbide of Inconel
To: With HVOF Coating—Chrome Carbide or Nickel-Chromium Self-Fluxing
       Alloy

2.0   Associated Documents

From: HVOF Coating Parameter (FI-3-50 Procedure) Inconel
To: HVOF Coating Parameter (FI-3-110 Procedure) Nickel-Chromium Self-
       Fluxing Alloy

4.7   Note:

From: Note: An ABS inspector may elect to be present during any or all of the
       preinspection steps. Their purpose will be as follows:
To: Note: When required, ABS inspector may elect to be present during any or
       all of the preinspection steps. Their purpose will be as follows:

8.3

From: Journals will be sprayed using an HVOF Chrome Carbide coating. An HVOF
       Inconel coating will be sprayed on any keyed fit, such as coupling area. The
       coating, as sprayed, will allow for finish grind stock. Use FI-3-50 or FI-3-60
       Parameter Procedure for this purpose.
To: Journals will be sprayed using either HVOF Chrome Carbide coating or HVOF
       Nickel-Chromium Self-Fluxing Alloy coating. Keyed fits, such as coupling
       areas, will be sprayed with HVOF Nickel-Chromium Self-Fluxing Alloy
       coating. The coating, as sprayed, will allow for finish grind stock. Use FI-3-60
       or FI-3-110 Parameter Procedure for this purpose.

Source: Praxair Surface Technologies, Houston, Texas.
Author's Note: Revised procedures must document changes that enhance component
reliability.

Table 10-A-6
Repair Procedure for Piston Rods and Plungers

High Velocity Liquid Fuel (HVLF) Tungsten Carbide Procedure for Tafa JP-5000 Repair of Piston Rods and Plungers

1. Rod will be checked for straightness, amount of wear, thread damage, piston fit size, etc., and findings documented in as "received condition." Hardness of packing/wiper section will also be documented.
2. Unless specified differently by customer, packing and wiper ring section plus at least ½" at each end of area shall be ground undersize to remove damage and wear. Edge of undercut will have a radius to prevent possible stress riser from occurring. Document hardness after undercut. If previously coated, all old coating will be removed.
   Note: On rods previously coated, there is an option to chemically remove the old coating without grinding. This is done in house.
3. Rod will be magnetic particle inspected to check for cracks throughout shaft. Rod shall be demagnetized after inspection to a residual level of 2 gauss or less. As an option, a customer may also request ultrasonic inspection of a rod.
4. Rod will then be masked and taped off with surface protection for all surfaces except those to be grit blasted.
5. An aluminum oxide grit will be used in grit blasting to provide a surface finish of 200 to 350 rms for coating. We use only new grit for this purpose.
6. Within 2 hours of the grit blast operation, an HVLF (3,300–3,900 feet per second [FPS] particle velocity) Tafa JP-5000 tungsten carbide overlay shall be used on packing/wiper sections. The rod temperature during coating application shall not exceed 350°F. We verify this using an infrared gun directed at point of jet stream at rod impact zone.
   Note: There are several tungsten carbide chemical compositions available. Depending on the type of service rod will see, we will determine, with customer approval, the proper composition of tungsten carbide and bonding matrix best suited for this particular application.
7. Coating shall be applied on the diameter 0.010" to 0.015" greater than the specified finish diameter. As an option, and when required, an application of a proprietary UCAR 100 sealer will immediately follow the coating process. Please note that rods that are in oxygen service should never be sealed!
   Note: If required, at this step, we can attach a coupon of like material substrate to be sprayed with the rod. This coupon will be supplied to the customer for any metallurgical examination you may wish to perform.
8. Rod will be finish ground with a diamond wheel, diamond honed, and super-finished to the desired rms required. This is determined by discharge pressure and will be discussed with customer prior to beginning repair. All reliefs and radii will be polished.
9. Dye penetrant check will be performed for final inspection against any indication of blisters, spalling, flaking, cracking, or pits.
10. Unless otherwise specified, rod shall be preserved and crated with proper supports in place, per customer requirements.

(Text continued on next page)

Table 10-A-6
Repair Procedure for Piston Rods and Plungers

High Velocity Liquid Fuel (HVLF) Tungsten Carbide Procedure for Tafa JP-5000 Repair of Piston Rods and Plungers—cont'd

11. PST will record work performed on rod which includes:
    a. Rod will be stamped on one end with PST job number
    b. All dimensional checks
    c. RC hardness
    d. Magnaflux results
    e. Conditions "as received" and corrections made
    f. Coating lengths and thickness of coating
    g. Type of coating and lot number used for repair
    h. Final rms finish documented with profilometer tape
    Above information will be made available upon request.

The Tafa JP-5000 HVLF system provides a dense wear-resistant sprayed coating with very good surface smoothness (1–2 rms attainable due to density of coatings) and bond strength greater than 10,000 lb, per ASTM C-633 tests.
Source: Praxair Surface Technologies, Houston, Texas.
Author's Note: Good repair shops will document
(a) what they will do ("necessary steps")
(b) how they have carried out the necessary steps
(c) inspection results (testing and/or measurements).

Table 10-A-7
Specification for HVOF: Repair of Centrifugal and Axial Compressors, Steam Turbines, High Speed Gear Shafts, Generator Rotors, Pump Elements, Etc.

Requirements

1. Base material shall be 4140, 4330, 4340, 400 series stainless steel, or 17-4 ASTM alloy steel composition. For other base alloys, ask Praxair for details and comments.
2. Base material shall not exceed 40 RC hardness.
3. Total coating dimension of restored area shall not exceed parameters established for each coating type. It is recommended that buildup on bearing fits and other impact areas be limited to 0.025" maximum diametral thickness.
4. When possible, supply information regarding type of service, operating speed, temperature and pressure to which coating will be subjected; also size, tolerance, TIR, and out-of-round limitation.

Process

Recommended process(es) for repair of centrifugal gas compressor components and for gas or steam turbine components shall be the Tafa JP 5000 HVOF coating system. Note that no "equivalent" is acceptable. Parameters set forth by manufacturer and spin test results conducted by Praxair will determine coating requirements.

Procedure

1. Part will be visually inspected and indicated for run-out and eccentricity condition upon receiving.
2. If required, a magnetic particle inspection will be performed to check the part for cracks. The part shall be demagnetized after inspection to a residual level not to exceed 2 gauss. Stainless parts can be dye penetrant inspected.
3. Part will be ground undersize, 0.005" minimum per side, to the maximum limit established per PST Coating Guideline 97 for each coating type, and for each area designated to be repaired. Each edge of undercut region shall be provided with a suitable radius. When possible, a substrate shoulder will be left at end of coated area.

Author's Note: Procedures may be generic or equipment-specific. Choose whichever ones give greatest assurance of achieving long-term component life!

Index

Page numbers followed by *f* indicate figures; those followed by *t* indicate tabular material.

617

CPSIA information can be obtained at www.ICGtesting.com
Printed in the USA
BVOW051711050912

299508BV00002B/4/P